Mammalian Cell Membranes
VOLUME ONE

Mammalian Cell Membranes

VOLUME ONE

General Concepts

Edited by

G. A. Jamieson Ph.D., D.Sc.
Research Director
American Red Cross Blood Research Laboratory
Bethesda, Maryland, USA
and
Adjunct Professor of Biochemistry
Georgetown University Schools of Medicine and Dentistry
Washington, DC, USA

and

D. M. Robinson Ph.D.
Professor of Biology, Georgetown University
and
Member, Vincent T. Lombardi Cancer Research Center
Georgetown University Schools of Medicine and Dentistry
Washington, DC, USA

BUTTERWORTHS
LONDON · BOSTON
Sydney · Wellington · Durban · Toronto

THE BUTTERWORTH GROUP

ENGLAND
Butterworth & Co (Publishers) Ltd
London: 88 Kingsway, WC2B 6AB

AUSTRALIA
Butterworths Pty Ltd
Sydney: 586 Pacific Highway, NSW 2067
Also at Melbourne, Brisbane,
Chatswood, Adelaide and Perth.

CANADA
Butterworth & Co (Canada) Ltd
Toronto: 2265 Midland Avenue,
Scarborough, Ontario, M1P 4S1

NEW ZEALAND
Butterworths of New Zealand Ltd
Wellington: 26–28 Waring Taylor Street, 1

SOUTH AFRICA
Butterworth & Co (South Africa) (Pty) Ltd
Durban: 152–154 Gale Street

USA
Butterworth (Publishers) Inc.
Boston: 19, Cummings Park,
Woburn, Mass. 01801

All rights reserved. No part of this publication may be reproduced or transmitted in any form or by any means, including photocopying and recording, without the written permission of the copyright holder, application for which should be addressed to the publisher. Such written permission must also be obtained before any part of this publication is stored in a retrieval system of any nature.

This book is sold subject to the Standard Conditions of Sale of Net Books and may not be resold in the UK below the net price given by Butterworths in their current price list.

First published 1976

© Butterworth & Co (Publishers) Ltd 1976

ISBN 0 408 70722 4

Library of Congress Cataloging in Publication Data
Main entry under title:

Mammalian cell membranes.

Includes bibliographical references and index.
CONTENTS: v. 1. General concepts.
1. Mammals—Cytology. 2. Cell membranes.
I. Jamieson, Graham A., 1929–   II. Robinson,
David Mason, 1932–   [DNLM: 1. Cell membrane.
2. Mammals. QH601 M265]
QL739.15.M35   599'.08'75   75-33317
ISBN 0-408-70722-4 (v. 1)

Filmset and printed Offset Litho in Great Britain by
Cox & Wyman Ltd, London, Fakenham and Reading

Contents

1. **A CURRENT INTERPRETATION OF THE ANATOMY OF THE MAMMALIAN CELL** *George G. Rose* (Houston) ... 1
   1.1 Introduction ... 1
   1.2 Membranous components ... 2
   1.3 Nonmembranous components ... 25

2. **THE SEPARATION AND CULTIVATION OF MAMMALIAN CELLS** ... 31
   D. M. Robinson (Washington, DC)
   2.1 Introduction ... 31
   2.2 Cell separation and fractionation ... 32
   2.3 Cell cultivation ... 36

3. **ISOLATION OF SURFACE MEMBRANES FROM MAMMALIAN CELLS** ... 45
   Mary Catherine Glick (Philadelphia)
   3.1 Introduction ... 45
   3.2 Preparation ... 46
   3.3 Isolation ... 55
   3.4 Markers ... 58
   3.5 Contaminants ... 61
   3.6 Procedural artifacts ... 64
   3.7 Comments on isolated membranes ... 66
   3.8 Conclusions ... 73

4. **PHYSICAL STUDIES OF MEMBRANES** *Y. K. Levine* (Leeds) ... 78

5. **PHYSICOCHEMICAL STUDIES OF CELLULAR MEMBRANES** ... 97
   D. Chapman (London)
   5.1 Introduction ... 97
   5.2 The constituents of cell membranes ... 98
   5.3 Physical properties of the membrane constituents ... 106
   5.4 Physicochemical properties of cell membranes ... 125

6. **MECHANICAL PROPERTIES OF CELLULAR MEMBRANES** ... 138
   Peter B. Canham (London, Ontario)
   6.1 Introduction ... 138
   6.2 Mechanical properties of the red cell membrane ... 139
   6.3 Mechanical properties of other membranes ... 157
   6.4 Summary and conclusions ... 158

CONTENTS

7	ENZYME DISTRIBUTION IN MAMMALIAN MEMBRANES	161
	R. H. Hinton and E. Reid (Guildford)	
7.1	Membranes to be discussed as membrane sites	161
7.2	Isolation of membrane fragments	164
7.3	Methodology for assigning enzymes to particular membranes	165
7.4	Methods for the study of membrane-associated enzymes	170
7.5	Enzymes of particular membrane systems	172
7.6	Concluding comments	191
8	ELECTROSTATIC CONTROL OF MEMBRANE PERMEABILITY VIA INTRAMEMBRANOUS PARTICLE AGGREGATION	198
	David Gingell (London)	
8.1	Introduction	198
8.2	Glycoprotein nature of intramembranous particles	200
8.3	Mobility of intramembranous particles	201
8.4	Forces between intramembranous particles	202
8.5	Results and discussion	205
8.6	Biological evidence	208
8.7	Synopsis and conclusion	217
	Appendix	220
9	BIOGENESIS OF MAMMALIAN MEMBRANES	224
	S. K. Malhotra (Edmonton)	
9.1	Introduction	224
9.2	Examples of model systems for membrane biogenesis	225
9.3	General remarks on the biogenesis of membranes	229
9.4	Evidence for mosaicism in membranes	234
9.5	Role of mitochondria in protein synthesis	238
9.6	Evidence for division of mitochondria in mammalian cells	239
9.7	Conclusions	240
10	A PERSPECTIVE ON MODELS OF MEMBRANE STRUCTURE	244
	John Lenard and Frank R. Landsberger (Piscataway, NJ and New York)	
10.1	Introduction	244
10.2	Some basic considerations	245
10.3	Dynamic structure of the lipid bilayer	248
10.4	Interactions between lipids and proteins	251
10.5	Membrane asymmetry	253
10.6	Arrangement and motion of membrane proteins	257
INDEX		265

Contributors

PETER B. CANHAM
Department of Biophysics, Health Sciences Centre, The University of Western Ontario, London, Ontario N6A 3K7, Canada

D. CHAPMAN
Department of Chemistry, Chelsea College, University of London, Manresa Road, London SW3 6LX, England

DAVID GINGELL
Department of Biology as Applied to Medicine, The Middlesex Hospital Medical School, London W1P 6DB, England

MARY CATHERINE GLICK
Department of Pediatrics, University of Pennsylvania School of Medicine, Children's Hospital of Philadelphia, One Children's Center, 34th Street and Civic Center Boulevard, Philadelphia, Pennsylvania 19104, USA

R. H. HINTON
Wolfson Bioanalytical Centre, University of Surrey, Leapale Lane, Guildford GU2 5XH, England

FRANK R. LANDSBERGER
The Rockefeller University, New York, NY 10021, USA

JOHN LENARD
Department of Physiology, College of Medicine and Dentistry of New Jersey, Rutgers Medical School, University Heights, Piscataway, New Jersey 08854, USA

Y. K. LEVINE
Department of Physical Chemistry, University of Leeds, Leeds LS2 9JT, England

S. K. MALHOTRA
Electron Microscope Laboratory, Biological Sciences, The University of Alberta, Edmonton, Alberta T6G 2E1, Canada

E. REID
Wolfson Bioanalytical Centre, University of Surrey, Leapale Lane, Guildford GU2 5XH, England

LIST OF CONTRIBUTORS

D. M. ROBINSON
Biology Department, Georgetown University, Washington, DC 20057, USA

GEORGE G. ROSE
Department of Internal Medicine, The University of Texas Health Science Center at Houston, Dental Branch, 6516 John Freeman Avenue, PO Box 20068, Houston, Texas 77025, USA

Preface

This series on 'MAMMALIAN CELL MEMBRANES' represents an attempt to bring together broadly based reviews of specific areas so as to provide as comprehensive a treatment of the subject as possible. We sought to avoid producing another collection of raw experimental data on membranes, rather have we encouraged authors to attempt interpretation, where possible, and to express freely their views on controversial topics. Again, we have suggested that authors not pay too much attention to attempts to avoid all overlap with fellow contributors in the hope that different points of view will provide greater illumination of controversial topics. In these ways, we hope that the series will prove readable for specialists and generalists alike.

This first volume, entitled *General Concepts*, serves to introduce the subject and covers the most essential aspects of physical and chemical studies which have contributed to our present knowledge of membrane structure and function. The second volume, *The Diversity of Membranes*, addresses itself to specific types of intra- and extracellular membranes, while the third volume, *Surface Membranes of Specific Cell Types*, as its title indicates, will review the knowledge that we have of the surface membranes of the various cell types which have been studied in any detail to this time. *Membranes and Cellular Functions* will be covered in Volume 4 and will concern ultrastructural, biochemical and physiological aspects. Since the cell surface represents the point of interaction with the cellular environment, Volume 5 addresses itself to *Responses to External Influences* and the way in which these are mediated by the plasma membrane.

As editors, our approach to our responsibilities has been rather permissive. With regard to nomenclature and useful abbreviations, we have used 'cell surfaces' and 'plasma membranes' where appropriate rather than 'cell membranes' since this last is nonspecific. Both British and American usage and spelling have been utilized depending upon personal preference of the authors and editors with, again, no attempt at rigid adherence to a particular style.

While the title of the series is 'MAMMALIAN CELL MEMBRANES', we have encouraged authors to introduce concepts and techniques from non-mammalian systems which may be useful in their application to eukaryotic cells. The aim of this series is to provide a background of information and, hopefully, a stimulation of interest to those investigators working in, or about to enter, this burgeoning field.

Finally, the editors would like to acknowledge the dedication and resourcefulness of their secretary and editorial assistant, Mrs Alice R. Scipio, in the coordination and preparation of these volumes.

G. A. JAMIESON
D. M. ROBINSON

1

A current interpretation of the anatomy of the mammalian cell

George G. Rose
Department of Internal Medicine, The University of Texas Health Science Center at Houston, Dental Branch

1.1 INTRODUCTION

Our appreciation of the anatomy of mammalian cells depends on the type of microscope through which we look at these tiny units of life and is conditioned by the state of the cells at the time of our observation. Perhaps it is unfair to give credit to one form of microscopy more than to others, but there is no doubt that the studies with the transmission electron microscope have yielded the greatest wealth of structural detail available for the construction of our present models and concepts of the anatomy of mammalian cells. Yet, we cannot fully comprehend a cell by using just one type of microscope any more than we can fully interpret another human being with just our eyes or our sense of touch. Our current concept of cellular anatomy, then, is very much dependent upon the types of sensors (microscopes) we use, the state of the cell(s) observed, and how we interpret and synthesize the stimuli we receive.

General textbooks of biology and chapters on the general description of cells abound with diagrams of the basic cell and its organelles. Such schema are designed to aid and assist in understanding the very complex. Unfortunately, they occasionally result in a myth, and the diagrams may be the total visual exposure some readers will have of cells and their components. That a myth may evolve from such schema seems all the more probable to those who have studied living cells, especially by time-lapse cinemicrography. Schema impose an inordinate rigidity to cells which in their natural environment undergo massive form changes at the same time they are maintaining specific anatomic structure. The myth of cellular anatomy developed years ago along with conventional histologic techniques, but in recent times has been perpetuated by the beautiful techniques of ultrathin sectioning and electron microscopy. On the other side of the coin is the living cell, so dynamic that to comprehend its true and everchanging structure we have to stop its activities by fixation, section it to see its insides, stain it, digest it, label it radioactively, send it through various counters, scan it, make it fluorescent and much more. In the interest of clarity, this retrievable information has been schematized, for only in this way have our finite minds been able to build a picture, bit by bit, of those complexities within membranes which we call cells. A true exegesis of the anatomy of cells requires, then, a blending of the actual with the factual, that is, of the cell as we are able to visualize it in its natural, or near natural, living microenvironment with information which specific techniques yield about its whole or parts in unnatural (fixed) states.

In the past, it was convenient to describe a cell as having two parts, nucleus and cytoplasm, presumably surrounded by a cell membrane thought by many to be only an interface. We now know that cell membranes are real structures intricately distributed throughout the cytoplasm as well, that those surrounding the nucleus occasionally are found to be continuous with those of the cytoplasm, that what was once thought to be only cytoplasmic may also be found in the nucleus, and that what was once thought to be only nuclear is now found in the cytoplasm. For these reasons this conventionally used topographic divisioning now seems antiquated. A structural classification based upon electron microscopic findings is used in this chapter to divide the cell into membranous and nonmembranous parts.

1.2   MEMBRANOUS COMPONENTS

1.2.1   Cell membranes (plasmalemmae)

It is now indisputable that cells are enclosed in membranes. In a living cell observed in tissue culture (*Figure 1.1*), a limiting membrane may appear to

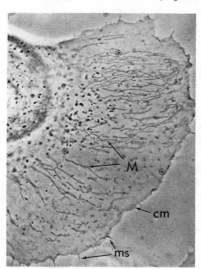

Figure 1.1   Portion of fetal human oral epithelial cell growing in tissue culture, showing the irregularity of the surrounding cell membrane (cm) from which protrude small probing and waving microspikes (ms). M, mitochondria. Phase contrast. ×2000. (From Rose, 1970, courtesy of Academic Press)

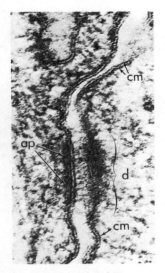

Figure 1.2 High-power view of the periphery of two opposing human gingiva epithelial cells in vitro, showing the unit membrane structure of their limiting cell membranes (cm). A poorly defined desmosome (d) is shown in the bracketed area with a dense attachment plaque (ap) just below the inner leaflet of the cell membrane on the left. Electron micrograph. × 220 000. (Courtesy of A. Sugimoto)

isolate fully the cell's content from its immediate surroundings. We know that this is an illusion for, at various levels of magnitude, a cell must communicate with its environment and undergo fundamental exchanges or perish. The plasmalemma of a living cell cannot, then, be the complete barrier that superficially it appears to be. With the electron microscope (EM), these surface cell membranes are viewed in cross section as trilaminar structures (unit membranes) of relatively uniform width (*Figure 1.2*). They vary in different cells from 6 to 10 nm thick and appear as two dark lines interpreted as protein on either side of a light intermediate bilayer of lipid.

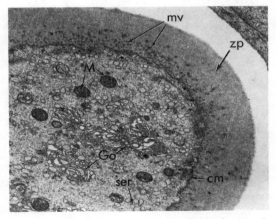

Figure 1.3 Fetal rat ova in tissue culture, showing the thickened glycocalyx (zona pellucida, zp) into which microvilli (mv) project from the cell membrane (cm). There is a large accumulation of Golgi saccules and vesicles (Go) in the center of the cell as well as typically granular mitochondria (M). There are a great many images of smooth endoplasmic reticulum (ser). Electron micrograph. × 10 000. (From Rose et al., 1970, courtesy of Academic Press)

To obtain the best view of this laminated structure requires sectioning it at a right angle, but since a cell is not a perfect cube or sphere, many areas of a membrane will be sectioned obliquely and therefore appear fuzzy (*Figure 1.2*). Outside of this definite trilaminar barrier, there is an indefinite fuzzy layer, which is fuzzy regardless of the angle of sectioning. It is not necessarily observed in every cell, and its thickness varies among the different

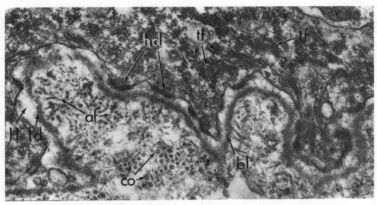

Figure 1.4 Section through the epithelial–connective tissue junction of human gingiva, showing the lamina lucida (ll) and lamina densa (ld) components of the basal lamina (bl). Anchoring fibrils (af) which project down the lower aspect of the lamina densa occasionally encircle fibrils of collagen (co) in the connective tissue. The hemidesmosomes (hd) are seen as areas of increased density between the basal lamina and the lower aspect of the basal epithelial cell. Electron micrograph. tf, tonofibrils. × 52 000. (Courtesy of N. Ijuhin)

cell types. This is sometimes referred to as the glycocalyx because of its glycoprotein–polysaccharide constituents and is especially thick around ova, where it forms the zona pellucida (*Figure 1.3*). Where cells closely abut one another, a condensation of this amorphous layer is sometimes present. It may be equivalent to the basal lamina (*Figure 1.4*) which separates epithelium from connective tissue.

1.2.1.1 ATTACHMENT DEVICES

In the oral epithelium of the human gingiva, as well as in many similar epithelial areas and in cardiac muscle (intercalated discs), three types of abutments or cell contacts are found: tight junctions, intermediate junctions and desmosomes. The basal cells also unite to the basal lamina with half-desmosomes (hemidesmosomes) rather than to cells of the underlying connective tissue.

In the tight junction the two outer leaflets of the cell membranes of opposing cells appear to have fused and thus form a pentalaminar contact zone (*Figure 1.5*), whereas in the intermediate junction there is a uniform space separating the two opposing membranes, and often a few tonofilaments are concentrated in the underlying cytoplasm as well. The desmosome appears the more complicated junction (*Figures 1.5 and 1.6*). The two opposing cell membranes are separated by an intermediate layer (presumably a condensation of the outer fuzzy layer), and there is a great concentration of tonofilaments in the underlying cytoplasm. The tonofilaments do not cross the cell membrane, as light-microscopic observations suggest, but are reflected back into the cell through an electron-dense attachment plaque

Figure 1.5 High-power view of attaching devices between upper epithelial cells of human gingiva in vivo. In the lower left corner, two profiles of fused cell membranes which form tight junctions (tj) are seen close together. Two desmosomes (d) and their associated tonofibrils (tf) are also shown. Electron micrograph. × 115 200. (Courtesy of H. Takarada)

(*Figure 1.6*) just inside the cell membrane. This attachment plaque is largely a concentration of these loops, but enzymatic studies have shown it to be digested by protease, whereas the tonofilaments are protease-resistant (Douglas, Ripley and Ellis, 1970). These three contacts are sometimes found together between cells as a 'junctional complex' with the tight junction being

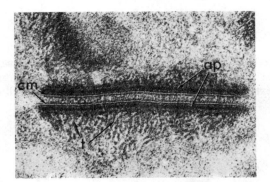

Figure 1.6 High-power view of a large desmosome between upper spinous epithelial cells of human gingiva in vivo. The desmosome is sectioned so that the intermediate plate between the two cell membranes (cm) is visualized as a dashed center line and the density of the attachment plaques (ap) internal to the cell membranes on both sides of the desmosome are noted as electron-dense areas which run parallel to the surfaces of the cell membranes. The tonofilaments (t) are sectioned either across their axes or tangentially. Electron micrograph. × 123 000. (Courtesy of H. Takarada)

most superficial, i.e. toward a lumen, the desmosome occupying the basal position, and the intermediate junction between them. This complex is seen in many places where epithelial cells form watertight barriers, such as throughout the gastrointestinal tract, the integument, and in gland acini.

Tight junctions are sometimes observed between fibroblasts. There are other types of junctions between highly specialized cells.

At the boundary of stromal tissue with epithelial tissue, there is the thickened extracellular basal lamina (*Figure 1.4*). It is composed of two layers: the lamina lucida is the relatively clear zone next to the basal cell, and the lamina densa is the fine granular and filamentous layer between the lamina lucida and the connective tissue. It has been proved experimentally that the basal lamina is of epithelial origin (Briggaman, Dalldorf and Wheeler, 1971). Normally, the lower portion or lamina densa is 40–80 nm thick and is a single layer, although in certain situations it may be thinner or thicker and multilayered (Takarada *et al.*, 1974c). The epithelial cells external to the basal lamina are associated with it by hemidesmosomes (*Figure 1.4*). These are similar to the desmosomes between cells, but since there is no confronting epithelial cell, there is only half a desmosome. The extracellular components of the underlying connective tissue, namely collagen fibrils, are loosely attached to the underside of the lamina densa. They do so indirectly through their association with the small anchoring fibrils which project down from the basal lamina (Susi, Belt and Kelly, 1961) and which have been shown to be of connective tissue origin (Briggaman, Dalldorf and Wheeler, 1971). Additionally, the microfibrils of elastic fibers have also been found blending with the basal lamina of human gingiva (Takarada, Cattoni and Rose, 1974a). In histological sections, this entire area is called the *basement membrane*.

1.2.1.2 PROJECTIONS

In epithelial cells of the oral cavity, the desmosomes between basal cells (i.e. the lowest layer of cells attached to the basal lamina) and between those in the next several upper layers are often found on cellular extensions

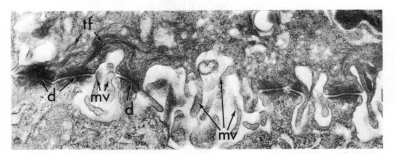

Figure 1.7 Section through two opposing epithelial cells of human gingiva, showing desmosomal attachment devices between which project the slender microvilli (mv). tf, tonofibrils; d, desmosome. Electron micrograph. × 27 760. (Courtesy of H. Takarada)

(*Figure 1.7*) which protrude into the intercellular spaces between the more slender microvilli. As viewed in the electron microscope, microvilli are finger-like extensions which also project from many epithelial cells into lumina. They are seen in great abundance in the surface cells of the gastrointestinal tract (the brush border of light microscopy, *Figure 1.8*), cells

A CURRENT INTERPRETATION OF THE ANATOMY OF THE MAMMALIAN CELL

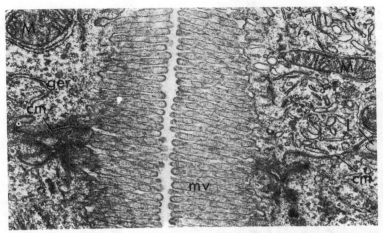

Figure 1.8 Section through opposing luminal walls of fetal mouse intestine, showing the parallel alignment of the microvilli (mv) which, with the light microscope, are seen as the brush border. ger, granular endoplasmic reticulum; cm, cell membrane; M, mitochondrion; L, lipid. Electron micrograph. ×27 300. (Courtesy of A. Sugimoto)

lining cystic spaces, thyroid acinar cells lining the colloid-containing thyroid follicles (Rose et al., 1970), and the epithelial lining of bile canaliculi (Rose, Kumegawa and Cattoni, 1968), to name a few sites. They were also shown projecting from the ova into the glycocalyx in Figure 1.3. Long axonal extensions of nerve cells, cilia and dendritic extensions of macrophages (Figure 1.9) and clear cells (melanocytes, Langerhans cells and Merkel cells)

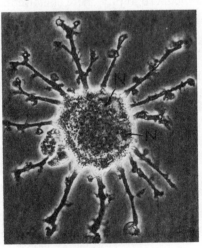

Figure 1.9 Binucleate macrophage cultivated from fetal mouse salivary gland. The long dendritic processes are engaged in membrane ruffling, pinocytosis and phagocytosis. Phase contrast. N, nucleus. ×950. (From Rose, 1970, courtesy of Academic Press)

are extremely specialized cellular projections (Nikai, Rose and Cattoni, 1970, 1971).

Both normally and pathologically the cell membrane may undergo blebbing or 'blistering' deformations (Costero and Pomerat, 1951). In tissue culture this blebbing (zeiosis) may appear in some cells as a reversible process without known provoking stimuli. Zeiotic blebs may be either single or multiple and either submicroscopic or large enough to contain more than half of the cell's contents (Rose, Cattoni and Pomerat, 1963). Any constituent of the cell may flow into these zeiotic blebs, including nuclei (*Figure 1.10*). Occasionally, a bleb may become detached from a cell and, in this

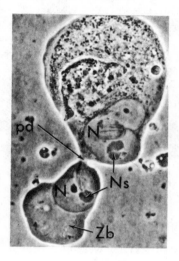

Figure 1.10 Cell from a human epidermoid carcinoma of the nasal pharynx (KB line) undergoing zeiosis and the enucleation of one of two nuclei (N) into a zeiotic bleb (Zb). Ultimately, the bleb was reduced, and the nucleus was reincorporated into the cell mass. Ns, nucleolus; pd, pedicle. Phase contrast. ×1400. (From Rose, 1970, courtesy of Academic Press)

way, the cell may lose some of its organelles (Rose, 1966). Large blebs of this kind have points of constriction, but when viewed with time-lapse cinemicrography, these constrictions are observed to originate as peristaltic-like waves at the distal end of the bleb which move toward the cell. A series of tight constrictions often occurs at the bleb–cell junction, forming an extended pedicle which prevents further transit of cytoplasmic structures in or out of the bleb. The bleb rarely separates from the cell, but most often the constriction forming the pedicle relaxes abruptly, and the bleb is reduced as its content flows back into the cell. Some electron microscopic images of cells *in vivo* occasionally show zeiotic-like blebs. Other blebs are irreversible and are due to noxious agents which disrupt the cell membrane.

1.2.1.3   INCLUSIONS

The contours of cells and their membranes are not smooth and regular, although free cells may appear without projections or indentations under certain forms of microscopy. For instance, most normal cells in an air-dried blood smear appear rounded. The same cells in a physiologic wet mount may be round, but as the leukocytes attach to the glass walls of the vessels supporting them, they develop many structural surface irregularities. Some of these irregularities are due to endocytosis, i.e. the ingestion of particulate matter (phagocytosis) or liquids (pinocytosis), while others are simply due to the cells' attachments and mobility. The sectioning of such free cells for electron microscopy displays the tortuosity of their membranes, even in cells which appear smooth-surfaced with the light microscope.

A CURRENT INTERPRETATION OF THE ANATOMY OF THE MAMMALIAN CELL

Cells may be observed in tissue culture to have peripheral ruffled membranes and phase-white droplets in the cytoplasm (*Figure 1.11*). These membranes appear to engulf or surround droplets of fluid. It takes 10 to 20 minutes for these droplets to coalesce inside of the cell and then to decrease in size and become rounder in shape as they move across the cytoplasm toward the nucleus. This, of course, is more easily viewed by time-lapse cinemicrography, but once the structural changes are viewed and the

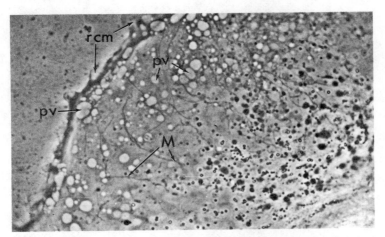

Figure 1.11 Periphery of a human gingival epithelial cell undergoing pinocytosis in tissue culture. At the ruffled cell membrane border (rcm), the enclosed pinocytotic vesicles (pv) are larger, since this is where they are coalescing and are flatter. Filamentous mitochondria (M) are also shown. Phase contrast. ×2100. (From Rose, 1970, courtesy of Academic Press)

activity ascertained, it is not difficult to diagnose pinocytosis with just a single observation of a living cell. With the electron microscope, pinocytosis of this type is confirmed by the appearance of large droplets interiorized close to the cell membrane (*Figure 1.12*). A folding of the membrane around new droplets may be seen at the cell surface.

The basal cells of the human gingiva and the endothelial cells of the underlying connective tissue *in vivo*, as well as a great many other cell types, exhibit a micropinocytosis observable only with the electron microscope. This micropinocytosis was first described in the endothelial cells of blood vessels (Palade, 1960) and is recorded as an invagination of the cell membrane with the engulfment of a fluid droplet. In the capillary endothelial cells of the lamina propria, the droplet apparently moves across the cell and is discharged by a reverse process into the connective tissue spaces. The same invaginatory process is evident in fibroblasts (*Figure 1.13*) and in epithelial cells. In the basal cells of the gingiva, small invaginations are seen on the surface that embraces the basal lamina (*Figure 1.14*). It has always been assumed that these invaginations of the cell membrane in this latter area represent a micropinocytosis or an interiorization of fluids (from the connective tissues), but it has not been determined whether they may, in fact, represent secretory processes. Since the epithelial cells have been proved to be responsible for the elaboration of the basal lamina, and since these

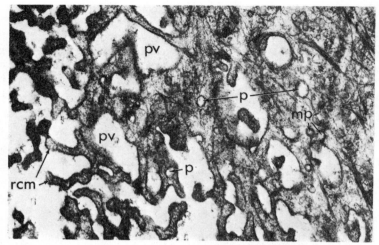

Figure 1.12 High-power view of a section of a human gingival epithelial cell in tissue culture, showing the periphery and the ruffled cell membrane (rcm). This is similar to the area shown in Figure 1.11. The pinocytotic vesicles (pv) are now seen to provide areas of micropinocytosis (mp) and the inclusion of pinosomes (p). rcm, ruffled cell membrane. Electron micrograph. ×40 000. (Courtesy of K. Kobayashi)

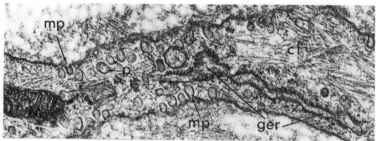

Figure 1.13 High-power view of section through a fibroblast in the lamina propria of human gingiva, showing micropinocytosis (mp) by the cell membrane. ger, granular endoplasmic reticulum; cf, cytofilaments; p, pinosome. Electron micrograph. ×52 000. (Courtesy of H. Takarada)

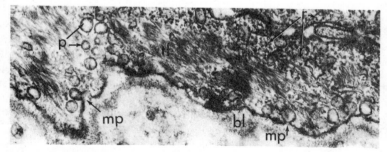

Figure 1.14 High-power view of section of the basal portion of a basal cell from human gingiva, showing micropinocytosis (mp) of the cell membrane adjacent to the basal lamina (bl). Interiorized pinosomes (p) are seen deeper in the cell amongst the tonofibrils (tf). Free ribosomes (r) are also dispersed throughout this cell. Electron micrograph. ×60 000. (Courtesy of H. Takarada)

invaginations are also seen in basal cells adjacent to teeth from which apparently they could derive no fluid nutrients to pinocytose, their secretory role in the elaboration of the basal lamina appears a distinct possibility.

Unique inclusions, particularly in basal cells of the gingiva, skin and other similar areas, are the melanosome complexes (*Figure 1.15*). These are membrane-bound inclusions containing melanosomes injected from the tips of melanocytic processes.

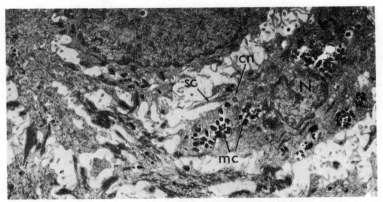

Figure 1.15 Basal area cells of the human gingival epithelium, showing accumulations of melanosomes in melanosome complexes (mc). The section shows the cell to the right with a centriole (cn) and a solitary cilium (sc). N, nucleus. Electron micrograph. × 12 000. (From Nikai, Rose and Cattoni, 1970. Copyright by the American Dental Association. Reprinted by permission)

Interiorization of cell debris and in fact whole living cells (cannibalism) by phagocytosis has been observed in tissue cultures (Rose, 1970, p. 208). The mechanism is basically similar to pinocytosis.

1.2.2   Organelles

1.2.2.1   MITOCHONDRIA

When the phase-contrast microscope appeared in the laboratory soon after World War II, students and skeptics alike began to believe in the validity of these evasive organelles. Numerous mitochondria (up to 1600 in a normal liver cell and 300 000 in some oocytes) occupy the cytoplasm of most cell types, but their number, shape and distribution vary widely, especially among different cell types. Generally, they are described as elongated, 'worm-like' bodies approximately 0.5 μm in diameter and a few micrometers in length, although they may be 8–10 μm in length, several times wider, or even spheroidal. A general characterization on the basis of size is hardly possible as any ten randomly selected mitochondria will generally have ten different dimensions. In living cells in tissue cultures, they are best observed with high-power phase-contrast microscopy (*Figures 1.1* and *1.11*). In this circumstance they appear in the cytoplasm as black, worm-like bodies which move discernibly and undergo contour changes. Their movements, of course, are best observed with the aid of time-lapse, phase-contrast cinemicrography.

Projected time-lapse movie sequences show not only bizarre conformational changes but also fissions and fusions, all of which seemingly occur randomly.

Mitochondria sectioned and viewed by the techniques of transmission electron microscopy reveal two characteristic components: (a) a double-walled enclosure composed of inner and outer unit membranes, and (b) a semipartitioning by invaginations of the inner membrane to form a variable number of cristae (*Figures 1.3, 1.8, 1.13* and *1.16*). A mitochondrion, then,

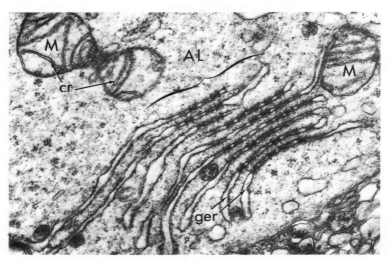

*Figure 1.16 High-power view of human gingival epithelial cell line (Smulow, J. B. and Glickman, I., Proc. Soc. exp. Biol. Med., **121**: 1294 (1966)), showing a profile of the annulate lamellae (AL) and its attachment to the granular endoplasmic reticulum (ger). In the upper portion of the micrograph, there are two dilated mitochondria (M) with widely spaced cristae (cr) projecting into the internal compartment of the mitochondria. Electron micrograph. × 40 000. (Courtesy of A. Sugimoto)*

is divided into two compartments: (a) the lesser compartment, which lies between the two parts of the double membrane and includes the area between the two leaves of the cristae projecting into the mitochondrion, and (b) the greater compartment, which is the area internal to both double membranes and between the cristae. The thickness of mitochondrial unit membranes (approximately 6 nm) is not as great as that of the unit membranes of the plasmalemma. The space of the lesser compartment is only about 6–8 nm in width and appears to be relatively clear, whereas the greater compartment bounded by the internal membranes is filled with a relatively dense material, the mitochondrial matrix. In addition, some mitochondria have osmiophilic matrix granules (approximately 30 nm in diameter) within the greater compartment. These are now concluded to be mainly lipid (Barnard and Afzelius, 1972) and are prominent in mitochondria of kidney tubule cells, epithelial cells of the small intestine, and osteoclasts. Their number has been reported to vary from 10 to 110 per mitochondrion. Since the cristae do not project completely across a mitochondrion except on its attached side, the matrix of the greater compartment is continuous. After special negative staining techniques, the surfaces of cristae are shown to be covered by spherical particles (8–10 nm) that project from the cristae on stems at regularly spaced intervals of 10 nm (Fernández-Morán, 1963). In usual preparations, these 'elementary' particles, as they are called, are not seen on the cristae but pop out after treatment with a hypotonic medium.

The variations in the structure of mitochondria seen by transmission electron microscopy after ultrathin sectioning are many. Occasionally, the cristae project longitudinally instead of exhibiting the more commonly observed lateral orientation. In the adrenal cells cristae may appear as tubules rather than as lamellae. The number of cristae will vary among mitochondria from different cell types; in liver and germinal cells there are few cristae but a great abundance of matrix, whereas in muscle cells the reverse is seen. In some situations, true ribosomes have been found in the matrix of mitochondria (Rabinowitz and Swift, 1970). It is also known that DNA is a component of this organelle (Nass, Nass and Afzelius, 1965), although its structure is not easily demonstrated. In rare situations, the outer membranes of mitochondria are continuous with either the outer nuclear envelope, the endoplasmic reticulum, or the plasmalemma.

1.2.2.2 ENDOPLASMIC RETICULUM (GER AND SER)

Although this is now seldom viewed as a reticulum, early electron microscopic observations of unsectioned cultured fibroblasts (Porter, Calude and Fullman, 1945) showed a reticulated material in the more central portion of the cytoplasm, hence the term endoplasmic reticulum. In rare situations, this reticulum has been observed in fresh tissue cells by phase-contrast microscopy (Fawcett and Ito, 1958) and in tissue-cultured cells (Rose, 1961b, 1970 (p. 297); Rose and Pomerat, 1960). However, the current sectioning techniques for transmission electron microscopy reveal the endoplasmic reticulum to be a part of the cellular vacuolar system, which comprises the granular endoplasmic reticulum (GER), the smooth endoplasmic reticulum (SER) and the Golgi complex.

The endoplasmic reticulum is a cytoplasmic system of closed unit membranes surrounding flattened or dilated cisternae. Those membranes which are studded with ribosomes on their external surfaces are called the granular endoplasmic reticulum (*Figures 1.8*, *1.13*, *1.16* and *1.17*), while those which are free of ribosomes are called the smooth endoplasmic reticulum (*Figure 1.3*). However, continuities exist between these two types of endoplasmic reticulum and occasionally other continuities with the unit membranes of the Golgi complex, the nuclear envelope, mitochondria and the plasmalemma are noted. As in mitochondria, the unit membrane forming the endoplasmic reticulum is thinner (5–6 nm) than that of the plasmalemma (6–10 nm). The cisternae or spaces between the membranes of the endoplasmic reticulum may be compressed and contain little if any opacity, or they may be quite distended and show varying degrees of density. Cells undergoing active protein synthesis, such as osteoblasts and fibroblasts, often show greatly distended cisternae containing a dense material (*Figure 1.17*). The amount of GER present in a cell type depends upon the amount of protein produced by the cell for release into the surrounding microenvironment. In the basal cells of the human gingiva, as in the skin, only a few images of the GER

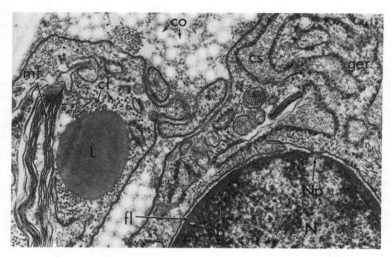

Figure 1.17 High-power view of portions of two fibroblasts in the lamina propria of the human gingiva. The cells are surrounded by collagen fibrils (co). The nucleus (N) of the cell on the right has a prominent nuclear pore (Np). Immediately inside the nuclear envelope, there is a striated layer, the fibrous lamina (fl), which is studded with chromatin particles (ch). A granular endoplasmic reticulum (ger), a large lipid granule (L), and a myelin figure (mf) are also seen in this micrograph. Cross sections of cytoplasmic filaments (cf) are seen in the cell to the left as prominent black dots somewhat smaller and more sharply delineated than the ribosomes attached to the ger. Electron micrograph. ×65 000. (Courtesy of H. Takarada)

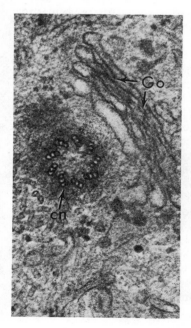

Figure 1.18 The centriole (cn) and Golgi complex (Go) in a hematopoietic tissue culture cell line from a patient with infectious mononucleosis. The perfect transverse section of the centriole reveals the microtubule components as nine sets of triplets, and the Golgi complex is sectioned so that four parallel lamellar units are seen, each composed of flattened sacs with terminal budding dilations. Electron micrograph. ×40 000. (From Moore, Kitamura and Toshimo, 1968, courtesy of the American Cancer Society)

A CURRENT INTERPRETATION OF THE ANATOMY OF THE MAMMALIAN CELL

will be seen, whereas there are dense parallel arrays of GER in pancreatic acinar cells and in plasma cells actively engaged in the secretion of zymogen granules and antibody protein.

In general, it may be said that the granular endoplasmic reticulum is involved in the synthesis of proteins, whereas the agranular, or smooth, endoplasmic reticulum is involved in the synthesis of lipids. The SER has also been implicated in the role of detoxification and in the synthesis of glycogen. In muscle cells, the SER is organized as the sarcoplasmic reticulum, and in nerve cells the GER is the Nissl substance.

1.2.2.3 GOLGI COMPLEX

Like the endoplasmic reticulum, the Golgi complex is part of the cytoplasmic vacuolar system and, like the smooth endoplasmic reticulum, it is a zone from which ribosomes are excluded and, therefore, it is not directly engaged in protein synthesis. In EM sections, it is composed of smooth-surfaced unit membranes approximately 6 nm thick arrayed as a series of parallel closed loops which form flattened sacs (*Figures 1.3* and *1.18*). In some cells, the Golgi complex is found adjacent to the nucleus only on one side, whereas in other cells it may be seen in various areas surrounding the nucleus.

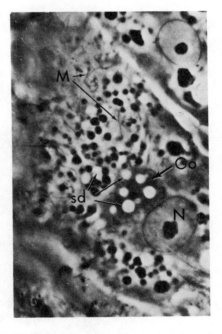

Figure 1.19 Embryonic chick osteoblasts in tissue culture, showing the juxtanuclear positioning of the phase-gray poorly defined Golgi complex (Go). Emanating from the Golgi complex are phase-white secretory droplets (sd). M, mitochondria; N, nucleus. (From Rose, 1970, courtesy of Academic Press)

Usually, the ends of these curved stacks of flattened sacs are dilated and adjacent to them are small round vesicles of varying sizes (40–80 nm). The flattened sacs (cisternae) appear as dense parallel membranes only if the cell is sectioned in the proper direction; otherwise, they appear as amorphous and intercommunicating vesicles. The interior of the cisternae may appear

16 A CURRENT INTERPRETATION OF THE ANATOMY OF THE MAMMALIAN CELL

to be quite clear, whereas in cells in which lipoprotein particles accumulate, there are varying degrees of density. Direct communications with the endoplasmic reticulum and the nuclear envelope have been observed.

When viewed by phase-contrast microscopy, the Golgi complex in living cells has been recorded as a juxtanuclear phase-gray body, often with phase-white droplets emanating from it (*Figure 1.19*; Rose, 1961a, b). It now appears certain that the general role of the Golgi complex is one of secretion and of carbohydrate synthesis. It is generally believed that the Golgi complex receives and packages secretory products by surrounding material from the endoplasmic reticulum with a limiting membrane.

The Golgi complex also gives a positive reaction with the periodic acid–Schiff stain for carbohydrate and it has been suggested that this carbohydrate material may migrate to the cell surface by way of the small vesicles and there contribute to the 'fuzzy' cell coat.

One of the chief products of the Golgi complex is the primary lysosome. Both lysosomes and certain vacuoles of the Golgi complex give the acid phosphatase reaction, and it is for this basic reason that the two have been equated.

1.2.2.4   LYSOSOMES

Lysosomes are round or oval small bodies (0.2–0.8 μm) which are enclosed by a single unit membrane (*Figure 1.20*). They contain large amounts of acid phosphatase and other hydrolytic enzymes. Although multi-enzyme particles may exist, it is possible that the hydrolytic enzymes are carried in specific lysosomes so that there is a great array of different lysosomes within a given cell. The numerous granules of polymorphonuclear leukocytes and the crystal-containing granules of eosinophils are notable examples

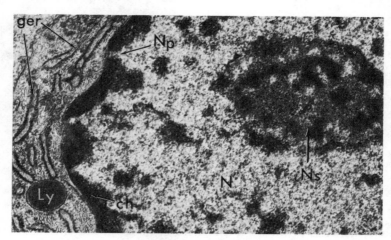

Figure 1.20   Portion of rat chondrocyte, showing a segment of the nucleus (N), the nucleolus (Ns), elements of the granular endoplasmic reticulum (ger), and a primary lysosome encircled in a unit membrane. Several nuclear pores (Np) are also seen passing between the dense accumulations of chromatin (ch) in the periphery of the nucleus. Ly, lysosome. Electron micrograph. ×40 000. (Courtesy of A. Sugimoto)

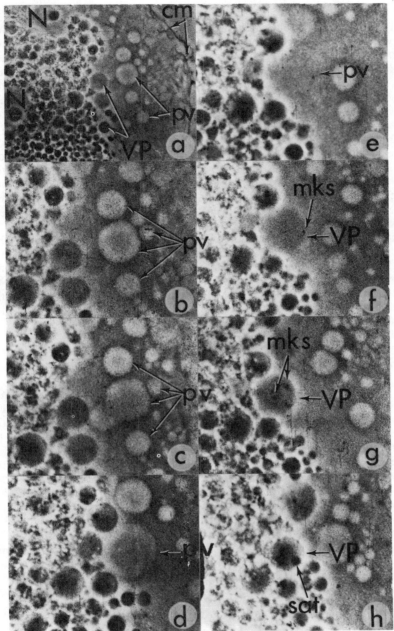

Figure 1.21 Series of micrographs extracted from a time-lapse motion picture film of a HeLa cell undergoing pinocytosis. In the low-power view (a), the limiting cell membrane (cm) and inrushing pinocytotic vesicles (pv) are shown. Their transformation into darker and rounder cytoplasmic spheres (VP) closer to the nucleus (N) is also shown. In the succeeding micrographs, a transformation and coalescing of several pinocytotic vesicles into a larger vesicle, which has its phase display shifted (light to dark) after encirclement by rapidly moving microkinetospheres (mks, lysosomes), is shown. sat, satellite body. Phase contrast. (a), ×940; (b)–(h), ×3000. (From Rose, 1970, courtesy of Academic Press)

of lysosomes. Lysosomes vary greatly in size, form and density. Basically, they are divided into primary and secondary types. Enzymes are not activated until the phagocytosed particles or pinocytosed droplets become associated with the lysosomes. The enzymes of the primary lysosome are synthesized by the ribosomes, accumulated in the endoplasmic reticulum, and then penetrate into the Golgi region where they are packaged and released into the cytoplasm. When the primary lysosome fuses with a pinosome or a phagosome, they become heteropinosomes or heterophagosomes.

In tissue culture, pinosomes surrounded by several lysosomes (microkinetospheres) which then fuse to form a satellite body on the surface of the pinosome have been observed (*Figure 1.21*; Rose, 1957). Ultimately, the pinosomes shrink as the enzymes commingle with the fluid droplets. The residual bodies observed under phase microscopy and in time-lapse cinemicrography eventually (1–3 days) become lost in a tightly compacted cluster of granules beside the nucleus.

In some EM sections of pathologic tissues, in completely digested residual bodies, phagosomes and pinosomes with attached lysosomes, may be found in the cytoplasm. On other occasions, autophagic vacuoles or lysosomes surrounding cell organelles (cytolysosomes) may be found in those parts of the cell involved in lysosomal digestive processes. These are frequently seen during starvation in liver cells and are the mechanism for autodegradation of a cell's constituents.

1.2.2.5   THE NUCLEAR ENVELOPE

The nucleus is surrounded by a double layering of unit membranes which are separated by the perinuclear space, 10–30 nm wide (*Figure 1.17*). This double-layered membrane is penetrated by 'pores' which, in mammalia, occupy about 10 percent of the surface area (*Figures 1.17* and *1.20*). It has been estimated that there are 35–65 pores per square micrometer of surface (Franke, 1966). The pores are approximately 60 nm in diameter and contain an electron-dense material in their centers. Considerable evidence has shown that macromolecules can pass from the nucleus through these pore regions, although it is not concluded that they are freely communicating orifices. The external surface of the outer membrane is usually studded with ribosomes. On occasion it can be seen merging with membranes of the GER and, in rare instances, with the membranes of mitochondria, the Golgi complex and the plasmalemma. Just inside, and apposed to, the inner membrane in some cells (notably melanocytes, some fibroblasts, and endothelial cells) is an electron-dense amorphous layer, the fibrous lamina (*Figure 1.17*; Fawcett, 1966).

1.2.3   **Special organelles, granules and inclusions**

1.2.3.1   ANNULATE LAMELLAE

In the cytoplasm of embryonic, neoplastic, germinal and other cells which are generally undergoing rapid proliferation, there are multiple layers of double unit membrane structures similar to the GER but with periodic interruptions, pores or annuli, which are identical with those of the nuclear envelope (*Figure 1.16*). These stacks of membranes are called the *annulate lamellae* and are seen around the nucleus in human oocytes where they may be continuous with the outer layer of the nuclear envelope. The unit membranes forming these lamellae are 7–9 nm in thickness and enclose a space about 30–50 nm across. Where the pairs of membranes of each lamella join, they form a pore. The diameter of the pore varies between 40 and 60 nm and the spaces between pores are about 100–200 nm. In some sections, as in the nuclear envelope, there are structures termed *annular diaphragms* which appear to bridge the pore. These appear to be membranes of greater density than either of the two membranes composing the nonannulate portion of the lamellae. The presence or absence of the diaphragm is explained in terms of the thickness of the sections and appears to be due to an inclusion of a portion of the wall or rim of the annulus in thicker sections. There is, without doubt, a homogeneous matrix of considerable electron density which forms a collar inside the pores and which extends a short distance from the pores. The annulate lamellae are also generally found to be continuous with the GER but do not themselves contain ribosomes except in the regions of continuity with it.

1.2.3.2   PEROXISOMES (MICROBODIES)

Earlier, these granules were thought to be lysosomes, but more recently they have been found to be rich in peroxidase, catalase, D-amino acid oxidase and to a lesser extent urate oxidase; they correspond morphologically to the microbodies found in kidney and liver cells. They are ovoid granules, less than a micrometer in diameter, are limited by a single unit membrane, and contain a fine granular substance which may condense and form an opaque or homogeneous crystalline core (*Figure 1.22*). Peroxisomes apparently originate in the endoplasmic reticulum of the cell, as do the lysosomes, and are packaged and released from the membranes of the Golgi complex. They are also present in macrophages and possibly in polymorphonuclear heterophils as well. The multivesicular bodies are thought to be related either to lysosomes or peroxisomes, but their direct relationship is unknown at this time.

1.2.3.3   GRANULES AND INCLUSIONS

Space will not permit a description of all the known granules and inclusions. Those cited, however, are representative of the great variety of types found throughout mammalian cells which are surrounded by unit membranes.

(a)   *Melanin*. Melanin granules are produced by melanocytes, but they are also found in Langerhans cells, melanophages and keratinocytes. The granule is ovoid or rod-shaped and varies from 0.4 to 1.0 μm in length and from 0.1 to 0.5 μm in width (*Figure 1.23*). In EM sections, the melanin granules

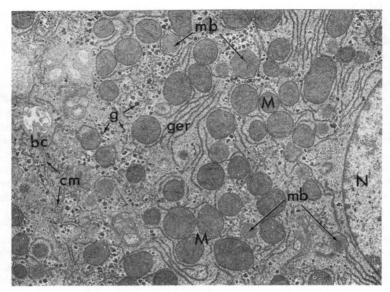

Figure 1.22 Adult rat liver, showing microbodies (mb) bound by a single membrane, elements of the granular endoplasmic reticulum (ger), spotty accumulations of glycogen (g), and granule-shaped mitochondria (M). The two opposing cell membranes (cm) to the left are sectioned through a dilation, the bile canaliculus (bc). N, nucleus. Electron micrograph. ×18 000. (Courtesy of M. Kumegawa)

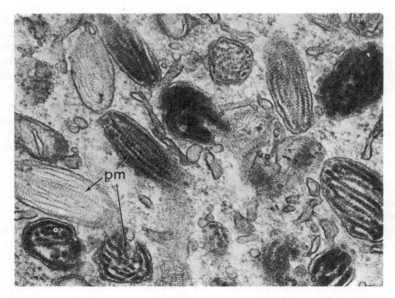

Figure 1.23 High-power view of premelanosomes (pm) in a human melanoma tissue culture cell line. The longitudinal arrangement of the helical internal structure and the variations of pigment deposited on the helices are seen in these granules which are bound to unit membranes. Electron micrograph. ×89 000. (Courtesy of G. Maul)

are enclosed in a unit membrane and have a specific internal structure which is seen as a series of parallel and folded or helicoid fibrils which run the length of the ovoid body. A variety of stages may be seen, commencing with ovoid bodies apparently devoid of this material and progressing through successive stages of increased fibrillar density to the last stage, in which there is a solid dense body without discernible fibrillar structure. The premelanosome is the early stage of this organelle and contains a protein matrix, on which melanin may or may not be deposited; the premelanosome, which contains increasing amounts of deposited melanin, ultimately becomes the mature melanosome. In the oral mucosa and integument, the melanocyte apparently injects the melanosomes into the basal keratinocytes where they accumulate in membrane-bound melanosomal complexes (*Figure 1.15*). An EM section through the integument or pigmented oral epithelium shows the greatest number of melanosomal complexes in the basal cells and a smaller number in the spinous cells.

(b)   *Lipofuchsin*. In cells of older animals, golden-brown pigment granules which fluoresce under UV light appear. They are regarded as residue bodies of insoluble degenerated organelles and may be derivatives of lysosomes. They are commonly seen in the cells of the myocardium while those of neurones have a tendency to grow myelin figures in their interior.

(c)   *Langerhans granules*. Like melanocytes, Langerhans cells (*Figure 1.24*) are dendritic cells of the integument and oral mucous membranes. Neither contains desmosomes, but both are found among the keratinocytes with long dendritic processes which reach across many cells. The Langerhans granule is characterized by a tennis racket-shape when sectioned in one plane. The handle of this racket-shaped granule has a central longitudinal line and cross striations (*Figure 1.24*, insert) about 9 nm apart. Most sections show only the profiles of the handle portion of the tennis racket-shaped granule. The dilated end of the granule is thought to be a Golgi vesicle from which the granule buds off. Other images show the rod portion of the granule continuous with the plasmalemma, so that the content of the granule appears to be ejected into the intercellular space. The size of the granule is variable, but at times may be 2 µm in length. Its function is thought to involve the extracellular space, as is that of the similarly formed membrane-coating granules (*see below*) of the keratinocytes. Eosinophilic granulomas (one form of histiocytosis X) have been observed in which histocytes are packed with typical Langerhans granules (Morales *et al.*, 1969). Numerous other studies of histiocytosis X have presented convincing evidence that the Langerhans granules are related to macrophagic activity and the Langerhans cell is in fact an epithelial macrophage.

(d)   *Secretory granules*. Cells have a great many varieties of secretory activities and some involve the discharge of large secretory granules from the cell. The production of collagen by fibroblasts, chondroblasts and osteoblasts occurs after synthesis in the GER with a concomitant dilation

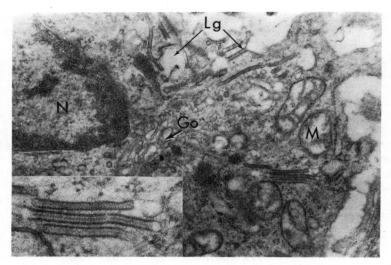

Figure 1.24 Langerhans cell in the epithelium of human gingiva, showing the special type of Golgi-derived Langerhans granule (Lg). Their characteristic fine structure reveals them as rod-shaped bodies with rounded ends and striated lamellae running down their centers. M, mitochondrion; N, nucleus; Go, Golgi complex. Electron micrograph. ×42 000 (insert ×112 000). (Courtesy of H. Nikai)

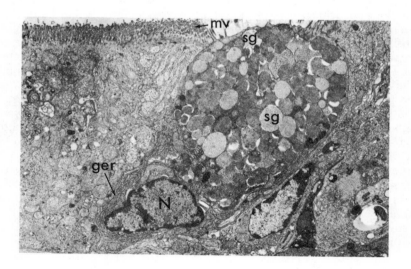

Figure 1.25 Goblet cell in adult rat intestine, showing numerous accumulations of secretory granules (sg) at various stages within the cytoplasm. The lumen of the intestine is at the top of the micrograph, where the brush border of parallel units of microvilli (mv) may be seen. N, nucleus; ger, granular endoplasmic reticulum. Electron micrograph. ×8000. (Courtesy of A. Sugimoto)

of the cisternae (*Figure 1.17*), but its transfer to the intercellular space does not appear to involve the formation of inclusion bodies. On the other hand, the cells of exocrine and endocrine glands do form membrane-bound inclusions in their secretory processes (*Figure 1.25*). Generally, these protein-rich secretions are synthesized on the ribosomes of the GER; the products are then directed through the Golgi complex where enclosing unit membranes are acquired and, in the case of mucopolysaccharide and glycoprotein secretions, the carbohydrate component is probably also synthesized by the Golgi complex. The zymogen granules of acinar cells are often so electron-dense that the limiting membrane may be undetectable; although it is always visible in low-density granules. Generally, the membrane of the zymogen granule of exocrine glands fuses with the plasmalemma in such a way that there is no break in membrane continuity during the discharge of the granule. On the other hand, endocrine gland cells have granules which are not so clearly polarized toward a luminal surface as are those of exocrine gland cells. Granules of the endocrine gland cells are often very dense and are surrounded by a clearly defined unit membrane which is not tightly bound to the dense proteinaceous core of the granule, i.e. there is a space between the density of the granule and the limiting membrane; these granules are also found emerging from the Golgi complex. Examples of these endocrine cells are the argentaffin cells of the gastrointestinal tract and the chromaffin cells of the adrenal medulla. Similar, but distinguishable from chromaffin granules, are lipid droplets surrounded by the membranes of the SER, like those in intestinal epithelial cells which have been fixed during absorption of fat. However, lipid bodies in cells are typically nonmembranous.

(e) *Membrane-coating granules* (MCG). MCG are submicroscopic cytoplasmic granules in the granular and spinous cells of the epidermis (Selby, 1957) and oral mucosa (Nikai and Takarada, 1966). They probably originate in the Golgi complex and move toward the cell periphery during the late stages of cell differentiation. There they fuse with the plasmalemma and empty their contents into the intercellular space. Generally, MCG are ovoid

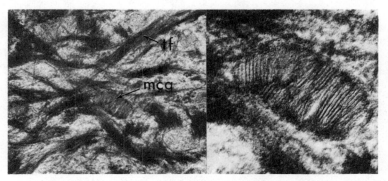

Figure 1.26 *Portion of upper epithelial cell of human gingiva, showing a small membrane-coating granule (mcg) surrounded by a dense array of tonofibrils (tf). The fine structure of this lamellated cytoplasmic granule, which is bound to the unit membrane, is shown to the right. Electron micrograph.* ×60 000; ×210 000. *(Courtesy of H. Takarada)*

and vary between 0.1 and 0.5 µm (*Figure 1.26*). They are enclosed in a unit membrane, the outer layer of which is about 3 nm thick and the inner layer somewhat thicker. There are alternately dense and less dense lamellae embedded in an amorphous matrix. MCG have been shown to accept a label with the acid phosphatase reaction product (Silverman and Kearns, 1970; Luzardo-Baptista and Garcia-Tamayo, 1971) and therefore bear some relationship to lysosomes. Similarly, Hashimoto (1971) found that the MCG were amenable to phospholipase C digestion. Since it is believed that phospholipids are part of the cement in the intercellular spaces, the role of the MCG as a cementing substance is suspected not only on the morphological basis of its evagination and ejection of its content into the intercellular space but on a biochemical basis as well.

(f)   *Myelin figures.* Stacks of unit membranes in whorling forms are occasionally found in the cytoplasm (*Figure 1.17*). Since they resemble the myelin sheaths of nerves, they are termed *myelin figures*. No functions have been ascribed to them, although they are most likely expressions of degeneration. They are often found as parts of secondary lysosomes, where they have been termed *residual bodies*. In the inflamed gingiva, they may be seen within and attached to mitochondria (Takarada *et al.*, 1974b), and these are interpreted as degenerative changes.

(g)   *Mast cell granules.* The granules of mast cells are limited by a single unit membrane. Their internal structures display considerable variability

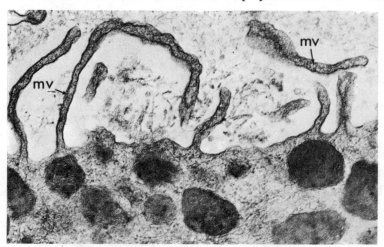

Figure 1.27   The periphery of a human mast cell in human gingiva, showing the special form of mast cell granules of crystal-like appearance and bound to the unit membrane. Microvilli (mv) are seen projecting into the connective tissue. Electron micrograph. ×65 000. (Courtesy of H. Nikai)

and may be observed as lamellar whorls of coarse subunits or crystalline-like forms (*Figure 1.27*). Pharmacological agents inducing a release of histamine in the tissues also cause a degranulation of mast cells. Besides histamine, mast cells also contain the potent anticoagulant, heparin.

1.3 NONMEMBRANOUS COMPONENTS

1.3.1 General

1.3.1.1 FILAMENTS

The cytoplasmic matrix of most cells contains fine proteinaceous filaments 4–5 nm in thickness and of varying lengths up to several micrometers (*Figure 1.13*). They are most common in stratified squamous epithelia, where their banded forms are recognized as the tonofibrils associated with desmosomes (*Figures 1.6, 1.7* and *1.26*) and hemidesmosomes (*Figure 1.14*). It is not known whether the chemical natures of filaments are the same among the various forms of cells, but it is likely that they have a common structural protein and that they function in reinforcing the cytoskeleton. Two special filaments are seen in muscle, the thicker myosin filaments (10 nm in diameter) and the thinner actin filaments (6 nm). Smooth muscle also contains actin filaments, and it is now thought that actin is also responsible for the ruffling of the plasmalemma in pinocytosis, phagocytosis and locomotion. Actin filaments are also attached to the inside of the tips of microvilli.

1.3.1.2 MICROTUBULES

Microtubules are widely occurring cytoplasmic constituents having an outside diameter of 20–27 nm and a wall thickness of 5–7 nm (*Figure 1.28*).

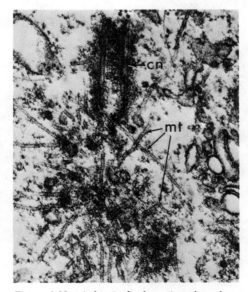

Figure 1.28 A longitudinal section through a centriole (cn) and its associated microtubules (mt) in a lymphoblastoid cell of a human hematopoietic cell line. Electron micrograph. ×65 000. (From Moore, Kitamura and Toshimo, 1968, courtesy of the American Cancer Society)

They appear to be hollow. The walls of microtubules are composed of approximately 13 filamentous subunits arranged in a helix and with a center-to-center spacing of 5.5–6.0 nm. In most cells microtubules have a random orientation, but in dividing cells they form the spindle apparatus, the caudal sheath of spermatids, the marginal band around nucleated erythrocytes, and they appear in longitudinal array in the axoplasm of neurones. They are intimately related to the structure of the centriole. Like the filaments, their function is most likely related to the integrity of the cytoskeleton. Their relation to cilia indicates that they are definitely involved in a motor apparatus function as well. Since the microtubules are involved in the vibratile cell processes, it is conceivable that microtubules of the cell's cytoplasm may have an involvement in the movements of the cytoplasm as well. They are also thought to be important in intracellular transport and give direction to organelle movement.

1.3.1.3 CENTRIOLES

A pair of minute granules may be seen in the juxtanuclear region of living cells. They are sometimes called the diplosome in histological sections, but in electron microscopy they are usually known as centrioles (*Figures 1.18* and *1.28*). Normally, they reside in the cell center (centrosome). In glandular

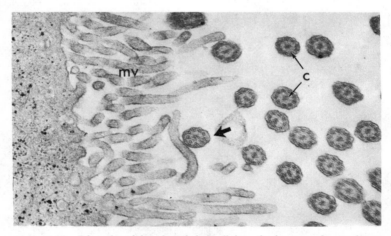

Figure 1.29 Edge of epididymal epithelial cell from fetal mouse, showing longitudinal sectioning of the microvilli (mv) and transverse sectioning of the cilia (c). The microfibril arrangement in the cilia is seen as nine peripherally oriented doublets and one center pair. Electron micrograph. × 56 670. (From Rose, 1970, courtesy of Academic Press)

epithelial cells, the centrosome is often partially surrounded by the Golgi complex at the apical pole of the nucleus, but in other cells, centrioles may be situated immediately beneath the free surface of the cell (*Figure 1.15*). They are self-replicating organelles, and they reproduce themselves in the early phases of mitosis. In the differentiation of ciliated epithelial cells, their reduplication gives rise to the formation of the basal bodies of the cilia which serve as their kinetic center. They apparently function in cyclosis during cell division and in ciliary activity.

Centrioles are tiny cylinders of varying lengths (0.25–2.0 µm) and diameters (0.1–0.2 µm). Primarily, their walls are composed of 27 microtubules equally spaced as nine triplets embedded in a dense amorphous matrix (*Figure 1.18*). Their centers are of a low-density homogeneous cytoplasmic material which may contain a few small dense granules. However, there is a 7.5 nm helical filament which hugs the inside surface of the wall. One end of the centriole appears to be closed, while the other is open. In some cells various ill-defined satellite structures, 50–70 nm in diameter, are spaced around two or more centrioles. The paired centrioles are arranged with their long axes perpendicular to one another. Transverse sectioning of the longitudinally oriented triplet fibers gives a pinwheel characteristic to the sectioned centriole. The dimensions of the nine tubule triplets are approximately the same as for microtubules. In ciliated cells, the base of the cilium is termed the basal corpuscle or basal centriole, as it has the same structure as a centriole. Transverse sections of cilia differ slightly from that of the centriole in having two central microfibrils and with the groups of outer fibrils as doublets (subfibrils A and B) rather than triplets (*Figure 1.29*). Subfibril A has a darker interior than that of subfibril B and has short arms that project toward subfibril B of the next doublet. Fibrous rootlets, which may be striated, extend from the basal body to the cytoplasm. In multiciliated cells, the centrioles, which give rise to the cilia, are duplicated many times. Usually more microtubules are observed around the basal bodies than in other parts of the cell's cytoplasm.

1.3.1.4   RIBOSOMES

With the EM, ribosomes are seen as flattened or spheroidal granules, or as star-shaped bodies with four to six arms implanted on a dense axis (*Figures 1.13, 1.14* and *1.17*). They occur primarily in the cytoplasm and measure approximately 15×25 nm. They are the synthesizers of protein and consist of ribosomal RNA and protein, and give cells their basic staining qualities. They contain 85 percent of the cellular RNA. The ribosome is composed of two major components of differing size, which may be visualized by negative staining. Where ribosomes are attached to the membranes of the endoplasmic reticulum, the larger unit makes the attachment. The protein synthesis taking place in the ribosomes apparently does not require a cytoplasmic membrane. Rapidly proliferating cells synthesizing protein for growth do not necessarily have a well-developed GER, but they do have a high concentration of free ribosomes in the cytoplasm. Similarly, cells producing tonofilaments, such as the integumentary keratinocytes or those of the gingiva, have very little need for the membranes of the GER, as their protein product is largely used within the cell. It seems that a close association of the ribosomes with the membranes of the GER is necessary only when the product being synthesized is going to be exported as a cell secretion.

1.3.1.5 LIPID

Lipid droplets commonly found in cells are triglycerides of fatty acids. They are largest in the adipose cells. Many cell types normally contain a few small droplets or may develop them under certain physiological conditions. They are a local energy store for lipid-containing structural components, such as the cellular membranes. With osmium fixation, the droplets of triglycerides are blackened by the reduction of the osmium and they are insoluble in alcohol and other agents used in dehydration (*Figure 1.17*). The lipid accumulates in the cytoplasmic matrix but is not limited by a unit membrane. Although a more intense reduction of osmium at the interface between the lipid and the surrounding cytoplasm often leads to the presumption that there is a limiting membrane, careful inspection will fail to reveal the trilaminar structure of lipoprotein membranes. An exception is the surrounding of lipid droplets by unit membranes in the absorptive intestinal epithelial cells. During the absorption of fat, the droplets appear in large numbers and rather uniform size within the lumina of the agranular and granular endoplasmic reticula and the Golgi complex, and also as bizarrely shaped lipid masses lying free in the cytoplasmic matrix.

1.3.1.6 GLYCOGEN

Glycogen is found in a great many cells besides the major carbohydrate storage cells of the liver (*Figure 1.22*) and skeletal muscle. In EM specimens it is usually in the form of roughly isodiametric particles measuring 15–30 nm in diameter, with slightly irregular outlines. Single particles are referred to as β-particles, while aggregates of larger size are commonly referred to as α-particles or rosettes, and may reach a diameter of 0.1 μm. The reason for the occurrence of either type of particle in any given cell is unknown, nor is it known whether their chemical or metabolic properties differ. The glycogen particles are larger than ribosomes but they are not easily differentiated where they are closely commingled.

1.3.1.7 CHROMATIN

For ultramicroscopy, the nuclear chromatin is best observed after fixation with glutaraldehyde and a postfixation with osmium tetroxide. The chromatin is then in the form of coarser and more discrete particles than when it is observed in the nucleoplasm after osmium fixation alone. Further staining with uranyl acetate and lead citrate reveals chromatin as a deeply stained and boldly contrasted, dense granular substance against the relatively low density of the nuclear matrix. The pattern of distribution of chromatin in EM sections appears to be quite similar to that seen with the light microscope. The chromatin is often found distributed around the nuclear periphery (*Figure 1.20*), where it is interrupted by nuclear pores (*Figure 1.17*). This dense granular pattern may be seen about the nucleolus as well. The term *chromatin* implies DNA but, since free DNA does not occur, chromatin includes protein in combination with it. In the electron microscopy of cells in interphase, those portions of chromosomes that remain deeply stained and condensed are referred to as *heterochromatin* and are thought to be metabolically inert, whereas the *euchromatin*, which is a lightly staining dispersed form, is thought to be the genetically active form.

1.3.1.8 NUCLEOLUS

The nucleolus is the rounded body, or bodies, in the nucleus and is generally basophilic and eccentrically placed; it has been shown to be rich in RNA. In the living cell (*Figures 1.10* and *1.19*), it is a sharply demarcated refractile body in the nucleoplasm under bright field microscopy but is black under phase contrast. It may be single or multiple and may vary widely in size. The nucleolar apparatus, or complex, is the nucleolus proper (RNA) and the associated bodies of heterochromatin (DNA), which together appear to comprise the nucleolus (*Figure 1.20*) but which special staining has shown to be separate entities. With the light microscope, the components of the nucleolus proper in some cells are observable as (a) the pars amorpha, a structureless component, and (b) the nucleolonema, a thread-like element generally tangled around the pars amorpha. With electron microscopy, the reality of the nucleolonema can be observed as a coarse dense branching strand which forms a network about the compact, spherical pars amorpha, which is fine granular material of a somewhat lower density. The nucleolonema consists of an extremely fine-textured matrix with dense granules (diameter 15 nm) embedded within it. They are similar to ribosomes of the cytoplasm and have been considered to be the ribonucleoprotein component of the nucleolus.

1.3.2 Special

1.3.2.1 KERATOHYALIN GRANULES

With the light microscope, keratohyalin granules are observed as refractile cytoplasmic bodies occurring in keratinizing epithelium. With the polarizing

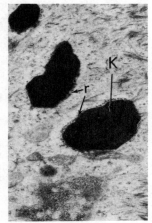

Figure 1.30 Keratohyalin granules (K) in a human gingival epithelial cell, showing their variable densities, the absence of a detectable unit membrane and their close association with ribosomes (r). Electron micrograph. × 38 000. (Courtesy of H. Nikai)

microscope, most of the granules are isotropic, but some reveal a weak birefringence indicating the presence of an oriented molecular structure. Keratohyalin granules vary from a few nanometers to several micrometers in size and can have many forms and textures: they may be round, ovoid or irregular and amorphous, granular or fibrous. In the lower epithelial strata, the granules are smaller than those in the upper layers, indicating that they do grow in size. They have not been demonstrated to be limited by a unit membrane, and their free surfaces are in direct contact with the cytoplasmic matrix. Bundles of filaments and ribosomal clusters often appear in the immediate vicinity of small keratohyalin granules (*Figure 1.30*). As the granules grow in size, these structures may become included in the content of the granule, resulting in granules which have granular and fibrous structure admixed in the amorphous matrix.

REFERENCES

BRIGGAMAN, R. A., DALLDORF, F. G. and WHEELER, C. E., JR. (1971). *J. Cell Biol.*, **51**:384.
BARNARD, T. and AFZELIUS, B. A. (1972). *Sub-Cell. Biochem.*, **1**:375.
COSTERO, I. and POMERAT, C. M. (1951). *Am. J. Anat.*, **89**:405.
DOUGLAS, W. H. J., RIPLEY, R. C. and ELLIS, R. A. (1970). *J. Cell Biol.*, **44**:211.
FAWCETT, D. W. (1966). *Am. J. Anat.*, **119**:129.
FAWCETT, D. W. and ITO, S. (1958). *J. biophys. biochem. Cytol.*, **4**:135.
FERNÁNDEZ-MORÁN, H. (1963). *Science, N.Y.*, **140**:381.
FRANKE, W. W. (1966). *J. Cell Biol.*, **31**:619.
HASHIMOTO, K. (1971). *Arch. derm. Forsch.*, **240**:349.
LUZARDO-BAPTISTA, M. J. and GARCIA-TAMAYO, J. (1971). *Paradontologie*, No. 2, 49.
MOORE, G. E., KITAMURA, H. and TOSHIMO, S. (1968). *Cancer, N.Y.*, **22**:245.
MORALES, A. R., FINE, G., HORN, R. C., JR. and WATSON, J. H. L. (1969). *Lab. Invest.*, **20**:412.
NASS, M. M. K., NASS, S. and AFZELIUS, B. A. (1965). *Expl Cell Res.*, **37**:516.
NIKAI, H. and TAKARADA, H. (1966). *6th International Congress of Electron Microscopy*, p. 581. Tokyo: Maruzen Company.
NIKAI, H., ROSE, G. G. and CATTONI, M. (1970). *J. dent. Res.*, **49**:1141.
NIKAI, H., ROSE, G. G. and CATTONI, M. (1971). *Archs oral Biol.*, **16**:835.
PALADE, G. E. (1960). *Anat. Rec.*, **136**:254.
PORTER, K. R., CALUDE, A. and FULLMAN, E. F. (1945). *J. exp. Med.*, **81**:233.
RABINOWITZ, M. and SWIFT, H. (1970). *Physiol. Rev.*, **50**:376.
ROSE, G. G. (1957). *J. biophys. biochem. Cytol.*, **3**:697.
ROSE, G. G. (1961a). *J. biophys. biochem. Cytol.*, **9**:463.
ROSE, G. G. (1961b). *Cancer Res.*, **21**:706.
ROSE, G. G. (1966). *Jl R. microsc. Soc.*, **86**:87.
ROSE, G. G. (1970) (Ed.). *Atlas of Vertebrate Cells in Tissue Culture*, New York; Academic Press.
ROSE, G. G., CATTONI, M. and POMERAT, C. M. (1963). *J. dent. Res.*, **42**:38.
ROSE, G. G., KUMEGAWA, M. and CATTONI, M. (1968). *J. Cell Biol.*, **39**:430.
ROSE, G. G. and POMERAT, C. M. (1960). *J. biophys. biochem. Cytol.*, **8**:423.
ROSE, G. G., KUMEGAWA, M., NIKAI, H., BRACHO, M. and CATTONI, M. (1970). *Microvascular Res.*, **2**:24.
SELBY, C. C. (1957). *J. invest. Derm.*, **29**:131.
SILVERMAN, S. and KEARNS, G. (1970). *Archs oral biol.*, **15**:169.
SUSI, F. R., BELT, W. D. and KELLY, J. W. (1961). *J. Cell Biol.*, **34**:686.
TAKARADA, H., CATTONI, M. and ROSE, G. G. (1974a). *J. Periodont.*, **45**:288.
TAKARADA, H., CATTONI, M., SUGIMOTO, A. and ROSE, G. G. (1974b). *J. Periodont.*, **45**:30.
TAKARADA, H., CATTONI, M., SUGIMOTO, A. and ROSE, G. G. (1974c). *J. Periodont.*, **45**:155.

2

The separation and cultivation of mammalian cells

D. M. Robinson
Biology Department, Georgetown University, Washington, DC

2.1 INTRODUCTION

This account provides information for workers unfamiliar with the techniques and methods of cell culture, who may wish to engage in the separation and cultivation of cells from mammalian tissue as a prelude to membrane studies. It is by no means a comprehensive or detailed account, but one which seeks to highlight the more recent advances and the more appropriate methods available, within the context of the present series. With the plasma membrane in particular, it is enormously important to consider the consequences of cell handling. Trypsin treatment alone may alter the chemical and physiological characteristics of the cell surface in profound ways (*vide infra*). Low-temperature storage methods may have similar effects, although to all intents and purposes the cells are unaffected as judged by viability. The approaches included here may help overcome some of these more obvious disadvantages.

Despite this consideration it is undoubtedly true that cells in culture, by providing well defined and homogeneous populations under precise control and without the inherent complexities of the whole animal, have been essential to the more recent advances in understanding the nature of mammalian cell membranes. Future studies, using media which are defined chemically, will undoubtedly serve to improve this understanding.

For full and detailed reviews of the subject of cell and tissue culture, the reader is directed to the works of Willmer (1965a, b, 1966), Rothblat and Cristofalo (1972), and Kruse and Patterson (1973).

2.2 CELL SEPARATION AND FRACTIONATION

2.2.1 Enzymatic treatment

The lysis of the extracellular matrix by degradative enzymes has been used extensively for the release of cells from solid tissue. Generally, exposure to the appropriate enzyme(s) is accompanied by restrained mechanical agitation, such as stirring or aspiration; however, careful application of a sequence of different enzymes may itself be sufficient to produce viable single-cell suspensions.

2.2.1.1 TRYPSIN

The most widely used enzyme for solid tissue dissociation is trypsin. Crystalline trypsin is available and effective, but the commercial preparation of the enzyme most frequently used is a crude pancreatic extract and amounts to a completely undefined product to which the tissue is exposed. The resulting viability of cells released by such an extract may vary with the batch used. The tryptic component catalyzes the hydrolysis of peptide bonds between the carboxy group of arginine or lysine and the amino group of an adjacent amino acid (Desnuelle, 1960) and the release of cells from tissue is assumed to be a consequence of breakage of the protein links in the intercellular matrix. The released cells must be recovered before irreversible injury is sustained; reduction of trypsin-induced damage is usually achieved by provision of a trypsin inhibitor such as serum. Solutions of trypsin are usually prepared in balanced salt solutions at pH 7.2 (37 °C) and tissue is exposed in stirred suspension with removal of released cells into serum-containing medium at 15 to 20 minute intervals. Detailed methods have recently been published for mammalian tissues in general (Shipman, 1973) and kidney in particular (Montes de Oca, 1973). Specific disadvantages of exposure to trypsin are discussed later in this chapter.

2.2.1.2 PRONASE

This enzyme has proved particularly useful for specific problems such as the removal of the zona pellucida from the mammalian egg (Mintz, 1963; Gwatkin, 1964, 1965). The commercial preparation is derived from cultures of *Streptomyces griseus* and contains protease, amidopeptidase, carboxypeptidase and lipase activity (Narahashi, Shibuya and Yanagita, 1968). It is used much in the same fashion as trypsin and in some circumstances produces suspensions of cells from tissue which have less tendency to clump than do those released by trypsin (Gwatkin and Thomson, 1964). For solid tissue dissociation, pronase acts quickly and effectively, although the author has found it to be particularly damaging to some cells, possibly owing to its rather wide range of enzymatic activities and fast action (Kahn, Ashwood Smith and Robinson, 1965). Serum contains no pronase-inhibitory activity, so that released cells must be washed free of the enzyme without delay, for they will continue to be degraded if simply suspended directly in growth medium. A recent, detailed method for the use of pronase has been provided by Gwatkin (1973).

2.2.1.3 COLLAGENASE

As a principal component of the intercellular matrix of many tissues, being essentially extracellular and presumably not extensively bonded to the plasma membrane, collagen seems an obvious target in tissue dissociation where intact, undamaged cells are required to be released. Perhaps more for traditional rather than rational reasons, collagenase has received relatively little attention by comparison with trypsin and pronase. However, heart muscle and thyroid, especially from the mammalian embryo, are very successfully dissociated by this enzyme (Hilfer, 1973). The commercial preparations are relatively pure, but some residual protease (Hilfer and Brown, 1971), polysaccharidase and esterase (Peterkofsky and Diegelmann, 1971; Bernfield, Banerjee and Cohn, 1972) activities are found. Indeed these contaminants may themselves cause tissue dissociation in some circumstances—such as the salivary gland—where little or no collagen may be present (Bernfield, Banerjee and Cohn, 1972). A recent account of the method of Cahn, Coon and Cahn (1967) has been published by Hilfer (1973).

2.2.1.4 SEQUENTIAL ENZYME TREATMENT

Such techniques as this are used where conventional methods like trypsinization, or nonenzymatic treatments, fail to produce a sufficiently high yield of cells from solid tissue. One example is the adult mammary gland of the mouse, where a sequence of collagenase and hyaluronidase for 'structure weakening', followed by pronase and DNase, produces a maximal attainable cell yield (Wiepjes and Prop, 1970).

2.2.2 Nonenzymatic treatment

Cutting, stirring and aspiration of tissues and cells are normal and obvious attributes of most dissociation methods and will receive no further mention here. So far solid tissue has been considered, but blood provides a special situation, where tissue dissociation becomes synonymous with cell fractionation. Most of the appropriate techniques for cell fractionation are included in this section, but for a comprehensive treatment of blood, the reader should consult Cutts (1970).

2.2.2.1 CHELATING AGENTS

Calcium and magnesium play an important role in the attachment of cells to each other and removal of these cations by chelation, using the disodium salt of ethylenediaminetetraacetic acid (EDTA) or ethylene glycol-bis(β-aminoethyl ether)-N,N'-tetraacetic acid (EGTA), is often employed in tissue dissociation, frequently in conjunction with enzymes such as trypsin (Montes de Oca, 1973).

2.2.2.2 VELOCITY SEDIMENTATION

Spherical particles that are suspended in a fluid of lower specific gravity follow Stokes' Law, which may be symbolized

$$v = \frac{2ga^2(d_1 - d_2)}{9n} \qquad (2.1)$$

where v = final velocity, g = acceleration due to gravity, a = radius of the particle, d_1 = specific gravity of the particle, d_2 = specific gravity of the fluid medium and n = coefficient of viscosity of the fluid medium.

This provides the theoretical basis for the observation that cells fall out of suspension in medium under the influence of gravity at a rate which depends upon their size. The method has been found to be extremely sensitive and highly reproducible in the quantitative fractionation of viable cells such as bone marrow cells (Peterson and Evans, 1967), spleen cells (Miller and Phillips, 1969) and immune-competent cells (Miller and Phillips, 1970). The spleen cells, fractionated across a shallow gradient of calf serum, were separated virtually independently of cell shape (Miller and Phillips, 1969). The sensitivity of the method permits separation of an asynchronous population of mammalian cells into fractions synchronized in all phases of the cell cycle (MacDonald and Miller, 1970). Because of the virtual absence of trauma for the cells being separated, velocity sedimentation seems an obvious method of choice when membrane studies are planned for fractionated cells. Recent reviews of the technique are provided by Cutts (1970) and Miller (1973). The use of this technique in combination with standing ultrasonic waves may have special value for blood cell separation (Baker, 1972).

2.2.2.3 CENTRIFUGATION

Centrifugation offers speed of separation of subpopulations of cells, based upon differences in specific gravity, as its main advantage. Methods range from simple centrifugation in isotonic suspending fluids, through equilibration centrifugation on gradients of various types, to the use of special machines such as the elutriator rotor and the Rastgeldi threshold centrifuge.

Differential centrifugation at high speed has been shown to be particularly effective and reproducible, especially when Ficoll gradients are used. It has thus been employed successfully in concentration of the parietal cells of the gastric mucosa (Walder and Lunseth, 1963) and its sensitivity has been demonstrated by its use in the identification of antigen-specific surface receptors on the progenitors of antibody-producing cells (Gorczynski, Miller and Phillips, 1970).

Using a colloidal silica density gradient, Wolff and Pertoft (1972) were able to separate mitotic from interphase HeLa cells by centrifugation. These cells were fully viable and capable of producing synchronous cultures on harvesting.

Techniques not yet fully developed for general mammalian cell fractionation, yet showing promise for the future, involve the elutriator rotor, which has been used for leukocyte separation from malaria-infected blood (McEwen *et al.*, 1971) and the separation of mast cells from mixed populations in rat peritoneal washings (Glick *et al.*, 1971). The great advantages of the elutriator rotor are that no gradients are used, no pellets of cells are produced, the suspending medium is isotonic at all stages, continuous processing is feasible and separations of quantities of cells from the milligram to the gram range are possible. One other special centrifuge technique with promise is the Rastgeldi threshold system, which has been used to select cells of *Chlamydomonas reinhardii* for synchronous culture (Knutsen *et al.*, 1973). The Rastgeldi centrifuge may be used to avoid long isopycnic centrifugations of the type described by Hopkins, Sietz and Schmidt (1970) and the use of gradients. The principle of the machine is described by Rastgeldi (1958).

2.2.2.4   CHEMICALLY DERIVATIZED SURFACES

The principle employed here involves the use of mechanical supports coupled to proteins capable of binding to specific cell surface components. In this way the isolation of antibody-forming cells on columns of derivatized beads has been developed (Wigzell and Andersson, 1969; Truffa-Bachi and Wofsy, 1970; Davie and Paul, 1970). Separation of SV_{40}-transformed 3T3 cells from normal 3T3 cells has been achieved by Sephadex beads derivatized with D-galactose (Chipowsky, Lee and Roseman, 1973). Removal of cells from such supports presents problems, however, since surface structures involved are rarely available as competitive inhibitors of cell binding. An improvement has been the development of derivatized fibers (Edelman, Rutishauser and Millette, 1971; Rutishauser, Millette and Edelman, 1972) where such molecules as lectins or antibodies are covalently bonded to the fibers to give binding specificity for the cells. Adsorbed cells are removed by competitive means, or by mechanical or enzymatic cleavage. The major disadvantage lies in the possible effects of the ligand on the structure or function of the cell. A recent account of the current methods has been published by Edelman (1973).

2.2.2.5   BEHAVIORAL CHARACTERISTICS OF CELLS

It is possible to take advantage of the fact that some cells adhere more readily and tenaciously to a substrate like glass than do others, and to make separations accordingly. Rabinowitz (1964, 1965) showed that the cells which adhere to glass in the presence of fresh serum and which require EDTA for release include: neutrophilic, eosinophilic and basophilic granulocytes; metamyelocytes and monocytes. On the other hand the cells which fail to adhere include: normal and leukemic lymphocytes; lymphosarcoma cells; blasts; promyelocytes and erythrocytes. The procedure, reviewed by Cutts (1970), permits the preparation of lymphocyte-rich suspensions contaminated by 0–2 percent of granulocytes and not subject to trauma.

Use has been made of the phagocytic properties of some leukocytes in separating them from mixed populations, for example by allowing them to ingest iron particles and then immobilizing them in the field of a strong electromagnet (Rous and Beard, 1934). St George, Friedman and Byers (1954) also used iron ingestion to recover reticuloendothelial cells from liver. The problem here is that almost any type of cell will become phagocytic under appropriate conditions *in vitro* (Cameron, 1952).

2.2.2.6 FREEZING AND THAWING

An ingenious technique has been applied to the separation of different populations of lymphocytes, which takes advantage of the differential response of stimulated as opposed to unstimulated cells to various cooling rates on freezing (Knight, Farrant and Morris, 1972). The method allows recovery of a chosen population of cells in a mixture. Cells undergoing blast transformation are recovered at the expense of residual, unstimulated small lymphocytes by a slow cooling rate detrimental to the latter. Alternatively a fast cooling rate can be used to preserve unstimulated cells and destroy those in process of transformation. With further development this technique might well be used to separate B and T lymphocytes.

2.3 CELL CULTIVATION

2.3.1 **Medium requirements**

Surprisingly little is known about the nutritional requirements of mammalian cells. As a consequence, the formulation of nutrient media for growth *in vitro*, to replace the intercellular fluids derived from blood *in vivo*, remains empirical and to a large extent imprecise. A few cell lines have been 'trained' to grow in fully chemically defined synthetic media, but for the most part the addition of serum is essential. A recent review of the role of serum has been provided by Temin, Pierson and Dulak (1972). Apart from providing nutrients, growth media provide a protective environment for the cell. The advent of broad-spectrum antibiotics, especially those with anti-*Mycoplasma* activity, such as kanamycin and chlortetracycline, has allowed cell culture to proceed with neither covert nor overt contamination.

A troublesome attribute of growth media, at least until recently, has been the ubiquitous use of sodium bicarbonate and CO_2 as the buffering system used to maintain a pH of about 7.2 (37 °C). The maintenance of a suitable partial pressure of CO_2 in the gas space of culture vessels required permanent airtight stoppering. The introduction by Good *et al.* (1966) of certain synthetic amino and sulfonic acids as nonvolatile buffers which appear to be well tolerated by most cells has largely overcome this inconvenience. However, it appears that some of these buffers may have effects on cell surfaces produced by mechanisms other than the control of pH (Morris, 1971; Robinson and Meryman, 1972) and care should be exercised in their use in membrane-related studies.

A detailed account of the various growth media is inappropriate here, especially since a complete but succinct account has recently been published by Weymouth (1972).

2.3.2 Primary cultures

Primary cultures are those relatively recently derived from organized tissue, with or without the capacity for serial culture, but, in the former case, having a finite lifetime. No standard procedure exists for the preparation of primary cultures, which may be derived from tissues of diverse age and origin. Three different cell types will be considered here, but for a complete account the reader should consult the book by Kruse and Patterson (1973).

2.3.2.1 HUMAN DIPLOID FIBROBLASTS

Such cells as these are easily derived from a variety of human tissues and when in culture retain chromosomal stability, demonstrate heritable biochemical properties and replicate in a predictable fashion over a period of several months. Human diploid cells have been used extensively in genetic studies and have provided substrates for human virus vaccines. The essential stability and homogeneity of these cells, coupled with the possibility of producing them in large numbers, make them ideal candidates for biochemical and physiological studies of cell membranes. Perhaps the best known such cell line is WI-38, derived from fetal lung (Hayflick and Moorhead, 1961) which has become highly characterized, especially with regard to amino acid requirements (Griffiths, 1970), glycolysis (Cristofalo and Kritchevsky, 1966), enzyme content (Cristofalo, Kabakjian and Kritchevsky, 1971; Wang et al., 1970), lipid content (Howard and Kritchevsky, 1969; Kritchevsky and Howard, 1966), cholesterol biosynthesis (Rothblat, Boyd and Deal, 1971), and population dynamics in culture (Hayflick, 1965). This line, with a useful body of knowledge already available, clearly provides excellent material for the study of membranes of 'normal' human cells under carefully controlled conditions. A recent and detailed review of human diploid cells has been published by Mellman and Cristofalo (1972). A good, simple method for culturing fetal human diploid cells is provided by Hayflick (1973).

2.3.2.2 HUMAN BLOOD LEUKOCYTES

Although leukocyte cultures have been used a great deal in human karyology, they are now becoming of equal interest in the study of the initiation of cell division, especially in the context of immune transformation. However, these cells provide a potentially valuable system for workers interested in cell surface changes accompanying malignant transformation. Several permanent cell lines are derived from blood leukocytes and appear to depend for their permanency upon the presence of a Herpes-like virus (Henle and Henle, 1968; Moore, Gerner and Franklin, 1967). The association of a similar virus particle with cell cultures derived from Burkitt's lymphoma, having similar proliferative capacity (Klein, 1973) may be more than a coincidence. The comparison of plasma membranes from normal and abnormal leukocytes in culture under similar circumstances of replication may therefore be illuminating.

Normal blood leukocyte culture is simple and highly reproducible, exsanguination being followed by stimulation of the white cells to undergo blast transformation and cell division. Plant lectins such as phytohemagglutinin (PHA) or Pokeweed mitogen (PWM) are used in this respect. A recent account of the standard method has been given by Moorhead (1973).

2.3.2.3 MUSCLE CELL CULTURES

The study of developmental biology in mammalian systems has largely been confined to organs and organisms; cellular or subcellular studies are comparatively few in number. This may be because differentiation by mammalian cells in culture is not a common phenomenon. An exception to this generalization appears to be muscle, at least in the hands of such investigators as Konigsberg (1965), Hauschka (1968), Holtzer (1970) and Fischman (1970). It should be made clear that primary muscle cell cultures are not homogeneous in the sense of fibroblasts or leukocytes, indeed most myoblasts are accompanied by fibroblasts in culture (Konigsberg, 1963). However, cloning techniques have been used (Konigsberg, 1963; Hauschka and Konigsberg, 1966) with considerable success to demonstrate that single myoblasts may give rise to differentiated muscle colonies. It is with this point in view that the writer suggests muscle cell cultures as suitable systems for the study of the role of membranes in development, albeit early development.

Dissociation of muscle tissue and preparation of primary cultures follow conventional lines, but in one respect in particular, muscle cells appear to be sensitive: the relationship to the substrate. In most cases, for example chick embryo myoblasts, a collagen substrate is required for colony differentiation (Konigsberg and Hauschka, 1965; Hauschka and Konigsberg, 1966). Gelatin (denatured collagen) may substitute for collagen but other possible substrates such as glycyl-L-hydroxyproline, poly(L-Pro–Gly-L-Pro), copolymer (Gly-L-Pro) and various polyamines and proteins, do not work (Hauschka, 1972). A comprehensive review of muscle cell culture with a full discussion of the substrate problem has been published by Hauschka (1972).

2.3.3 Permanent cultures

These comprise cell lines having an indefinite lifetime, capable of continuous serial propagation and usually characterized by heteroploidy and oncogenicity. Several methods of culture will be described here, all of which have been used successfully with permanent cultures, but some of which may be applicable to primary cultures.

2.3.3.1 ANCHORAGE-DEPENDENT CELLS

Anchorage-dependent cells (Stoker et al., 1968) require a substrate on which to grow and cannot usually be grown in suspension. Normally such cells are grown attached to the surface of glass or plastic dishes or flasks, overlaid by medium, but this imposes severe limitations on the economical production of large numbers of cells. Accordingly, several devices which enable mass culture to be achieved more easily have been produced, and some of these will be described.

One of the commonest improvements on static culture flasks is provided by the roller bottle system. Here, cylindrical flasks are turned slowly on their longitudinal axes and a small quantity of medium thereby washes over the whole of the inner surface. Cells grow over the whole of this surface. The addition of perfusion systems to roller bottles has considerably enhanced the process (Kruse, Keen and Whittle, 1970).

Large-scale production of monolayers of cells in a relatively small space has been made possible by the multiplate propagator, which is a glass or steel container within which are stacked glass or titanium discs, supported about 6.4 mm apart. The vessel can be perfused with growth medium and cells grow over the whole of the available surface area. A small multi-plate propagator of about 10 liters' capacity can provide a surface area for cell growth equivalent to that provided by more than 200 stationary 1 liter bottles (Schleicher, 1973). A variation on the same theme involves packed spiral films of plastic sheet within a glass container (House, Shearer and Maroudas, 1972).

Such attempts to approach the economy in space and medium afforded by suspension culture have been achieved in an ingenious fashion by the use of microcarriers. Microcarrier culture provides for the growth of anchorage-dependent cells on the surface of small spherical particles which may themselves be kept in suspension by stirring. This system offers possibilities like those of conventional suspension culture for precise measurement and control of such environmental variables as pH, CO_2 and O_2 tension, temperature, nutrient supply and waste removal. A major advantage of microcarrier culture is the opportunity afforded to grow primary as well as permanent cultures in this way. Porous silica beads have been used as microcarriers (Bizzini, Chermann and Raynaud, 1971), but are generally too heavy to be kept in suspension at low stirring speeds. By far the best available material is DEAE-Sephadex A50, a negatively charged, nontoxic, smooth, spherical particle of about 60–280 µm in diameter when wet. Pretreatment of the beads is fairly time-consuming, but treated and sterilized beads may be stored for long periods at 4 °C. The system offers an elegant method for the large-scale production of cells which will not multiply in free suspension. A good account has been given by Van Wezel (1973).

Perhaps the most exciting recent innovation in cell culture is the artificial capillary system first described by Knazeck *et al.* (1972). Here, commercially available hollow fibers (semipermeable cellulose acetate membranes fashioned into capillaries) are collected to form a capillary bed within a small glass tube. Cells grow within the tube on the outside surface of the capillaries, while medium is perfused through the bore of each of the capillaries. Silicone polycarbonate fibers, permeable to oxygen, are also included in the bed so as to avoid anoxia when necessary. Cells grown in such devices allow simulation of growth *in vivo* more closely than has ever been possible before. For example, the influence of one cell upon another reaches a maximum at cell densities approaching 10^8 cells/ml. Such concentrations are impossible to achieve by monolayer or suspension culture yet are actually achieved in artificial capillaries, where cells grow into solid tissue-like masses. It may well be possible to follow the cell surface changes occurring as a function of cell density in this type of culture, and humoral relationships between different cell types kept in separate units within a common perfusion pathway may clearly be investigated. A recent account of artificial capillary systems has been given by Knazeck and Gullino (1973).

2.3.3.2   CELLS IN SUSPENSION

Mammalian cells which grow in free suspension may be handled to all intents and purposes as bacteria and may therefore be grown in bulk as batch or continuous-flow cultures. Batch cultures range from the simple stirred glass vessel to highly sophisticated contrivances such as the spin filter chamber of Himmelfarb, Thayer and Martin (1969). Continuous-flow cultures demand equally complex machinery in the form of chemostats with precise control of environmental variables (Pirt and Callow, 1964). Suspension cultures provide cells in a highly reproducible metabolic state and cell harvesting is a simple matter of centrifugation. Large quantities of cells, by comparison with anchorage-dependent cultures, may be produced from small volumes of growth medium in small containers. It is the case, however, that cells capable of growth in suspension are 'transformed' or 'malignant' by nature, often directly derived from tumors *in vivo*. With this reservation, and as knowledge of the nutrient requirements of mammalian cells increases, suspension cultures are likely to become especially important, particularly since the technique opens the way to 'steady-state' conditions which should enable the relationship of the cell and its membranes to the environment to be explored fully. For further reading, the following are selected from a voluminous literature: Moore and Ulrich (1965); Telling and Elsworth (1965); McLimans *et al.* (1966), and Peraino, Bacchetti and Eisler (1970).

2.3.4   Harvesting cells from culture

The ease of harvesting cells from suspension culture by simple centrifugation has been mentioned above. However, the situation with anchorage-dependent cells is considerably different; removal of cells from surfaces usually requires enzyme treatment, often in combination with chelation. Although much the same circumstances apply here as were described above for the enzymatic dissociation of solid tissues, a good deal more is known about the damaging effects of these treatments on cells harvested from culture. This section will therefore emphasize the dangers inherent in harvesting anchorage-dependent cells which might be required for membrane, especially plasma membrane, studies.

Even the briefest possible exposure to trypsin (several minutes), necessary to harvest cells from monolayers, can remove as much as 30 percent of the plasma membrane glycoproteins (Buck, Glick and Warren, 1970), and EDTA has been shown to have a similar effect (Glick, 1974). Trypsin has also been shown to remove acetylcholinesterase from cells (Hall and Kelley, 1971). The surface of Chinese hamster cells harvested by 0.25 percent trypsin is damaged to such an extent that the cells can no longer take up [^{3}H]uridine, although it is interesting that this damage is repaired after several hours following attachment of the cell to a substrate (Robinson, 1973). This damage is repaired much more quickly when the cells are placed in 'conditioned' medium taken from old cultures of similar cells. Harvesting cells by pronase can be even more disastrous (Kahn, Ashwood Smith and Robinson, 1965). Although removal of cells may be achieved by such mechanical methods as shaking or scraping, even this is not without hazard, as shown by Lutz (1973), who has demonstrated the loss of selected pieces of cell surface by homogenization procedures.

Altogether the process of harvesting cells from culture may have effects likely seriously to influence subsequent investigations, particularly investigations of the plasma membrane.

2.3.5 Single cell isolation and cloning

In any study where the homogeneity of cell cultures is of concern, colonies derived from single cells may be required. Several techniques are available for the isolation of single cells, but the subsequent growth of colonies from them often requires unusual attention to medium components, since single cells in relatively large volumes of medium cannot readily 'condition' their surroundings in the same way that a high density culture can.

A review of earlier techniques by Sanford *et al.* (1961) includes an account of the highly reliable capillary method. Here cells are dispersed by trypsin or mechanical means (interestingly Sanford recommends avoiding the use of trypsin because of its damaging effects on cells), then taken up in a capillary tube to be viewed microscopically. Sections of tube in which single cells may be seen are broken off and placed in flasks containing growth medium. Cells derived from the single progenitor eventually spill out from the tube and colonize the flask. A similar technique developed by Macpherson (1973) involves aspiration of a single cell from a suspension into a fine pipette, by means of which the cell is transferred to its own cup in a multiple-chamber plastic tray. Here it can be fed and observed by inverted microscope. Cells developing in the cup are subsequently transferred to a conventional culture vessel for propagation. A simpler but more equivocal method involves inoculation of a very dilute suspension of cells into a culture vessel and leaving them undisturbed to produce colonies which are then mechanically removed (Puck, Marcus and Ciecura, 1956). Several techniques of cloning have been discussed by Kruse and Patterson (1973).

2.3.6 Cell preservation

By use of penetrating cryoprotective agents (CPA) and suitable rates of cooling and warming it is possible to recover 100 percent of most mammalian cells frozen in suspension from storage in liquid nitrogen at $-196\,°C$. Storage at temperatures above $-130\,°C$ for prolonged periods of time may produce loss of some cells.

Generally glycerol or dimethylsulfoxide (DMSO) are used as CPA; the former penetrates cells only slowly and is therefore best introduced in a stepwise or gradual fashion so as to avoid osmotic stress, while the latter penetrates quickly but has toxic properties which may not supervene if cooling is initiated as soon as possible after exposure to the DMSO. The possibility exists that sublethal damage, which has been demonstrated in unprotected cells on freezing (Robinson, 1972, 1973), may be sustained by cells apparently surviving freeze-preservation using CPA. Such damage is likely to affect the plasma membrane in particular, so that a recovery period in culture would be a desirable safeguard if membrane studies are planned for frozen stocks of cells. The theory of cryoprotection and cell survival has been covered well by Mazur (1970) and a suitable account of practical methods for the freeze-preservation of mammalian cells is included by Shannon and Macy (1973).

REFERENCES

BAKER, N. V. (1972). *Nature, Lond.*, **239**:398.
BERNFIELD, M. R., BANERJEE, S. D. and COHN, R. H. (1972). *J. Cell Biol.*, **52**:647.
BIZZINI, B., CHERMANN, C. J. and RAYNAUD, M. (1971). Dutch Pat. NL-7010119Q, Fr. Pat. 2053565.
BUCK, C. A., GLICK, M. C. and WARREN, L. A. (1970). *Biochemistry*, **9**:4567.
CAHN, R. D., COON, H. G. and CAHN, M. B. (1967). *Methods in Developmental Biology*, p. 493. Ed. F. H. WILT and N. K. WESSELLS. New York; T. Y. Crowell.
CAMERON, G. C. (1952). *Pathology of the Cell*, p. 110. Edinburgh; Oliver & Boyd.
CHIPOWSKY, S., LEE, Y. C. and ROSEMAN, S. (1973). *Proc. natn. Acad. Sci. U.S.A.*, **70**:2309.
CRISTOFALO, V. J. and KRITCHEVSKY, D. (1966). *J. cell comp. Physiol.*, **67**:125.
CRISTOFALO, V. J., KABAKJIAN, J. R. and KRITCHEVSKY, D. (1967). *Proc. Soc. exp. Biol. Med.*, **126**:649.
CUTTS, J. H. (1970). *Cell Separation Methods in Hematology*, p. 228. New York; Academic Press.
DAVIE, J. M. and PAUL, W. E. (1970). *Cell. Immun.*, **1**:404.
DESNUELLE, P. (1960). *The Enzymes*, Vol. 4, p. 119. Ed. P. D. BOYER, H. LARDY and K. MYRBACK. New York; Academic Press.
EDELMAN, G. M. (1973). *Tissue Culture, Methods and Applications*, p. 29. Ed. P. F. KRUSE and M. K. PATTERSON. New York; Academic Press.
EDELMAN, G. M., RUTISHAUSER, U. and MILLETTE, C. F. (1971). *Proc. natn. Acad. Sci. U.S.A.*, **68**:2153.
FISCHMAN, D. A. (1970). *Curr. Topics dev. Biol.*, **5**:235
GLICK, D., VON REDLICH, D., JUHOS, E. T. and MCEWEN, C. R. (1971). *Expl Cell Res.*, **65**:23.
GLICK, M. C. (1974). *Biology and Chemistry of Eukaryotic Cell Surfaces*, p. 213. Ed. E. Y. C. LEE and E. E. SMITH. New York; Academic Press.
GOOD, N. E., WINGET, G. D., WINTER, W., CONNOLLY, T. N., IZAWA, S. and SINGH, R. M. M. (1966). *Biochemistry*, **5**:467.
GORCZYNSKI, R. M., MILLER, R. G. and PHILLIPS, R. A. (1970). *Immunology*, **19**:817.
GRIFFITHS, J. B. (1970). *J. Cell Sci.*, **6**:739.
GWATKIN, R. B. L. (1964). *J. Reprod. Fert.*, **6**:325.
GWATKIN, R. B. L. (1965). *J. Reprod. Fert.*, **7**:99.
GWATKIN, R. B. L. (1973). *Tissue Culture, Methods and Applications*, p. 3. Ed. P. F. KRUSE and M. K. PATTERSON. New York; Academic Press.
GWATKIN, R. B. L. and THOMSON, J. L. (1964). *Nature, Lond.*, **201**:1242.
HALL, Z. W. and KELLEY, R. B. (1971). *Nature, New Biol.*, **232**:62.
HAUSCHKA, S. D. (1968). *Results and Problems in Muscle Differentiation*. Vol. 1, p. 37. Ed. H. URSPRUNG. Berlin; Springer-Verlag.
HAUSCHKA, S. D. (1972). *Growth, Nutrition and Metabolism of Cells in Culture*, Vol. 1, p. 67. Ed. G. M. ROTHBLAT and V. J. CRISTOFALO. New York; Academic Press.
HAUSCHKA, S. D. and KONIGSBERG, I. R. (1966). *Proc. natn. Acad. Sci. U.S.A.*, **55**:119.
HAYFLICK, L. (1965). *Expl Cell Res.*, **37**:614.

HAYFLICK, L. (1973). *Tissue Culture, Methods and Applications*, p. 43. Ed. P. F. KRUSE and M. K. PATTERSON. New York; Academic Press.
HAYFLICK, L. and MOORHEAD, P. S. (1961). *Expl Cell Res.*, **25**:585.
HENLE, W. and HENLE, G. (1968). *Perspectives in Virology*, Vol. 4. Ed. M. POLLARD. New York; Academic Press.
HILFER, S. R. (1973). *Tissue Culture, Methods and Applications*, p. 16. Ed. P. F. KRUSE and M. K. PATTERSON. New York; Academic Press.
HILFER, S. R. and BROWN, J. M. (1971). *Expl Cell Res.*, **65**:246.
HIMMELFARB, P., THAYER, P. S. and MARTIN, H. E. (1969). *Science, N.Y.*, **164**:555.
HOLTZER, H. (1970). *Cell Differentiation*, p. 476. Ed. O. SCHJEIDE and J. DE VELLIS. Princeton, N.J.; Van Nostrand–Reinhold.
HOPKINS, H. A., SITZ, T. O. and SCHMIDT, R. R. (1970). *J. Cell Physiol.*, **76**:231.
HOUSE, W., SHEARER, M. and MAROUDAS, N. G. (1972). *Expl Cell Res.*, **71**:293.
HOWARD, B. V. and KRITCHEVSKY, D. (1969). *Biochim. biophys. Acta*, **187**:293.
KAHN, J., ASHWOOD SMITH, M. J. and ROBINSON, D. M. (1965). *Expl Cell Res.*, **40**:445.
KLEIN, G. (1973). *Natn. Cancer Inst. Monogr.*, **37**:110.
KNAZECK, R. A. and GULLINO, P. M. (1973). *Tissue Culture, Methods and Applications*, p. 321. Ed. P. F. KRUSE and M. K. PATTERSON. New York; Academic Press.
KNAZECK, R. A., GULLINO, P. M., KOHLER, P. O. and DEDRICK, R. L. (1972). *Science, N.Y.*, **178**:65.
KNIGHT, S. C., FARRANT, J. and MORRIS, G. J. (1972). *Nature, New Biol.*, **239**:88.
KNUTSEN, G., LIEN, T., SCHREINER, O. and VAAGE, R. (1973). *Expl Cell Res.*, **81**:26.
KONIGSBERG, I. R. (1963). *Science, N.Y.*, **140**:1273.
KONIGSBERG, I. R. (1965). *Organogenesis*, p. 337. Ed. R. C. DEHAAN and H. URSPRUNG. New York; Holt.
KONIGSBERG, I. R. and HAUSCHKA, S. D. (1965). *Reproduction: Molecular, Subcellular and Cellular*, p. 243. Ed. M. LOCKE. New York; Academic Press.
KRITCHEVSKY, D. and HOWARD, B. V. (1966). *Annls Med. exp. Biol. Fenn.*, **44**:343.
KRUSE, P. F., KEEN, L. N. and WHITTLE, W. L. (1970). *In Vitro*, **6**:75.
KRUSE, P. and PATTERSON, M. K. (1973) (Eds). *Tissue Culture, Methods and Applications*, p. 868. New York; Academic Press.
LUTZ, F. (1973). *Comp. biochem. Physiol.*, **45B**:805.
MACDONALD, H. R. and MILLER, R. G. (1970). *Biophys. J.*, **10**:834.
MACPHERSON, I. (1973). *Tissue Culture, Methods and Applications*, p. 241. Ed. P. KRUSE and M. K. PATTERSON. New York; Academic Press.
MAZUR, P. (1970). *Science, N.Y.*, **168**:939.
MCEWEN, C. R., JUHOS, E. T., STALLARD, R. W., SCHNELL, J. V., SIDDIQUI, W. A. and GEIMAN, Q. M. (1971). *J. Parasit.*, **57**:887.
MCLIMANS, W. F., MOUNT, D. T., BOGDITCH, S., CROUSE, E. J., HARRIS, G. and MOORE, G. E. (1966). *Ann. N.Y. Acad. Sci.*, **139**:190.
MELLMAN, W. J. and CRISTOFALO, V. J. (1972). *Growth, Nutrition and Metabolism of Cells in Culture*, Vol. 1, p. 327. Ed. G. H. ROTHBLAT and V. J. CRISTOFALO. New York; Academic Press.
MILLER, R. G. (1973). *New Techniques in Biophysics and Cell Biology*, p. 87. Ed. R. H. PAIN and B. J. SMITH. London; John Wiley.
MILLER, R. G. and PHILLIPS, R. A. (1969). *J. Cell Physiol.*, **73**:191.
MILLER, R. G. and PHILLIPS, R. A. (1970). *Proc. Soc. exp. Biol. Med.*, **135**:63.
MINTZ, B. (1963). *Science, N.Y.*, **138**:594.
MONTES DE OCA, H. (1973). *Tissue Culture, Methods and Applications*, p. 8. Ed. P. F. KRUSE and M. K. PATTERSON. New York; Academic Press.
MOORE, G. E. and ULRICH, K. (1965). *J. surg. Res.*, **5**:270.
MOORE, G. E., GERNER, R. F. and FRANKLIN, H. A. (1967). *J. Am. med. Ass.*, **199**:519.
MOORHEAD, P. S. (1973). *Tissue Culture, Methods and Applications*, p. 56. Ed. P. F. KRUSE and M. K. PATTERSON. New York; Academic Press.
MORRIS, J. E. (1971). *Biochem. biophys. Res. Commun.*, **43**:1436.
NARAHASHI, Y., SHIBUYA, K. and YANAGITA, M. (1968). *J. Biochem., Tokyo*, **64**:427.
PERAINO, C., BACCHETTI, S. and EISLER, W. J. (1970). *Science, N.Y.*, **169**:647.
PETERKOFSKY, B. and DIEGELMANN, R. (1971). *Biochemistry*, **10**:988.
PETERSON, E. A. and EVANS, W. H. (1967). *Nature, Lond.*, **214**:824.
PIRT, S. J. and CALLOW, D. S. (1964). *Expl Cell Res.*, **33**:413.
PUCK, T. T., MARCUS, P. I. and CIECURA, S. J. (1956). *J. exp. Med.*, **103**:273.
RABINOWITZ, Y. (1964). *Blood*, **23**:811.

RABINOWITZ, Y. (1965). *Blood,* **26**:100.
RASTGELDI, S. (1958). *Acta physiol. scand.,* **44**:Suppl. 152.
ROBINSON, D. M. (1972). *Molecular and Cellular Repair Processes,* p. 104. Ed. R. F. BEERS, R. M. HERRIOT and R. CARMICHAEL TILGHMAN. Baltimore; Johns Hopkins University Press.
ROBINSON, D. M. (1973). *Cryobiology,* **10**:413.
ROBINSON, D. M. and MERYMAN, H. T. (1972). *In Vitro,* **7**:251.
ROTHBLAT, G. H., BOYD, R. and DEAL, C. (1971). *Expl Cell Res.,* **67**:436.
ROTHBLAT, G. H. and CRISTOFALO, V. J. (1972) (Eds). *Growth, Nutrition and Metabolism of Cells in Culture,* Vol. 1, p. 471, Vol. 2, p. 445. New York; Academic Press.
ROUS, P. and BEARD, J. W. (1934). *J. exp. Med.,* **59**:577.
RUTISHAUSER, U., MILLETTE, C. F. and EDELMAN, G. M. (1972). *Proc. natn. Acad. Sci, U.S.A.,* **69**:1596.
SANFORD, K. K., COVALESKY, A. B., DUPREF, L. T. and EARLE, W. R. (1961). *Expl Cell Res.,* **23**:261.
SCHLEICHER, J. B. (1973). *Tissue Culture, Methods and Applications,* p. 333. Ed. P. KRUSE and M. K. PATTERSON. New York; Academic Press.
SHANNON, J. E. and MACY, M. L. (1973). *Tissue Culture, Methods and Applications,* p. 712. Ed. P. KRUSE and M. K. PATTERSON. New York; Academic Press.
SHIPMAN, C. (1973). *Tissue Culture, Methods and Applications,* p. 5. Ed. P. KRUSE and M. K. PATTERSON. New York; Academic Press.
ST GEORGE, S., FRIEDMAN, M. and BYERS, S. O. (1954). *Science, N.Y.,* **120**:463.
STOKER, M., O'NEILL, C., BERRYMAN, S. and WAXMAN, V. (1968). *Int. J. Cancer,* **3**:683.
TELLING, R. C. and ELLSWORTH, R. (1965). *Biotechnol. Bioengng,* **7**:417.
TEMIN, H. M., PIERSON, R. W. and DULAK, N. C. (1972). *Growth, Nutrition and Metabolism of Cells in Culture,* Vol. 1., p. 50. Ed. G. H. ROTHBLAT and V. J. CRISTOFALO. New York, Academic Press.
TRUFFA-BACHI, P. and WOFSY, L. (1970). *Proc. natn. Acad. Sci. U.S.A.,* **66**:685.
VAN WEZEL, A. L. (1973). *Tissue Culture, Methods and Applications,* p. 372. Ed. P. F. KRUSE and M. K. PATTERSON. New York; Academic Press.
WALDER, A. I. and LUNSETH, J. B. (1963). *Proc. Soc. exp. Biol. Med.,* **112**:494.
WANG, K. M., ROSE, N. R., BARTHOLOMEW, E. A., BALZER, M., BERDE, K. and FOLDVARY, M. (1970). *Expl Cell Res.,* **61**:357.
WEYMOUTH, C. (1972). *Growth, Nutrition and Metabolism of Cells in Culture,* Vol. 1, p. 11. Ed. G. H. ROTHBLAT and V. J. CRISTOFALO. New York; Academic Press.
WIEPJES, G. J. and PROP, F. J. A. (1970). *Expl Cell Res.,* **61**:451.
WIGZELL, H. and ANDERSSON, B. (1969). *J. exp. Med.,* **129**:23.
WILLMER, E. N. (1965a) (Ed.). *Cells and Tissues in Culture; Methods, Biology and Physiology,* Vol. 1, p. 788. New York; Academic Press.
WILLMER, E. N. (1965b) (Ed.). *Cells and Tissues in Culture; Methods, Biology and Physiology,* Vol. 2, p. 809. New York; Academic Press.
WILLMER, E. N. (1966) (Ed.). *Cells and Tissues in Culture; Methods, Biology and Physiology,* Vol. 3, p. 826. New York; Academic Press.
WOLFF, D. A. and PERTOFT, H. (1972). *J. Cell Biol.,* **55**:579.

3
Isolation of surface membranes from mammalian cells

Mary Catherine Glick
Department of Pediatrics, University of Pennsylvania School of Medicine, Philadelphia

3.1 INTRODUCTION

With the study of surface membranes from mammalian cells new concepts have proliferated on cell interactions, differentiation, hormonal activity, oncogenic transformations and mitosis. At the time of writing of this manuscript many of these concepts are hazy and need continued definition but it is to be hoped that, as methods for the isolation of membranes improve and become more uniform, the problems will be unraveled.

This chapter is not meant as a laboratory protocol but as an attempt to evaluate various concepts rather than to criticize the individual methods, and to try to bring some perspective to the subject. For the most part, the literature to be reviewed will be from 1971 to 1974. Previous reviews have discussed the methods available prior to 1969 (*see* Steck and Wallach, 1970; Warren and Glick, 1971). Reviews on the isolation of plasma membranes (*see* DePierre and Karnovsky, 1973) and plasma membrane fragments (*see* Hinton, 1972) through 1971 have appeared. The latter review includes several tables of the methods used for liver and other cells, categorizing the preparations used previously. A negative approach to the problems of membrane isolation was assumed by DePierre and Karnovsky (1973) and again methods have been tabulated for the various tissues. Several other reviews on membranes discuss isolation procedures (*see* Oseroff, Robbins and Burger, 1973; Steck and Fox, 1972; Steck, 1972; Wallach and Lin, 1973). It is, therefore, not necessary to reiterate these points but an attempt will be made to summarize the more recent methods and discuss in detail those variations which are novel modifications or represent new approaches. A few methods which were reviewed previously will be discussed here as they are directly concerned with most of the more recent methods.

The methods of preparing surface membranes will be described on the basis of the final product. That is, some of the methods result in large fragments or whole membranes and can thus be identified by examination in the phase contrast microscope. These membranes are identified as surface membranes of the cell since without question they represent the outer layer —albeit the fractions can contain contaminants. Other methods result in small fragments or the vesiculation of the membranes and will be discussed separately as these methods require for identification the use of markers such as enzymes or membrane components. Membranes obtained by vesiculation are not completely identified as surface membranes and will be referred to as *plasma membranes*. Of course, the methods using morphological identification can be further substantiated by the use of enzyme or chemical markers and, in like manner, the vesicles which are isolated can be viewed in the electron microscope. Since some of the internal membranes are similar to surface membranes—in glycopeptide composition for example (Keshgegian and Glick, 1973)—one should exercise caution when the cell is broken and small membrane fragments or vesicles are isolated.

The procedures used to separate the membranes from the other cell contents will be described separately. Many variations of these separation procedures have been reported, and in practice it is necessary to work out methods for individual cells.

3.2 PREPARATION

3.2.1 Whole membranes

3.2.1.1 ENUCLEATED CELLS

The isolation of whole membranes from mammalian cells was first accomplished for erythrocytes. As early as 1952, Ponder was able to obtain red cell stroma, or ghosts as they are often called, in extremely high yields. The methods involve exposing the cell to hypotonic solutions so that the internal proteins pass out of the cell without apparent rupture of the stroma.

(a) *Hypotonic lysis.* The method which is most often used to obtain erythrocyte stroma is that of Dodge, Mitchell and Hanahan, 1963. Modifications of the method have been reviewed (Steck and Wallach, 1970; Warren and Glick, 1971). The erythrocytes are treated with 15 volumes of 20 milliosmolar sodium phosphate buffer, pH 7.4. The stroma are washed until they are free of hemoglobin. The absence of hemoglobin is considered the criterion of purity of these preparations, in addition to their appearance in the electron microscope.

Ghosts obtained by this procedure appear as shown in *Figure 3.1*. Using negative staining, the visual results depend upon the negative stain employed (McMillan and Luftig, 1973). A similar finding is true when the membranes are examined in thin section. In the latter case, the fixative determines the appearance of the final product in the electron microscope. In fact, 25–40 percent of the membrane protein from the red cell stroma was removed by

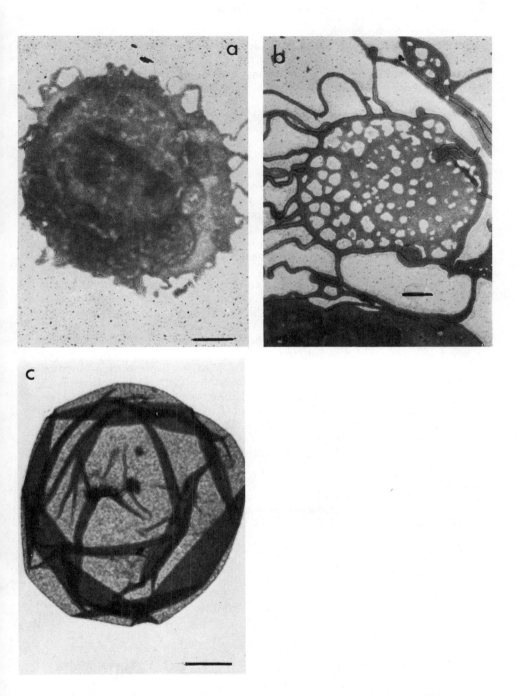

Figure 3.1 Electron micrographs of erythrocyte ghosts obtained after treatment with three negative stains: (a) 1.0% silicotungstate, (b) 1.0% phosphotungstate, and (c) 2.0% uranyl acetate. (a) ×11100, (b) ×6400, (c) ×11100. Scale represents 1.0 μm. (From McMillan and Luftig, 1973, reprinted with permission)

the ethanol washes employed for dehydration of the specimen when fixation with osmium was used, leaving not much to visualize apart from the phospholipid bilayer (McMillan and Luftig, 1973; Reynolds, 1973). Thus, if morphological criteria in the electron microscope are used to define the preparations, it is important to use conditions that are least deleterious to the membranes.

3.2.1.2 NUCLEATED CELLS

The isolation of whole surface membranes from nucleated mammalian cells was first described by Warren, Glick and Nass (1966). Among the advantages of having whole membranes is that they can be visualized in the phase contrast microscope so that there is no doubt that the external surface still remains at the exterior. In addition, separation of the membranes from the cytoplasmic debris using gradients of sucrose solutions, for example, is facilitated by observation in the phase contrast microscope. The final purified product can be counted in a hemocytometer and all chemical and enzymatic results, as well as yields, can be quoted on a per membrane basis and related back to the whole cell. There are a number of modifications of the original methods but essentially the basic reagents are as first described. Two of these methods, zinc ion and Tris procedures, are most widely used and will be discussed individually. The methods can be applied to any cells which are separated without membrane damage. When one is dealing with whole organs or tissue masses, it is very hard to disperse the individual cells without membrane damage, therefore these methods have been used, for the most part, for tissue culture cells either in suspension or monolayer. Contrary to comments on the procedures (Rothfield and Finkelstein, 1968; Atkinson and Summers, 1971), these methods allow the demonstration of enzyme activities. If the presence of $ZnCl_2$ inactivates any enzymes it can be removed with EDTA. Indeed, neither of these methods appears particularly toxic to the cells.

(a)   *Zinc ion method.* The cells, after having been gently washed and harvested, are swollen under hypotonic conditions in the presence of 1 mM $ZnCl_2$ (Warren and Glick, 1969). The membrane pulls away from the

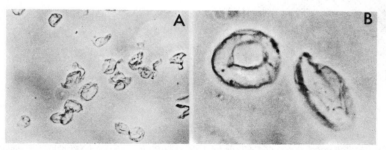

Figure 3.2   Surface membranes isolated from L cells by the zinc ion procedure. Phase contrast. (A) ×350, (B) ×1400. (M. C. Glick, unpublished)

cytoplasm and the nucleus and cytoplasmic contents are expelled through a fracture in the membrane by gentle homogenization in a Dounce homogenizer (Type B, loose). *Figure 3.2* shows the final product as seen in the phase-contrast microscope. The yields of purified membranes vary from 10 to 60 percent, depending on the starting cell type. The method has been used to isolate membranes from a variety of cells (Lerner, Meinke and Goldstein, 1971; Perdue and Sneider, 1970; Brunette and Till, 1971; Glick, Kimhi and Littauer, 1973; Charalampous, Gonatas and Melbourne, 1973; Shin and Carraway, 1973). For the most part the modifications involve the purification procedures or the production of membrane fragments and will be discussed later. This method is generally the easiest available for obtaining whole membranes. The membranes have enzyme activities, suggesting that $ZnCl_2$ does not interfere. However, if $ZnCl_2$ cannot be tolerated in the final preparation it can be removed with EDTA (Perdue and Sneider, 1970).

(b)   *Tris method.* Tris-HCl has been used to isolate membranes from a number of different cell types. The original method reported for use with L cells, mouse fibroblasts (Warren, Glick and Nass, 1966), or Ehrlich ascites cells (Warren and Glick, 1969) described conditions using 0.05 M Tris-HCl over a pH range of 7.2–8.2. The cells are washed and suspended in the Tris buffer with 5 mM $MgCl_2$ and gently broken in a Dounce homogenizer (Type B). The membranes are whole, as with the previous method, and the isolation procedure can be monitored by phase contrast microscopy. The yield of purified membranes is 10–30 percent.

A modification which involves a novel method to break the cells was used to isolate surface membranes from an L-asparaginase-sensitive cell line of leukemic cells (L5178Y/CA55). Hypotonic swelling of saline-washed cells in water for 2 min was followed by the addition of 0.01 M Tris, pH 7.2, 0.02 M KCl and sucrose to 3 percent (Dods, Essner and Barclay, 1972). Homogenization was carried out by use of a 35 M B-D Yale Precision syringe with a No. 21 needle (3.8 cm) at 10 strokes/min for approximately 1 min. This was followed by purification over glycerol gradients.

Table 3.1 summarizes some of the literature since 1970 where Tris buffer has been used as the homogenizing medium for particular cells. Frequently, the use of Tris buffer with little modification is reported as a new procedure (Bosmann, Hagopian and Eylar, 1968; Stonehill and Huppert, 1968; Boone *et al.*, 1969; Atkinson and Summers, 1971; Forte, Forte and Heinz, 1973; Smith and Crittenden, 1973). In the hands of most investigators the method yields not whole membranes but large fragments. Undoubtedly, overvigorous homogenization and somewhat careless handling of the fractions are the causes. In many cases it is not necessary to have whole membranes but the ease of identification and separation of the intact membranes are useful. Sonication can be used to obtain vesicles after the membranes have been purified (Boone *et al.*, 1969).

(c)   *Other homogenizing media to obtain whole membranes.* Rodbell (1967) obtained whole membranes from fat cells of rat epididymal adipose tissue by swelling the cells in hypotonic medium and lysing them with gentle

Table 3.1 USE OF TRIS BUFFER AS HOMOGENIZING MEDIUM SINCE 1970

Cell type	Breakage	Additions	Ref.
Lymphocytes (human)	Potter–Elvehjem	NaCl	Demus (1973)
Lymphocytes (pig)	Minced and stirred	NaCl	Allan and Crumpton (1970)
Macrophages (mouse)	Vortex	$MgCl_2$	Werb and Cohn (1972)
Ehrlich ascites	Dounce	NaCl, $CaCl_2$	Forte, Forte and Heinz (1973)
HeLa	Dounce	$MgCl_2$, NaCl	Atkinson and Summers (1971)
L5178Y/CA55	Syringe	KCl, sucrose	Dods, Essner and Barclay (1972)
Hamster kidney and rat embryo fibroblast	N_2 cavitation	$MgCl_2$, sucrose	Graham (1972)
Fat, white (rat or mouse)	Swinny filter	Sucrose, ATP, EDTA	Lauter, Solyom and Trams (1972)
Fat, white (rat)	Glass homogenizer	Sucrose, EDTA	Czech and Lynn (1973a)
Chick embryo fibroblasts	Dounce	$MgCl_2$	Smith and Crittenden (1973)
Thyroid (bovine)	Polytron and Dounce	Sucrose, EGTA, dithiothreitol	Wolff and Jones (1971)
Liver (rat)	Potter–Elvehjem	Sucrose	Touster et al. (1970)
Liver (rainbow trout)	Braun homogenizer	Sucrose	Lutz (1973)
Renal (rat)	Dounce	Sucrose, EDTA	Marx, Fedach and Aurbach (1972)
Neurones (rat)	Teflon glass	Sucrose, $MgCl_2$	Henn, Hansson and Hamberger (1972)
Neurones (rabbit)	Teflon glass	Sucrose	Karlsson, Hamberger and Henn (1973)
Glial (rabbit)	Teflon glass	Sucrose	Karlsson, Hamberger and Henn (1973)
Intestinal brush border (rat)	Teflon glass	HEPES, D-mannitol	Hopfer et al. (1973)
Kidney cortex (bovine)	Dounce	Sucrose, $CaCl_2$	Sutcliffe et al. (1973)

inversion. These membranes contained 26 percent of the mitochondria and endoplasmic reticulum of the whole cells, an unacceptable amount of contamination. Giacobino and Perrelet (1971) subsequently were able to isolate much cleaner preparations.

Fluorescein mercuric acetate (FMA) has been used to fix the cells before homogenization (Warren, Glick and Nass, 1966). This reagent inactivates enzymes and may in addition extract proteins and glycoproteins from the membrane. The carbohydrate composition of membranes of mouse fibroblasts obtained with FMA is quite different from those obtained with $ZnCl_2$ (*Table 3.2*).

Table 3.2 CARBOHYDRATE CONTENT OF SURFACE MEMBRANES AND WHOLE CELLS

L cell	Sialic acid	Hexosamines	Fucose	Mannose	Galactose
		(μmol $\times$ 10^{-10} per surface membrane or cell)			
Surface membranes					
Zn^{2+}	6.5$\pm$1.4 (73)	25.9$\pm$ 5.5 (44)	2.3$\pm$2.2 (36)	8.5$\pm$ 2.5 (16)	21.6$\pm$2.9 (66)
FMA	3.4$\pm$0.4 (38)	5.0$\pm$ 1.9 (9)	0.7$\pm$0.6 (11)	5.7$\pm$ 1.8 (11)	11.9$\pm$3.3 (36)
Whole cells	8.9$\pm$1.7 (100)	58.3$\pm$17.3 (100)	6.3$\pm$3.8 (100)	52.8$\pm$12 (100)	32.8$\pm$4.6 (100)

The values represent the mean values and the standard deviation for each carbohydrate. The numbers in parenthesis represent the percentage of the total carbohydrate of the cell found in the surface membrane (Glick, Comstock and Warren, 1970).

3.2.2 Large fragments

The first method of obtaining surface membranes, other than from erythrocytes, was devised by Neville (1960) for liver cells and is the most widely used of all methods and, indeed, liver itself is perhaps the target of more investigators than any other tissue. In a review article. DePierre and Karnovsky (1973) cited more than 35 references where different investigators have isolated membranes from liver cells from 1960 to 1971.

(a) *Hypotonic* $NaHCO_3$. In the method of Neville, minced liver is homogenized several times in a Dounce homogenizer with centrifugation and discarding of the supernatant material. The hypotonic homogenizing medium is 1 mM $NaHCO_3$, pH 7.5 (Neville, 1960). Neville and Glossmann (1971) have shown that the membranes recovered from rat liver by this method are bile canaliculi.

Many modifications of this method have been developed; one which is widely used is that of Emmelot *et al.* (1964). These investigators modified the sucrose gradients, making a sharper separation of the membranes from the cellular debris. Ray (1970) improved the yield of liver membranes by adding 0.5 mM $CaCl_2$ to the homogenizing medium, thereby preventing clumping of the nuclei. After extraction of the membranes with 4 percent N-laurylsarcosinate–Tris buffer, pH 7.8, Evans and Gurd (1972) isolated nexuses of the gap junction. Most of the other membrane proteins and enzyme activities were solubilized. Membranes have been prepared from a variety of cells including rat kidney brush border (Wilfong and Neville, 1970), human leukemic lymphocytes (Marique and Hildebrand, 1973), and murine plasmocytoma cells in culture (Lelievre, 1973). In general, most of the modifications involve either the washing procedure or the sucrose gradients, and the final results, including the yield, are again as in the case of the zinc ion or Tris methods, probably more as a consequence of the membrane handling and the particular cell type.

(b) *Hypotonic* $Na_2B_4O_7$. Large fragments have been prepared from liver cells by gentle homogenization in a Dounce homogenizer in the presence of 1 mM $Na_2B_4O_7$ and 0.5 mM $CaCl_2$, pH 7.5 (Dorling and Le Page, 1973). This method appears to yield bile canaliculi as well as most of the hepatocyte surface, as compared with the method of Neville, which yields mainly bile canaliculi (Neville and Glossmann, 1971). *Figure 3.3* shows the appearance of these membranes in the electron microscope. The authors suggest that the borate buffer combines with the surface carbohydrates and by increasing the net negative charge favors repulsion of cell particles, thus preventing clumping of nuclei. However, according to others, $CaCl_2$ prevents aggregation of nuclei (Hunter and Commerford, 1961; Ray, 1970).

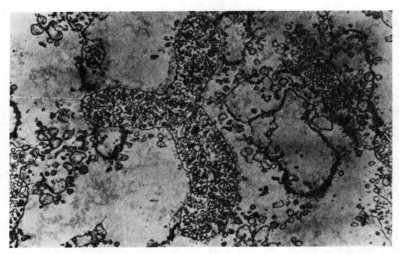

Figure 3.3 An electron micrograph of a liver plasma membrane preparation from rat. In the center of the field there is a large profile of a bile canaliculus containing many microvilli in section. The bile canaliculus is attached to pieces of double membrane at five different places. Other pieces of membrane and microvilli in section are scattered around the profile. × 6000. (From Dorling and Le Page, 1972, courtesy of Associated Scientific Publishers)

(c)   *Zinc ions.* $ZnCl_2$ with Tween-20 has been used to obtain membrane fragments (Warren, Glick and Nass, 1966; Perdue and Sneider, 1970). Omission of the Tween-20 yields membranes which are intact but with a few large fragments. Hemminki (1972, 1973) has obtained fragments from cells of newborn rat brain although the use of dimethylsulfoxide may have denatured the enzymes. Barland and Schroeder (1970) and Scher and Barland (1972) treated fibroblasts with 1 mM $ZnCl_2$ and dimethylsulfoxide while the cells were still attached to the culture dish. Subsequent treatment with saturated fluorescein mercuric acetate in 0.02 M Tris buffer, pH 8.1, and shaking of the cells removed the upper parts of the membranes, which were then decanted and isolated. The bottom parts of the membranes as well as much of the other cell particulate material remained attached to the culture dish. This could be a way to separate sections of the membranes with different functions. The membranes appear as large fragments.

(d)   *Tris.* Renal plasma membranes have been isolated as large fragments by gently homogenizing the cells in the presence of 0.01 M Tris, pH 7.5,

1 mM EDTA and sucrose (Marx, Fedak and Aurbach, 1972). These preparations probably are not brush border membranes but could be derived from the blood front of the renal tubular cells.

Preparing plasma membranes from bovine thyroid cells, Wolff and Jones (1971) found that homogenization in a Polytron homogenizer, in the presence of 3 mM Tris, pH 7.4, 0.25 M sucrose, 3 mM EGTA and 3 mM dithiothreitol, yielded larger fragments which had greater adenyl cyclase activity than membranes prepared with more vigorous homogenization. Recently, Czech and Lynn (1973a) have isolated membranes from fat cells by combining the methods of McKeel and Jarett (1970) and Laudat *et al.* (1972). The method involves homogenizing the fat cells in 5 mM Tris, pH 7.5, with 1 mM EDTA and 0.25 M sucrose in a glass homogenizer with a Teflon pestle.

(e) *Other procedures.* Hypotonic swelling of mouse leukemia cells in water for 6–10 min, followed by homogenization through a syringe in 0.25 M sucrose with 1 mM EDTA, pH 7.0, was used to prepare membrane fragments (Dods, Essner and Barclay, 1972). Glutaraldehyde (8%) was added to the homogenate before purification (Nachman, Ferris and Hirsch, 1971), which was accomplished by differential centrifugation, and centrifugation through gradients of sucrose solutions. The method has been used to isolate membranes from asparaginase-resistant leukemic cells (Kessel and Bosmann, 1970).

3.2.3 Vesicles

The method of Kamat and Wallach (1965) is used by many investigators to prepare membrane vesicles and has been reviewed in detail (Steck and Wallach, 1970). Unfortunately, methods using vesiculation depend largely on the use of enzyme markers, so discrepancies are prevalent in the preparations (Steck, 1972).

(a) *Nitrogen cavitation.* Cells are disrupted in 5 mM Tris buffer, pH 7.4, in 0.25 M sucrose containing 0.2 mM $MgSO_4$ by nitrogen cavitation. The vesicles formed from the surface membranes are separated from the other membranes of the cell by centrifugation on Ficoll gradients in the presence of divalent ions (Kamat and Wallach, 1965). A rather elaborate procedure was worked out by the authors and it appears to separate membrane vesicles with high specific activities of some marker enzymes (Wallach, 1967).

Ferber *et al.* (1972) have used this method to isolate plasma membranes from lymphocytes. The lymphocytes were freed from dead cells and phagocytes by passing them through a nylon column and subsequently suspending in 0.02 M HEPES buffer, pH 7.4, with 0.13 M NaCl and 0.5 mM $MgCl_2$. Sucrose was added after the cells had been suspended in the buffer, and they were then disrupted by nitrogen cavitation. The authors consider this procedure mild. The method has been used with very little modification to isolate vesicles from BHK_{21} cells (Gahmberg and Simons, 1970), hamster kidney and rat embryo fibroblasts (Graham, 1972) and thymocytes (van Blitterswijk, Emmelot and Feltkamp, 1973).

(b) *Glycerol lysis.* Barber, Pepper and Jamieson (1971) compared several methods for preparing membranes from human blood platelets and developed a new technique based on hypotonic lysis after loading the platelets intracellularly with glycerol (Barber and Jamieson, 1970). This method produced membrane vesicles of two different densities but of identical enzymatic composition. At least in the case of platelets, this method seems to separate the membranes with little damage to the intracellular organelles (Kaulen and Gross, 1973).

(c) *Sodium chloride.* Perdue, Kletzien and Miller (1971a) prepared membrane vesicles from fibroblasts harvested by treatment with collagenase (0.1%). The cells were homogenized in 0.16 M NaCl in a Potter–Elvehjem homogenizer. Membranes from other fibroblasts have been prepared by this method (Wickus and Robbins, 1973). The membranes appear in the electron microscope as small vesicles (*Figure 3.4*).

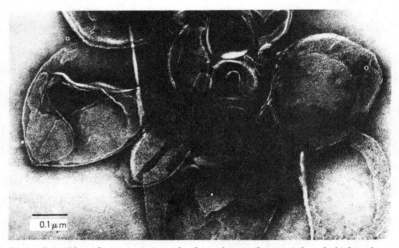

Figure 3.4 Photoelectron micrograph of membranes from uninfected chick embryo fibroblasts which have been stained with phosphotungstic acid, × 110 000. (From Perdue, Kletzien and Miller, 1971a, courtesy of Associated Scientific Publishers)

(d) *Other procedures.* Fibrosarcoma cells were homogenized in 0.25 M sucrose in a Potter–Elvehjem homogenizer to prepare vesicles (Perdue, Warner and Miller, 1973). Plasma membranes have been isolated as small fragments or vesicles from synaptosomes (Levitan, Mushynski and Ramirez, 1972; Breckenridge and Morgan, 1972; Morgan *et al.*, 1971), neurones (Henn, Hansson and Hamberger, 1972), and brush borders of intestinal epithelial cells (Forstner, Sabesin and Isselbacher, 1968; Hopfer *et al.*, 1973) or renal tubule cells (Berger and Sacktor, 1970; Thomas and Kinne, 1972; Chesney, Sacktor and Rowan, 1973).

Erythrocyte stroma have been vesiculated by homogenization in 0.5 mM sodium phosphate buffer, pH 8.0, with or without $MgSO_4$ (Kant and Steck, 1972).

3.3   ISOLATION

After the membranes are removed from the cells, whether intact or as fragments or vesicles, it is usually a problem to obtain purified fractions. Although no real innovations in methods of separating the membranes from the cells have been made in the past eight years, there has been some recent progress in ways of isolating these membranes from the other cell particulates.

3.3.1   Density gradient centrifugation

In the past, isolation has been accomplished by centrifugation over varying numbers of gradients of sucrose, glycerol or Ficoll; these are time-consuming and laborious procedures. In order to obtain membranes with a high degree of purity it is sometimes necessary to use as many as three different sucrose gradients (Warren and Glick, 1969) or manipulate the gradients with varying amounts of cations (Kamat and Wallach, 1965). The advantage of these methods, although cumbersome, is that the procedure is reasonbly reproducible when it is defined for the particular cell type. The use of zonal rotors makes it possible to perform these separation procedures on a relatively large scale (Anderson *et al.*, 1968; Hinton *et al.*, 1970; Evans, 1970; Evans and Gurd, 1973), and has been discussed in detail by Hinton (1972).

The separation of surface membranes into bands of two different densities with slightly different chemical composition is noted by many investigators. Suggestions have been made regarding the arrangement of proteins and glycoproteins within the membrane on the basis of these bands (Wray and Perdue, 1974). Although this may be true for preparations of vesicles, intact surface membranes also give two or three bands which probably reflect the heterogeneity of the starting cells due to nutritional differences or stages of the cell cycle. Functional differences between the two populations have been demonstrated in the case of blood platelets (Barber, Pepper and Jamieson, 1971).

3.3.2   Aqueous two-phase system

Taking advantage of the aqueous two-phase polymer systems developed by Albertsson (1970), Brunette and Till (1971) were able to separate surface membranes of L cells prepared by the zinc ion technique (Warren and Glick, 1969) from the other cell particulate material. A mixture of dextran 500, polyethylene glycol and $ZnCl_2$, buffered at pH 6.5 with 0.22 M phosphate, was used in their original method. The membranes came to the interface of this two-phase system and after several recycles were sedimented out of the suspension. This fraction contained about 6–7 percent of the total cell protein and electron micrographs showed the typical membrane appearance. The method appears to be much less cumbersome than gradient centrifugation.

This method was used in a comparative study of chorioallantoic cells and chick fibroblasts and yielded final products which required further purification on sucrose gradients (Israel, Vergus and Semmel, 1973), although modifications made to the original procedure may have influenced their results.

Using the method of Ray (1970), Lesko *et al.* (1973) modified this two-phase system by omitting $ZnCl_2$ and isolated membranes from rat liver. A comparison of the final products obtained by sucrose density gradients and the polymer system showed the two preparations to be remarkably similar. Enzyme markers were used to attest the presence of surface membranes and lack of other contaminants. The two-phase system yielded 49 percent more membrane material than the sucrose gradients. The authors suggested that for effective purification it is important not to overload the polymer system with excess starting material.

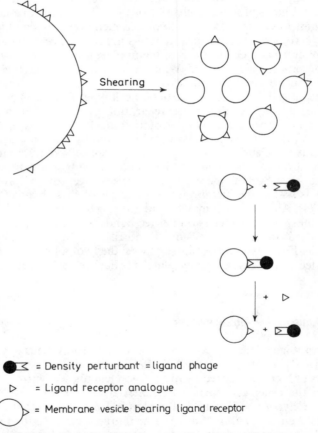

Figure 3.5 Schematic outline of membrane fractionation by affinity density perturbation. A plasma membrane bearing multiple receptors (▷) is sheared into membrane fragments carrying different numbers of receptors in varying distributions. These are reacted with the ligand (Σ) coupled to the density perturbant, i.e. K29 phage (●), producing a membrane–receptor–ligand–phage complex of higher density than the membrane itself and of lower density than the density perturbant. Addition of a low-molecular-weight dissociating agent (▶) returns the membrane and density perturbant to their original densities. (From Wallach et al., 1972, courtesy of the Federation of European Biochemical Societies)

3.3.3 Affinity density perturbation

This approach to the isolation of membrane fragments takes advantage of specific receptors on the membrane surface (Wallach et al., 1972). A specific ligand which combines with the receptor is covalently coupled to a labeled particle of very high density. The membrane fragments which are combined with the density-perturbing particles are then separated by ultracentrifugation. The membrane–ligand complex is dissociated by the addition of reagents that have higher affinity for the ligand than has the receptor. Figure 3.5 illustrates the fractionation procedures as described by Wallach et al. (1972) for lymphocyte membranes using concanavalin A, radioactively labeled with ^{125}I as ligand, which was converted into a density perturbant by coupling to coliphage K29 with glutaraldehyde. An excess of α,α-trehalose dissociated the complex after centrifugation. Under these conditions there was irregularity of membrane distribution, which probably points to heterogeneity in the number, or affinities, of the concanavalin A binding sites. The method does, however, hold promise for separation of specific fragments as any number of ligands, such as antibodies or peptide hormones, can be used.

3.3.4 Phagocytosis of latex beads

Wetzel and Korn (1969) described a method for isolating phagocytotic vacuoles from *Amoeba*. After having ingested latex particles, the amoebae were homogenized and centrifuged through solutions of sucrose to isolate those phagosomes which contained latex particles. Using this technique, Heine and Schnaitman (1971), and Werb and Cohn (1972) isolated phagolysosomal membranes from mouse fibroblasts and rat peripheral macrophages, respectively. The protein and acid hydrolases were leaked from the phagolysosomes by repeated freezing and thawing and diluting with water. The membranes remained on the particles. This procedure affords a method to study specialized fragments of the membrane and has been used to isolate membrane fragments from KB cells (Charalampous, Gonatas and Melbourne, 1973). In this study a comparison was made with membrane fragments isolated by the zinc ion technique (Warren and Glick, 1969). Using many criteria, the authors found the two preparations to be similar, suggesting that the isolated vesicles represent the whole membrane.

3.3.5 Free-flow electrophoresis

Taking advantage of the different electrical charges on each side of the cell membrane of the renal proximal tubule, Heidrich et al. (1972) separated two different fractions of membranes by free flow electrophoresis. Crude brush border fractions prepared by gentle homogenization of the renal cortex and centrifugations were subject to electrophoresis at a rate of 20 mg of protein per hour. Although the process requires specialized apparatus it affords separation of relatively large amounts of membrane material. The fraction nearest to the cathode contained morphologically intact microvilli (brush border) with a high specific activity of alkaline phosphatase. The second fraction, which moved toward the anode, contained junctional complexes, had a high specific activity of Na^+,K^+-ATPase and represented the plasma membrane from the basal infoldings. Thus, Heidrich *et al.* (1972) were able to separate the two sides of the membrane and also show that the alkaline phosphatase and Na^+,K^+-ATPase are on different sides of the cell. The presence of EDTA in the homogenizing medium lowered the electrophoretic mobility of the membranes.

3.3.6 Affinity chromatography

When a particular biological activity is sought on the membrane, affinity chromatography can be used to purify this particular area. It is quite possible that with the use of concanavalin A bound to Sepharose, for example, membrane fragments can be separated from all other particulates. There is always the concern, however, that nuclear fragments combine at the same time (Keshgegian and Glick, 1973). Membrane fragments have been passed over glass beads (Warren, Glick and Nass, 1966) so it is possible that they might pass over Sepharose beads. Concanavalin A receptors have been extracted from isolated and solubilized pig lymphocyte membranes (Allan, Auger and Crumpton, 1972) and proteins from erythrocyte stroma (Tarone, Prat and Comoglio, 1973). Insulin and glucagon receptors have been extracted from plasma membranes (Cuatrecasas, 1972; Rodbell *et al.*, 1971).

Although not reported in the literature, it may be possible to separate surface membranes by the fiber fractionation technique described for fractionation of whole cells (Edelman, Rutishauser and Millette, 1971). After separating the remaining whole cells and nuclei by a low-speed gradient, the membranes and smaller particles could be absorbed on the derivatized fibers for separation. The derivative does depend on the availability of a specific marker for the membrane such as antibodies or plant lectins.

3.4 MARKERS

In the present state of our knowledge, there are no universal chemical, biological or enzymatic markers for surface membranes. However, when vesicles or small fragments are prepared, markers have to be used for identification. There appear to be vast differences of opinion as well as misconceptions concerning their use and the discussion to follow will point this out.

3.4.1 Enzymes

Perhaps the most used markers for membranes have been enzymes. The subject has been thoroughly reviewed and the fallacies of their excessive use pertinently pointed out (Solyom and Trams, 1972). The presence, or absence for that matter, of an enzyme on the membrane is hard to establish. Indeed, if one asks whether or not all membranes have the same enzymatic activities, or if any activity is always associated with the same membrane, the answer is very likely *no*.

There is enough documentation at present to allow the following statements: (a) Enzymes characteristic of one membrane are not necessarily characteristic of others (Giacobino and Perrelet, 1971; Thomas and Kinne, 1972; Glossmann and Neville, 1972; Lelievre and Paraf, 1973). (b) Enzymes found on the cell surface under one condition are not there to the same extent under others (Lelievre, Prigent and Paraf, 1971; Graham, 1972; Evans and Gurd, 1973). (c) Artifacts in the method of breaking open the cells and purifying the membranes can add enzymes which are soluble *in vivo* or, conversely, break off small fragments of the membrane with certain enzymatic activities (Wolff and Jones, 1971). (d) The specific activities of the enzymes may be misleading in that they will vary with the amount of protein recovered in the membrane fraction and hence will depend on the purity of the preparation. (e) The activity of an enzyme such as 5′-nucleotidase can be confused with other phosphatase activities. (f) Very few authors report a complete cellular distribution and recovery of the enzymes. (g) The enzyme activity may be destroyed by the reagents used to prepare the membranes (Hemminki, 1973). (h) The presence of isoenzymes in subcellular fractions may complicate the situation (Weinhouse and Ono, 1973). (i) There are examples of species (Lauter, Solyom and Trams, 1972), tissue (Ray, 1970), and sex (Emmelot and Bos, 1971b; Solyom and Lauter, 1973) differences as well as those which occur in subcellular fractionations (Touster *et al.*, 1970; Morgan *et al.*, 1971) and distribution of specific enzymes on different parts of the membrane (Heidrich *et al.*, 1972; Weiser, 1973b). (j) No one enzyme activity has been exclusively proven to be a universal membrane marker.

Adenyl cyclase (Sutherland, Robison and Butcher, 1968), which catalyzes the reaction ATP→cyclic 3′,5′-AMP+PP_i, has been proposed as potentially the best membrane marker enzyme (Solyom and Trams, 1972). However, caution should be imposed since, to date, only a comparatively few well-defined membrane preparations have been assayed for this enzyme and most studies have been performed with liver (Marinetti, Ray and Tomasi, 1969; Emmelot and Bos, 1971a) or fat cells (Rodbell, 1971; McKeel and Jarett, 1970). The enzyme is not enriched in fractions of microvillus membranes from rat kidney (Wilfong and Neville, 1970). Indeed, not only has the enzyme been reported in sarcoplasmic reticulum (Entman, Levey and Epstein, 1969), but also more than one adenyl cyclase may exist (Reik *et al.*, 1970) in either inhibited or activated form (Constantopoulos and Najjar, 1973). A combination of hormone sensitivity and enzyme activity is possible using liver (Rodbell *et al.*, 1971) or kidney-cortex (Sutcliffe *et al.*, 1973) membranes. In assaying for an enzyme, it is important to standardize the assay procedure and the reader is referred to Rodbell (1971) for the adenyl cyclase assays. A competitive inhibitor such as α,β-methylene-adenosine-5′-triphosphate may also be useful (Krug *et al.*, 1973).

5′-Nucleotidase has been used most extensively as a plasma membrane marker for a variety of cells. If the data are examined, great variations in specific activities are seen and, indeed, few membrane preparations show a significant increase in specific activity over the homogenate. It may be a good marker for liver plasma membranes where bile front membranes are predominantly isolated but even here there are species (Lutz, 1973) or, perhaps, other differences. Changes in enzyme activity due to viral transformation (Perdue et al., 1971b) and contact inhibition of cultured cells (Lelievre, Prigent and Paraf, 1971) have been reported. Increased activity is noted in the presence of HEPES rather than Tris buffer (Ferber et al., 1972). Using muscle cells, Schimmel et al. (1973) separated membrane fragments containing increased activity for Na^+,K^+-ATPase but not 5'-nucleotidase.

Similar findings are seen with the use of Na^+,K^+-ATPase as a marker. This enzyme is distributed throughout the cell (unpublished observations) and is enriched only slightly in some membrane preparations (Berman, Gram and Spirtes, 1969). It is not found in the microvillus membranes of renal proximal tubule (Wilfong and Neville, 1970; Heidrich et al., 1972). Hemminki (1973) reported that ATPase is destroyed by $ZnCl_2$ in rat brain membranes, but others have found that Zn^{2+} does not interfere with this activity in cells (Warren, Glick and Nass, 1966). Hayden et al. (1973) found that pH affected the subcellular fractionation of neural tissue as well as the solubilization of acetylcholinesterase from the membrane fractions.

On the other side, it can be argued that the membrane preparations which do not show increased specific activities of enzymes (a) are not pure; (b) have lost activity due to inactivation or physical removal; or (c) show a distribution of the particular enzyme in all the cell fractions representing a fragmentation of the membranes. Indeed, valid but circular arguments lead to the conclusion that more than enzymes is needed for the validation of membrane fractions.

When enzyme markers are used it is suggested that the investigators help to standardize the assays by following the suggestions outlined by Solyom and Trams (1972).

3.4.2 Biological properties

If a specific biological activity is associated with the membranes of the cells which are being studied then it is possible to use this as a marker for the membrane fragments. It must be kept in mind, however, that the isolated fragments represent the particular receptor sites and not the whole membrane. ^{125}I-labeled glucagon (Rodbell et al., 1971) or insulin (Cuatrecasas, 1972) or plant lectin (Allan, Auger and Crumpton, 1972) or α-bungarotoxin (Schimmel et al., 1973) receptor sites are some examples. The distribution of antigens on some membranes has been defined (Vitetta, Uhr and Boyse, 1972; Nathenson and Muramatsu, 1971; Muramatsu et al., 1973; Konda, Stockert and Smith, 1973). The plant lectins (Sharon and Lis, 1972) may prove to be useful but, since most of them are specific not for membranes but for carbohydrates, one must be careful in the interpretation of the results.

3.4.3 Chemical components

The use of chemical markers to identify membranes has been reviewed (Glick, 1974b), as has the use of protein labels (Wallach, 1972) and characteristic membrane proteins (Guidotti, 1972). Surface membranes are enriched in carbohydrates (Glick, Comstock and Warren, 1970; Glick *et al.*, 1971). In these studies, isolated surface membranes were counted and related to the whole cells. The surface membranes contained only 70 percent of the sialic acid of the whole cell (*Table 3.2*). This amount varied according to the growth conditions and stage of the cell cycle (Glick, 1974a). Lysosomes are rich in sialic acid (Glick *et al.*, 1971) so it is impossible clearly to identify membrane fragments or vesicles on the basis of sialic acid. Gangliosides, which are sialic-acid-containing glycolipids, do not co-purify with 5'-nucleotidase in bovine mammary gland and rat liver membranes (Keenan, Huang and Morré, 1972), suggesting that one of these markers is not specific for the surface membrane. Only disialogangliosides are found in the surface membrane of the L cell (Weinstein *et al.*, 1970).

Fucose is found almost exclusively in glycoproteins and, again, the membrane is enriched in fucose (*Table 3.2*). However, a distribution throughout cellular fractions indicates that it is present in small amounts in other particulate matter (Chiarugi and Urbano, 1972; Keshgegian and Glick, 1973) and as fucolipids (Steiner and Melnick, 1974).

Surface membranes are also enriched in cholesterol (Weinstein *et al.*, 1969) and the cholesterol:phospholipid ratio is considered by some authors as characteristic of membranes. A brief summary of cholesterol:phospholipid ratios shows them to vary from 0.38 to 1.4 (*Table 3.3*). A ratio of 0.6 or above is considered to be characteristic of a surface membrane while a ratio of 0.1 is characteristic of endoplasmic reticulum.

3.4.4 Iodination as a marker

The technique of Phillips and Morrison (1971) for the catalytic iodination of surface proteins by lactoperoxidase has been used to label erythrocyte membrane proteins radioactively. Lactoperoxidase does not penetrate the cells under the conditions used so only the outer proteins are iodinated in the intact cell. The technique utilized in this way is not valid as a cell surface marker for the following reasons: (a) only one to three surface proteins are labeled in the intact cell (Phillips and Morrison, 1971; Barber and Jamieson, 1971; Poduslo, Greenberg and Glick, 1972; Nachman, Hubbard and Ferris, 1973; Czech and Lynn, 1973b) and these contain less than five percent of the radioactivity; (b) 90 percent or more of the radioactive iodine is rapidly taken up by whole cells (Poduslo, Greenberg and Glick, 1972); (c) chemical iodination of the lipids takes place (Poduslo, Greenberg and Glick, 1972). Thus, if one breaks open the cell under these conditions to prepare membrane fragments, it will be impossible to distinguish these from other cell parts in the presence of excess radioactivity. The possibility exists that other membrane probes may be useful to label the surface for isolation (Wallach, 1972).

3.5 CONTAMINANTS

A few enzymes which can be used to identify contaminating material in the membrane preparation are listed below. Some investigators report the

Table 3.3 SOME COMPONENTS FOUND IN SURFACE MEMBRANE FRACTIONS

Cell type	Homogenization	Cholesterol : phospholipid molar ratio	RNA (μg/mg of protein)	DNA	Ref.
Erythrocytes (human)	Hemoglobin-free	0.9–1.0			Steck, Fairbanks and Wallach (1971)
Erythrocytes (human)	Hemoglobin-free — normal target spur	0.8–0.9 1.0–1.2 1.1–1.4			Cooper et al. (1972)
Erythrocytes (rat)	Hemoglobin-free	0.74			Glossmann and Neville (1971)
Lymphocytes (human)	Tris, Potter-Elvehjem: light heavy	0.75 0.69	8 10	20 15	Demus (1973)
Lymphocytes, leukemic (human)	NaHCO₃, Dounce	0.38	10	0	Marique and Hildebrand (1973)
Lymphocytes (pig)	HEPES, sucrose, MgCl₂, N₂ cavitation	1.03	25	10	Ferber et al. (1972)
Lymphocytes (pig)	Tris, NaCl, stirring	1.01	28	0	Allan and Crumpton (1970)
Platelets (human)	Glycerol lysis	0.49	3.3	0	Barber and Jamieson (1970)
Thymocytes (calf)	Hanks, N₂ cavitation	0.61	30	0.2	Van Blitterswijk, Emmelot and Feltkamp (1973)
Macrophage (mouse)	Tris, latex bead, vortex	1.09			Werb and Cohn (1972)
Chick embryo fibroblasts	Imidazole, Thomas	1.05			Israel, Verjus and Semmel (1973)
L5178Y/CA55	Tris, syringe	0.95	1.5		Dods, Essner and Barclay (1972)
BRL rat epithelial	NaCl, Dounce	0.97	12		Perdue et al. (1971b)
BRL-MSV	NaCl, Dounce	0.88	9		Perdue et al. (1971b)
	ZnCl₂		60	0	
L cells	FMA, Dounce Tris	0.69 0.74	60	0	Glick and Warren (1969); Weinstein et al. (1969)

Ehrlich ascites	Tris, syringe	1.19	2		Dods, Essner and Barclay (1972)
CEF	NaCl, Potter–Elvehjem	0.6	36	7	Perdue, Kletzien and Miller (1971a)
CEF/RBA	NaCl, Potter–Elvehjem	0.6	23		Perdue, Kletzien and Miller (1971a)
Wing tumor	Sucrose		30		Perdue, Warner and Miller (1973)
Liver (mouse)	$NaHCO_3$, Dounce	0.46	3	4.5	Evans (1970)
Liver (rat)	$NaHCO_3$, Dounce	0.61			Glossmann and Neville (1971)
Liver (rat) male: female:	$NaHCO_3$, Dounce	1.6 2.1			Solyom and Lauter (1973)
Liver (rat)	$NaHCO_3$, $CaCl_2$, Dounce		10		Ray (1970)
Liver (rat)	Tris, sucrose, Potter–Elvehjem	0.83	71	11	Touster *et al.* (1970)
Liver (rat)	$Na_2B_4O_7$, $CaCl_2$, Dounce	0.83	15–18	0	Dorling and Le Page (1973)
Newborn brain (rat)	$ZnCl_2$, Dounce		39		Hemminki (1973)
Synaptosomal (rat)	Osmotic shock		5.7	2	Levitan, Mushynski and Ramirez (1972)
Synaptosomal (rat)	EDTA, osmotic shock	0.44	<1		Morgan *et al.* (1971); Breckenridge, Gombos and Morgan (1972)
Synaptosomal (rat)	Osmotic shock	0.46			Cotman, Mahler and Anderson (1969)
Intestinal microvillus	EDTA	1.2			Forstner, Sabesin and Isselbacher (1968)
Kidney brush border (rat)	$NaHCO_3$, Dounce	0.46			Glossmann and Neville (1971)
Smooth muscle (rat)	Polytron	0.82	70	20	Kidwai, Radcliffe and Daniel (1971)

enzymes as surface membrane markers, and perhaps they are under certain conditions.

Nucleus—NAD pyrophosphorylase
Mitochondria—succinate dehydrogenase
Rough endoplasmic reticulum—glucose-6-phosphatase
Lysosomes—acid phosphatase
Golgi—uridine diphosphatase
Unidentified structures as seen in the EM—enzymes not described
Soluble—glycolytic enzymes

Low temperature spectroscopy has been used to exclude low concentrations of mitochondrial or microsomal components (Weinstein *et al.*, 1969).

3.6 PROCEDURAL ARTIFACTS

3.6.1 Loss of components

Many of the variations in membrane preparations, even from the same tissue, may involve loss of components in preparing the tissue or cells prior to the membrane isolation procedures. Many tissues are digested with enzymes to disperse the cells and the cells are washed in a variety of buffers. It is reported that acetylcholinesterase, for example, is removed from the cell by trypsin (Hall and Kelly, 1971) or is extracted with Tris buffer (Hayden *et al.*, 1973). Approximately 20–30 percent of the surface membrane glycoproteins are removed by the brief trypsinization used in harvesting cells (Buck, Glick and Warren, 1970). Saline extracts of cells are used as membrane preparations (Jansons and Burger, 1973). EDTA, used in the preparation of membranes (*see Table 3.1*) removed surface glycoproteins from cells grown in culture (Glick, 1974a) and 25 percent of the protein from liver cell membrane was solubilized in alkaline EDTA (Gurd, Evans and Perkins, 1972). Physical force of the homogenizing procedure may remove tiny and specific pieces of the membrane (Wolff and Jones, 1971; Lutz, 1973). Proteins of erythrocyte stroma have been released at low ionic strength (Harris, 1971) or extracted at alkaline pH (Steck and Yu, 1973). The pH and type of buffer as well as the divalent cation concentration affected the presence of acetylcholinesterase activity in nerve cell membranes (Hayden *et al.*, 1973; Taylor, Nelson and Shirachi, 1973).

Table 3.2 points out some differences in the final membrane preparations obtained from the same cell line by two different methods. Membranes prepared by the Tris procedure (Warren and Glick, 1969) have protein, RNA and sialic acid values equivalent to those prepared by the zinc ion procedure (unpublished). Enzyme activities of membranes from KB cells are similar when prepared by either the zinc ion procedure or by latex bead ingestion (Charalampous, Gonatas and Melbourne, 1973). These findings suggest a loss of components from membranes prepared with fluorescein mercuric acetate. Organic mercurials have been shown to extract proteins from erythrocyte membranes (Steck and Yu, 1973) and solubilize the stroma (Cantrell, 1973).

3.6.2   Adsorption of components

As easily as components can be lost from the membranes during isolation procedures, so cell components can be adsorbed from the soluble cytoplasm of the cell. The pH of the homogenizing medium has been reported to cause proteins to bind to the membranes. Hexokinase isoenzymes, although cytoplasmic, can be adsorbed on membranes (Steck and Wallach, 1970).

3.6.3   Activation of endogenous enzymes

It is quite possible that enzymes such as proteases become activated during the homogenizing procedure. In methods which involve violent disruption of the cells it is hard to imagine that lysosomal membranes are not disrupted as well as other cell particles. With the activation of proteases, glycosidases and lipases the final membrane composition is quite different. Indeed, unless these effects and other artifacts are ruled out, comparisons of proteins of surface membranes as depicted by gel electrophoresis are hard to interpret (Greenberg and Glick, 1972; Wickus and Robbins, 1973).

3.6.4   Isolation of specific fragments

It has been shown that the preparation of liver cell membranes by the Neville procedure (1960) yields mostly membranes of bile front canaliculi (Neville and Glossmann, 1971). Distinctions of fragments from mixed cell populations are also made, such as between synaptosomal membranes and synaptic vesicles (Breckenridge and Morgan, 1972) as well as separations of synaptosomal and glial cell fragments (Karlsson, Hamberger and Henn, 1973). In an adaptation of the methods of Warren and Glick (1969), Scher and Barland (1972) have specifically removed the upper portion of the membranes from cells grown in monolayers. Renal cell membranes divide into specific fragments (Heidrich et al., 1972). The use of methods such as affinity density perturbations (Wallach et al., 1972) may isolate specific fragments.

Many investigators report two membrane fractions with somewhat different specificities when membrane fragments are isolated (Barber and Jamieson, 1970; Demus, 1973). Indeed, in isolating the membranes from any source, membranes of several densities are obtained even with relatively intact membranes (Warren and Glick, 1969; Perdue et al., 1971b). The presence of inside or outside vesicles also complicates the preparations (Kant and Steck, 1972). These vesicles have been discussed in detail (Steck, 1972).

3.6.5   Rearrangement of membrane proteins

With the use of dansyl chloride as a membrane probe it has been suggested that the erythrocyte membrane proteins reorganize during isolation. The argument is that dansyl chloride labeled the same proteins in the intact cell as in the membrane because of the ease of penetration of a small molecule (Schmidt-Ullrich, Knüfermann and Wallach, 1973). Carraway, Kobylka and Triplett (1971) came to a similar conclusion using another low-molecular-weight compound, acetic anhydride, as a probe. It is also argued that phosphatidylserine and phosphatidylethanolamine are labeled poorly in the whole cell but strongly in isolated membranes. Bretscher (1973) does not consider these findings to be due to membrane rearrangement but rather suggests that, because of pH differences, the inside membrane proteins are not labeled by small-molecule probes in the intact cell.

Iodination of the proteins of intact cells in the presence of lactoperoxidase does, however, appear to be vectorially oriented (Phillips and Morrison, 1971). However, when isolated membranes are iodinated most of the proteins are labeled, suggesting either a rearrangement during the isolation procedure or that most of the proteins are on the cytoplasmic side of the membrane (Phillips and Morrison, 1971; Poduslo, Greenberg and Glick, 1972; Nachman, Hubbard and Ferris, 1973). The possibility that rearrangement is not the answer comes from the fact that twice the amount of radioactivity was detected in the group of proteins exposed to the external environment (Poduslo, Greenberg and Glick, 1972), suggesting that these proteins could extend through the membrane in a manner analogous to that found by Bretscher (1971). Also suggesting the absence of any rearrangement is the fact that fat cell membrane proteins respond to trypsin in a manner similar to the proteins in the intact cell (Czech and Lynn, 1973a). Currently, however, this matter is not solved and evidence from these techniques favors rearrangement of the membrane proteins during the isolation procedures almost as strongly as it does the contrary.

3.7 COMMENTS ON ISOLATED MEMBRANES

For the most part, although a great deal of effort has been spent devising methods for preparing membranes, very few studies with these isolated membranes have explicitly proved many points. Studies aimed at elucidating the role of adenyl cyclase (Rodbell, 1974), the structure of erythrocyte stroma (Marchesi, Furthmayr and Tomita, 1974), or the specific recognition of embryonic (Merrell and Glaser, 1973) or liver cells (Pricer and Ashwell, 1971; Aronson, Tan and Peters, 1973) have proved fruitful. It may be that in most systems a good many preliminary data are required before pertinent questions can be asked. On the other hand, part of the difficulty has been in posing answerable questions and interpreting the results. Of course, the small yields and purity of the membrane preparations have slowed the progress.

There have been a number of studies on the chemical characterization of isolated surface membranes, or what the investigators call 'surface membranes'. It is clearly impossible to tabulate these results with any particular significance. Even preparations from similar cells made in different laboratories yield different analytical data. The following are discussed briefly in an attempt to point out the variety of results.

3.7.1 Gel electrophoresis

Gel electrophoresis appears to be the most used single technique although the data which have been generated are complicated not only by the differences in membrane preparations but also the procedural differences of the gels. Many artifacts are inherent; for example, the ratio of sodium dodecyl sulfate in the solubilizing buffer to the amount of protein can alter the migration rates of the polypeptides (Greenberg and Glick, 1972). Nevertheless, an attempted summary of selected results which appear to be comparable is given in *Table 3.4*, where some similarities can be observed in most preparations. The predominant and recurring proteins have molecular weights of 95 000, and with the exception of erythrocytes and lymphocytes, approximately 45 000. Actin has a molecular weight of 46 000, so perhaps an actin-like protein is present in all membranes (Perdue, 1973).

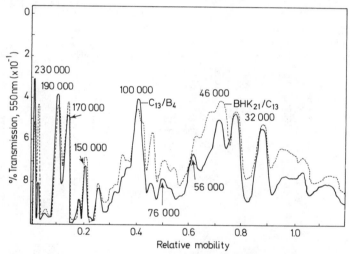

Figure 3.6 Comparison of the polypeptides from surface membranes of baby hamster kidney fibroblasts before (---; BHK_{21}/C_{13}) and after (——; C_{13}/B_4) transformation by the Bryan strain of Rous sarcoma virus. Protein (50 µg) from each membrane preparation was fractionated on sodium dodecyl sulfate gels of 5% polyacrylamide, stained with Coomazie blue and the absorbance scans were superimposed.

Wickus and Robbins (1973) compared the surface membrane proteins obtained in ghost (Brunette and Till, 1971) and vesicle forms (Perdue, Kletzien and Miller, 1971a). Most of the proteins obtained by the two methods were similar although at least one protein was lost in the preparation of the vesicles. Also, two proteins were present in reduced amounts in the ghosts and one protein was enriched. It is possible that the preparation of the ghosts keeps some of the external proteins associated with the membranes. Banker, Crain and Cotman (1972) compared the polypeptides from synaptosomal membranes prepared by three procedures and found only minor differences. Shin and Carraway (1973) compared the proteins of

Table 3.4 POLYPEPTIDES OF SURFACE MEMBRANES

Cell type	Membrane preparation	Polypeptides Range (molecular weight × 10^3)	Major	Ref.
Erythrocytes (human)	Dodge	210–25	210, 200, 93	Reynolds (1973)
Erythrocytes (human)	Dodge	250–30	250, 89	Fairbanks, Steck and Wallach (1971)
Erythrocytes (human)	Dodge	300–30	300, 270, 90	Schmidt-Ullrich, Knüfermann and Wallach (1973)
Erythrocytes (human)	Dodge	290–23	290, 260, 95	Schmidt-Ullrich, Knüfermann and Wallach (1973)
Erythrocytes (sheep)	Dodge	250–10	250, 96, 48	Neville and Glossmann (1971)
Erythrocytes (rat)	Tris–NaCl	>160–30	160, 69	Allan and Crumpton (1971)
Lymphocytes (pig)	Tris–NaCl	>160–20	170. 95. 67	Demus (1973)
Lymphocytes (human)	Glycerol lysis	200–13	118, 110, 93, 45	Phillips (1972)
Platelets (human)	Sucrose, EDTA, $Na_2S_2O_5$, protein inhibitors	300–10	300, 170, 150, 45	Nachman, Hubbard and Ferris (1973)
Platelets (human)	$ZnCl_2$	230–15	230, 190, 170, 100, 46–39	Greenberg and Glick (1972)
L cells	$ZnCl_2$	230–15	230, 190. 100. 96. 56–39	Greenberg and Glick (1972)
$BHK_{21}C_{13}$	$ZnCl_2$	230–15	230, 190, 100. 46–39	Greenberg and Glick (1972)
$C_{13}B_4$	N_2 cavitation	>200–24	96, 76	Gahmberg (1971)
BHK_{21}	$ZnCl_2$, FMA	250–30	80, 66, 57	Scher and Barland (1972)
3T3	$ZnCl_2$, FMA	250–30	80, 66, 57	Scher and Barland (1972)
3T6	NaCl	>220–20	175, 99, 69, 47–20	Wray and Perdue (1974)
CEF	NaCl	>200–15	95, 43, 30	Wickus and Robbins (1973)

Cell type	Conditions			Reference
CEF	36°	ZnCl$_2$	200–15	95, 30	Wickus and Robbins (1973)
Infected with Ts–68	41°				
CEF					
Avian leukosis virus					
susceptible		Tris	>130–<10	125–66, 42	Smith and Crittenden (1973)
resistant		Tris	>130–<10	125–66, 42	Smith and Crittenden (1973)
Sarcoma 180 (mouse)		ZnCl$_2$	>300–33	280, 237, 216, 42, 36	Shin and Carraway (1973)
		ZnCl$_2$	>300–33	280, 42	Shin and Carraway (1973)
		Tris	>300–33	78, 42, 36	Shin and Carraway (1973)
Liver (mouse)		NaHCO$_3$	>200–15	150, 90, 50	Evans and Gurd (1972)
Liver (rat)		NaHCO$_3$	360–10	250, 196, 96, 57, 48	Neville and Glossmann (1971)
Fat, white (rat)		Tris, sucrose	168–22	94, 78, 42	Czech and Lynn (1973b)
Immature brain cells (rabbit)		ZnCl$_2$	>160–15	>60, 55, 45, 40	Hemminki (1972)
Neuronal (rabbit)		Tris, sucrose	>150–10	95, 52, 44, 19	Karlsson, Hamberger and Henn (1973)
Glial (rabbit)		Tris, sucrose	>150–10	95, 52, 44, 19	Karlsson, Hamberger and Henn (1973)
Synaptosomal (rabbit)		Osmotic shock	>150–10	95, 68, 60, 52, 44, 19	Karlsson, Hamberger and Henn (1973)
Synaptosomal (rat)		Osmotic shock	>150–10	95, 52, 44	Levitan, Mushynski and Ramirez (1973)
Synaptosomal (rat)		Osmotic shock	>200–25	99, 53, 42	Banker, Crain and Cotman (1972)
Intestinal microvillus (rat)		EDTA	>200–20	150, 95, 45	Weiser (1973a)
Kidney microvillus (rat)		NaHCO$_3$	360–10	>250, 100, 48	Neville and Glossmann (1971)

Sarcoma 180 ascites cell membranes prepared by Tris and zinc ion procedures (Warren and Glick, 1969) with the slight modifications of Atkinson and Summers (1971) and the two-phase separation method of Brunette and Till (1971). Higher-molecular-weight components were found with the use of zinc ions (*see Table 3.4*).

Greenberg and Glick (1972) found that proteins of molecular weights of 210 000, 96 000, 82 000 and 56 000–45 000 were diminished on viral transformation of a number of different cell lines in culture. The most prominent area was 56 000–46 000 daltons (*Figure 3.6*). This has been confirmed. Using different procedures for membrane isolation from different cells, Wickus and Robbins (1973) found several protein contents to be reduced, particularly one of molecular weight 45 000. Bussell and Robinson (1973) reported a protein of molecular weight 142 000 to be lost altogether. Although variations in the purity of the membrane preparations could account for the observed findings, the fact that they are reproducible under a variety of conditions suggests a positive observation.

3.7.2 Cholesterol/phospholipid

The cholesterol:phospholipid ratio is suggested as characteristic for surface membranes and *Table 3.3* summarizes some of the reported molar ratios, which range from 0.38 to 1.4. Although not given in *Table 3.3*, the cholesterol:phospholipid (C:PL) ratios of rat liver membranes have been reported as being from 0.26 to 0.81 (Weinstein *et al.*, 1969). This is perhaps due to the fact that a variety of structures are isolated from the liver membranes with varying degrees of purity.

That the ratio is critical in maintaining the shape of erythrocytes is suggested by the studies of Cooper *et al.* (1972). Under certain pathological conditions the erythrocytes are deformed to target or spur cells. The latter are rigid and the membranes have a high C:PL ratio. Intestinal microvillus membranes which are more rigid than other membranes also have a high C:PL ratio (Forstner, Sabesin and Isselbacher, 1968). Increased phospholipid synthesis in the membranes of lymphocytes has been shown to accompany the transformation to blast cells after stimulation of lectins or antigens (Resch *et al.*, 1972) and human leukemic lymphocytes have a higher content of phospholipids (Marique and Hildebrand, 1973) than normal lymphocytes (Demus, 1973). On the other hand, a strain of mouse leukemia cells has a more usual phospholipid content (Dods, Essner and Barclay, 1972). It is increasingly important to have this ratio defined in the membrane preparations.

3.7.3 Sialic acid

Sialic acid, a terminal sugar residue of glycoproteins and a constituent of gangliosides, is considered to be a ubiquitous membrane marker. This has since been shown not to be the case (Glick, Comstock and Warren, 1970; Glick *et al.*, 1971), although surface membranes are indeed enriched in sialic acid. *Table 3.5* summarizes some of the sialic acid values reported for

Table 3.5 SIALIC ACID CONTENT OF MEMBRANES

Cell type	Morphology of membranes	Sialic acid (nmol/mg protein)	Reference
Erythrocytes (human)	Whole	89	Steck, Fairbanks and Wallach (1971)
Erythrocytes (human)	Vesicles	107–117	Steck, Fairbanks and Wallach (1971)
Erythrocytes (rat)	Vesicles	61	Glossmann and Neville (1971)
Lymphocytes (pig)	Vesicles	46	Allan, Auger and Crumpton (1972)
Platelets (human)	Vesicles	17	Barber and Jamieson (1970)
Thymocytes (calf)	Vesicles	66	Van Blitterswijk, Emmelot and Feltkamp (1973)
L51784Y/CA55 (mouse)	Fragments	33	Dods, Essner and Barclay (1972)
HeLa (human)	Vesicles	14	Bosmann, Hagopian and Eylar (1968)
L cells (mouse)	Whole— $ZnCl_2$ FMA	15 24	Glick et al. (1971)
BHK_{21}/C_{13} (hamster)	Whole	34	Glick, M. C., unpublished
C_{13}/B_4 (hamster—RSV transformed)	Whole	32	Glick, M. C. unpublished
C_{13}/SR_7 (hamster—RSV transformed)	Whole	27	Glick, M. C., unpublished
Ehrlich ascites (mouse)	Fragments	42	Dods, Essner and Barclay (1972)
Ehrlich ascites (mouse)	Vesicles	26	Wallach (1967)
Sarcoma 180 ascites (mouse)	Fragments	20	Shin and Carraway (1973)
Chick embryo fibroblasts	Vesicles	106	Perdue, Kletzien and Miller (1971a)
CEF/RAV–49	Vesicles	114	Perdue, Kletzien and Miller (1971a)
CEF/RBA	Vesicles	79	Perdue, Kletzien and Miller (1971a)
CEF/morph[f] (fusiform) Fujinami	Vesicles	77	Perdue, Kletzien and Wray (1972)
CEF/morph[r] (round) Fujinami	Vesicles	67	Perdue, Kletzien and Wray (1972)
BRL (rat epithelial)	Large fragments	75	Perdue et al. (1971b)
BRL	Vesicles	68	Perdue et al. (1971b)
MSV-converted BRL	Vesicles	63	Perdue et al. (1971b)
Wing tumor	Vesicles	83	Perdue, Warner and Miller (1973)
Breast tumor	Vesicles	72	Perdue, Warner and Miller (1973)
Fat (rat)	Fragments	16	Czech and Lynn (1973a)
Liver (mouse)	Vesicles	65	Evans (1970)
Liver (rat)	Vesicles	73	Evans (1970)
Liver (rat)	Vesicles	48	Touster et al. (1970)
Liver (rat)	Fragments	27	Glossmann and Neville (1971)
Liver (rat)	Fragments	50	Ray (1970)
Kidney brush border (rat)	Fragments	34	Glossmann and Neville (1971)

isolated membranes. The amounts appear to vary not only with different, but also with similar cell types.

Using different cell lines in culture, Perdue and his colleagues made a comprehensive study of the sialic acid content of membrane vesicles of several oncogenic transformed cells and compared them to their normal counterparts (Perdue, Kletzien and Miller, 1971a; Perdue et al., 1971b). Tumor cells were also examined (Perdue, Warner and Miller, 1973). Virus-infected cells, although not oncogenic, were controls for virus infection and morphological mutants were used to rule out fusiform versus round cells (Perdue, Kletzien and Wray, 1972). The two examples of normal cells which were examined were chick embryo fibroblasts and rat epithelial cells (BRL). The former as well as the non-oncogenic virus-infected chick embryo cells had 25 percent more sialic acid in their membranes than those of the other cells examined, including the epithelial BRL cells. It is important, however, to have rigidly defined growth and harvesting conditions for the cells before comparisons of this kind are made (Buck, Glick and Warren, 1971; Hartmann et al., 1972). The percentage of the total cell sialic acid found in the surface membrane varies by as much as 25–30 percent, depending on the state of growth of the cell (Glick, Comstock and Warren, 1970). In other studies, one cell line of RSV-transformed fibroblasts had a sialic acid content in the surface membranes equal to that of the normal counterpart while another cell line of RSV-transformed cells showed an increase in the sialic acid content when compared with the normal counterpart (Buck, Glick and Warren, 1970). This suggests that sialic acid is not necessarily decreased after viral transformation.

3.7.4 Ribonucleic acid

Most of the membrane preparations which have been examined for nucleic acids contain RNA (*Table 3.3*), and electron micrographs show ribosomes attached to the surface membranes (Warren, Glick and Nass, 1966). However, only a few investigators consider these as significant observations. The surface membranes of tissue culture cells contain a unique species of RNA (Glick, unpublished results) while RNA characterized in synaptosomal membranes resembles mitochondrial RNA although it is not considered a contaminant (Levitan, Mushynski and Ramirez, 1972). The RNA from surface membranes of Ehrlich ascites tumor cells has been characterized (Juliano et al., 1972). Human diploid lymphocyte membranes have been shown to contain RNA (Del Villano and Lerner, 1972) and DNA (Lerner, Meinke and Goldstein, 1971), and calf thymocyte membranes contain RNA (van Blitterswijk, Emmelot and Feltkamp, 1973). In membranes prepared from rat liver and hepatomas, the RNA may be integrated into lipoprotein material (Davidova and Shapot, 1970). Highly purified surface membranes of L cells, $BHK_{21}C_{13}$, $C_{13}B_4$ and Ehrlich ascites cells incorporate amino acids into large peptides which are different from those synthesized by the ribosomal fractions (Glick and Warren, 1969; Glick, unpublished). Purified synaptosomal membranes containing RNA and DNA (Levitan, Mushynski and Ramirez, 1972) incorporate amino acids into membrane proteins (Ramirez, Levitan and Mushynski, 1972), and preparations of platelet membranes have been shown to synthesize proteins (Booyse and Rafelson, 1969). All of these observations depend on the purity of the membranes and for the most part, the preparations have been well defined. In addition, the differences found in the isolated RNA and protein synthesizing systems preclude the results on the basis of contamination with other cell particulates. It is perhaps into these areas where more work should be channeled.

3.8 CONCLUSIONS

The preparation of purified surface membranes is an art not to be approached lightly. Methods are available to prepare membranes from most cell types. These methods produce whole membranes, large or small fragments or vesicles. Cells performing different functions require variations of the isolation procedures. Therefore, the method to be used should be selected on the basis of the cell type as well as on the ability of the final membrane product to answer the proposed question significantly. For example, to study biological activities, one should not use methods which inactivate the enzymes, and, in like manner, methods which vesiculate the membranes should be avoided in studies to determine the presence of a particular component on the outer surface of the cell.

Two kinds of restrictions are placed on the interpretation of any data on isolated membranes. First, the possibility exists that not all of the cytoplasmic contents have been expelled, particularly when whole membranes are prepared. Second, it may not be possible to separate completely all of the internal membranes from those of the surface, particularly when the method causes vesiculation of the membrane systems.

The field of surface membranes has developed rapidly, and borders on the chaotic. Perhaps it would be helpful if certain rigid conditions were set up. Since there are vast differences among cells, the criteria cannot be universal but should be categorized for particular cell types. In addition, when possible, similar methods should be used for a particular cell type or comparisons be made with the other methods if there is a conflict.

The following points are suggested as aids to standardize the preparations:

1. A moratorium should be imposed by editors on publishing methods for the isolation of surface membranes. Only methods which show real innovation should be published as separate papers.
2. Terminology should define as surface membranes only those preparations which meet a number of required criteria.
3. Standard criteria should be set up for the establishment of marker enzymes and chemical components. These will vary for different cell types but at least one marker in each category should be required.
4. Biological markers should be the definitive criteria whenever possible.
5. Standardized methods should be used to assay enzyme markers and chemical components as well as biological activities.
6. The degree of contamination by other cell organelles should be reported, again using standardized procedures.
7. Critical consideration should be given to the procedures used prior to

membrane isolation and the influence these procedures have on the final products.

8. The yield of membranes and the degree of purification should be clearly stated.

It is anticipated that when the preparations become standardized, membranes will serve as useful tools in defining some of the problems of molecular biology.

REFERENCES

ALBERTSSON, P. A. (1970). *Adv. Protein Chem.*, **24**:309.
ALLAN, D., AUGER, J. and CRUMPTON, M. J. (1972). *Nature, New Biol.*, **236**:23.
ALLAN, D. and CRUMPTON, M. J. (1970). *Biochem. J.*, **120**:133.
ALLAN, D. and CRUMPTON, M. J. (1971). *Biochem. J.*, **123**:967.
ANDERSON, N. G., LANSING, A., LIEBERMAN, I., BANKIN, C. T., JR. and ELROD, H. (1968). *Wistar Monograph No. 8*, p. 23.
ARONSON, N. N., JR., TAN, L. Y. and PETERS, B. P. (1973). *Biochem. biophys. Res. Commun.*, **53**:112.
ATKINSON, P. H. and SUMMERS, D. F. (1971). *J. biol. Chem.*, **246**:5162.
BANKER, G., CRAIN, B. and COTMAN, C. W. (1972). *Brain Res., Osaka*, **42**:508.
BARBER, A. J. and JAMIESON, G. A. (1970). *J. biol. Chem.*, **245**:6357.
BARBER, A. J. and JAMIESON, G. A. (1971). *Biochemistry*, **10**:4711.
BARBER, A. J., PEPPER, D. S. and JAMIESON, G. A. (1971). *Thromb. Diath, haemorrh.*, **26**:38.
BARLAND, P. and SCHROEDER, E. A. (1970). *J. Cell Biol.*, **47**:662.
BERGER, S. J. and SACKTOR, B. (1970). *J. Cell Biol.*, **47**:637.
BERMAN, H. M., GRAM, W. and SPIRTES, M. A. (1969). *Biochim. biophys. Acta*, **183**:10.
BOONE, C. W., FORD, L. E., BOND, H. E., STUART, D. C. and LORENZ, D. (1969). *J. Cell Biol.*, **41**:378.
BOOYSE, F. M. and RAFELSON, M. E. (1969). *Dynamics of Thrombus Formation and Dissolution*, p. 149. Ed. S. A. JOHNSON and M. M. GUEST. Philadelphia; J. B. Lippincott.
BOSMANN, H. B., HAGOPIAN, A. and EYLAR, E. H. (1968). *Archs Biochem. Biophys.*, **128**:51.
BRECKENRIDGE, W. C., GOMBOS, G. and MORGAN, I. C. (1972). *Biochim. biophys. Acta*, **266**:695.
BRECKENRIDGE, W. C. and MORGAN, I. G. (1972). *FEBS Lett.*, **22**:253.
BRETSCHER, M. S. (1971). *Nature, New Biol.*, **231**:229.
BRETSCHER, M. S. (1973). *Nature, New Biol.*, **246**:116.
BRUNETTE, D. M. and TILL, J. E. (1971). *J. Membrane Biol.*, **5**:215.
BUCK, C. A., GLICK, M. C. and WARREN, L. A. (1970). *Biochemistry*, **9**:4567.
BUCK, C. A., GLICK, M. C. and WARREN, L. A. (1971). *Biochemistry*, **10**:2176.
BUSSELL, R. H. and ROBINSON, W. S. (1973). *J. Virology*, **12**:322.
CANTRELL, A. C. (1973). *Biochim. biophys. Acta*, **311**:381.
CARRAWAY, K. L., KOBYLKA, D. and TRIPLETT, R. B. (1971). *Biochim. biophys. Acta*, **241**:934.
CHARALAMPOUS, F. C., GONATAS, N. K. and MELBOURNE, A. D. (1973). *J. Cell Biol.*, **59**:421.
CHESNEY, R. W., SACKTOR, B. and ROWEN, R. (1973). *J. biol. Chem.*, **248**:2182.
CHIARUGI, V. P. and URBANO, P. (1972). *J. gen. Virology*, **14**:133.
CONSTANTOPOULOS, A. and NAJJAR, V. A. (1973). *Biochem. biophys. Res. Commun.*, **53**:794.
COOPER, R. A., DILOY-PURAY, M., LANDO, P. and GREENBERG, M. S. (1972). *J. clin. Invest.*, **51**:3182.
COTMAN, C. W., MAHLER, H. R. and ANDERSON, N. G. (1969). *Biochim. biophys. Acta*, **163**:272.
CUATRECASAS, P. (1972). *Proc. natn. Acad. Sci. U.S.A.*, **69**:318.
CZECH, M. P. and LYNN, W. S. (1973a). *J. biol. Chem.*, **248**:5081.
CZECH, M. P. and LYNN, W. S. (1973b). *Biochemistry*, **12**:3597.
DAVIDOVA, S. YA. and SHAPOT, V. S. (1970). *FEBS Lett.*, **6**:349.
DEL VILLANO, B. C. and LERNER, R. A. (1972). *Fedn. Proc. Fedn. Am. Socs exp. Biol.*, **31**:804ab.
DEMUS, H. (1973). *Biochim. biophys. Acta*, **291**:93.
DEPIERRE, J. W. and KARNOVSKY, M. L. (1973). *J. Cell Biol.*, **56**:275.
DODGE, J. T., MITCHELL, C. and HANAHAN, D. J. (1963). *Archs Biochem. Biophys.*, **100**:119.
DODS, R. F., ESSNER, E. and BARCLAY, M. (1972). *Biochem. biophys. Res. Commun.*, **46**:1074.
DORLING, P. R. and LE PAGE, R. N. (1973). *Biochim. biophys. Acta*, **318**:33.
EDELMAN, G. M., RUTISHAUSER, U. and MILLETTE, C. F. (1971). *Proc. natn. Acad. Sci. U.S.A.*, **68**:2153.
EMMELOT, P., BOS, C. J., BENEDETTI, E. L. and RÜMKE, P. L. (1964). *Biochim. biophys. Acta*, **90**:126.

EMMELOT, P. and BOS, C. J. (1971a). *Biochim. biophys. Acta*, **249**:285.
EMMELOT, P. and BOS, C. J. (1971b). *Biochim. biophys. Acta*, **249**:293.
ENTMAN, M. L., LEVEY, G. S. and EPSTEIN, S. E. (1969). *Biochem, biophys. Res. Commun.*, **35**:728.
EVANS, W. H. (1970). *Biochem. J.*, **166**:833.
EVANS, W. H. and GURD, J. W. (1972). *Biochem. J.*, **128**:691.
EVANS, W. H. and GURD, J. W. (1973). *Biochem. J.*, **133**:189.
FAIRBANKS, G., STECK, T. L. and WALLACH, D. F. H. (1971). *Biochemistry*, **10**:2606.
FERBER, E., RESCH, K., WALLACH, D. F. H. and IMM, W. (1972). *Biochim. biophys. Acta*, **266**:494.
FORSTNER, G. G., SABESIN, S. M. and ISSELBACHER, K. J. (1968). *Biochem. J.*, **106**:381.
FORTE, J. G., FORTE, T. M. and HEINZ, E. (1973). *Biochim. biophys. Acta*, **298**:827.
GAHMBERG, C. G. (1971). *Biochim. biophys. Acta*, **249**:69.
GAHMBERG, C. G. and SIMONS, K. (1970). *Acta path. microbiol. scand.*, **78**:176.
GIACOBINO, J. P. and PERRELET, A. (1971). *Experientia*, **27**:259.
GLICK, M. C. (1974a). *Biology and Chemistry of Eucaryotic Cell Surfaces*, p. 213. Ed. E. Y. C. LEE and E. E. SMITH. New York, Academic Press.
GLICK, M. C. (1974b). *Methods in Membrane Biology*, p. 157. Ed. E. KORN. New York; Academic Press.
GLICK, M. C., COMSTOCK, C. A. and WARREN, L. (1970). *Biochim. biophys. Acta*, **219**:290.
GLICK, M. C. and WARREN, L. (1969). *Proc. natn. Acad. Sci. U.S.A.*, **63**:563.
GLICK, M. C., KIMHI, Y. and LITTAUER, U. Z. (1973). *Proc. natn. Acad. Sci. U.S.A.*, **70**:1682.
GLICK, M. C., COMSTOCK, C. A., COHEN, M. A. and WARREN, L. (1971). *Biochim. biophys. Acta*, **233**:247.
GLOSSMANN, H. and NEVILLE, D. M., JR. (1971). *J. biol. Chem.*, **246**:6339.
GLOSSMANN, H. and NEVILLE, D. M., JR. (1972). *FEBS Lett.*, **19**:340.
GRAHAM, J. M. (1972). *Biochem. J.*, **130**:1113.
GREENBERG, C. S. and GLICK, M. C. (1972). *Biochemistry*, **11**:3680.
GUIDOTTI, G. (1972). *A. Rev. Biochem.*, **41**:731.
GURD, J. W., EVANS, W. H. and PERKINS, H. R. (1972). *Biochem. J.*, **126**:459.
HALL, Z. W. and KELLY, R. B., (1971). *Nature, New. Biol.*, **232**:62.
HARRIS, J. R. (1971). *Biochem. J.*, **122**:38.
HARTMANN, J. F., BUCK, C. A., DEFENDI, V., GLICK, M. C. and WARREN, L. (1972). *J. cell Physiol.*, **80**:159.
HAYDEN, C. O., TAYLOR, J. E., FORREST, A. B. and SHIRACHI, D. Y. (1973). *Proc. west. pharmac. Soc.*, **16**:99.
HEIDRICH, H., KINNE, R., SAFFRAN, E. K. and HANNIG, K. (1972). *J. Cell Biol.*, **54**:232.
HEINE, J. W. and SCHNAITMAN, C. A. (1971). *J. Cell Biol.*, **48**:703.
HEMMINKI, K. (1972). *Life Sci.*, **2**:1173.
HEMMINKI, K. (1973). *Int. J. Neurosci.*, **5**:81.
HENN, F. A., HANSSON, H.-A. and HAMBERGER, A. (1972). *J. Cell Biol.*, **53**:654.
HINTON, R. H. (1972). *Subcellular Components: Preparation and Fractionation*, pp. 119–156. Ed. G. D. BIRNIE. London; Butterworths.
HINTON, R. D., DOBROTA, M., FITZSIMONS, J. T. R. and REID, E. (1970). *Eur. J. Biochem.*, **12**:349.
HOPFER, U., NELSON, K., PERROTTO, J. and ISSELBACHER, K. F. (1973). *J. biol. Chem.*, **248**:25.
HUNTER, M. J. and COMMERFORD, S. L. (1961). *Biochim. biophys. Acta*, **47**:580.
ISRAEL, A., VERJUS, M.-A. and SEMMEL, M. (1973). *Biochim. biophys. Acta*, **318**:155.
JANSONS, V. K. and BURGER, M. M. (1973). *Biochim. biophys. Acta*, **291**:127.
JULIANO, R., CISZKOWSKI, J., WAITE, D. and MAYHEW, E. (1972). *FEBS Lett.*, **22**:27.
KAMAT, V. B. and WALLACH, D. F. H. (1965). *Science, N.Y.*, **148**:1343.
KANT, J. A. and STECK, T. L. (1972). *Nature, New Biol.*, **240**:26.
KARLSSON, J.-O., HAMBERGER, A. and HENN, F. A. (1973). *Biochim. biophys. Acta*, **298**:219.
KAULEN, H. D. and GROSS, R. (1973). *Thromb. Diath. haemorrh*, **30**:199.
KEENAN, T. W., HUANG, C. M. and MORRÉ, D. J. (1972). *Biochem. biophys. Res. Commun.* **47**:1277.
KESHGEGIAN, A. A. and GLICK, M. C. (1973). *Biochemistry*, **12**:1221.
KESSEL, D. and BOSMANN, H. B. (1970). *FEBS Lett.*, **10**:85.
KIDWAI, A. M., RADCLIFFE, M. A. and DANIEL, E. E. (1971). *Biochim. biophys. Acta*, **233**:538.
KONDA, S., STOCKERT, E. and SMITH, R. T. (1973). *Cell. Immun.*, **7**:275.
KRUG, F., PARIKH, I., ILLIANO, G. and CUATRECASAS, P. (1973). *J. biol. Chem.*, **248**:1203.
LAUDAT, M. H., PAIRAULT, J., BAYER, P., MARTIN, M. and LAUDAT, PH. (1972). *Biochim. biophys. Acta*, **265**:1005.
LAUTER, C. J., SOLYOM, A. and TRAMS, E. G. (1972). *Biochim. biophys. Acta*, **266**:511.
LELIEVRE, L. (1973). *Biochim. biophys. Acta*, **291**:662.
LELIEVRE, L. and PARAF, A. (1973). *Biochim. biophys. Acta*, **291**:671.

LELIEVRE, L., PRIGENT, B. and PARAF, A. (1971). *Biochem, biophys. Res. Commun.*, **45**:637.
LERNER, R. A., MEINKE, W. and GOLDSTEIN, D. A. (1971). *Proc. natn. Acad. Sci. U.S.A.*, **68**:1212.
LESKO, L., DONLON, M., MARINETTI, G. V. and HARE, J. D. (1973). *Biochim. biophys. Acta*, **311**:173.
LEVITAN, I. B., MUSHYNSKI, W. E. and RAMIREZ, G. (1972). *J. biol. Chem.*, **247**:5376.
LUTZ, F. (1973). *Comp. biochem. Physiol.*, **45B**:805.
MARCHESI, V. T., FURTHMAYR, H. and TOMITA, M. (1974). *Biology and Chemistry of Eucaryotic Cell Surfaces*, p. 273. Ed. E. Y. C. LEE and E. E. SMITH. New York; Academic Press.
MARINETTI, G. V., RAY, T. K. and TOMASI, V. (1969). *Biochem. biophys. Res. Commun.*, **36**:185.
MARIQUE, D. and HILDEBRAND, J. (1973). *Cancer Res.*, **33**:2761.
MARX, S. J., FEDAK, S. A. and AURBACH, G. D. (1972). *J. biol. Chem.*, **247**:6913.
MCKEEL, D. W. and JARETT, L. J. (1970). *J. Cell Biol.*, **44**:417.
MCMILLAN, P. N. and LUFTIG, R. B. (1973). *Proc. natn. Acad. Sci. U.S.A.*, **70**:3060.
MERRELL, R. and GLASER, L. (1973). *Proc. natn. Acad. Sci. U.S.A.*, **70**:2794.
MORGAN, I. G., WOLFE, L. S., MANDEL, P. and GOMBOS, G. (1971). *Biochim. biophys. Acta*, **241**:737.
MURAMATSU, T., NATHENSON, S. G., BOYSE, E. A. and OLD, L. J. (1973). *J. exp. Med.*, **137**:1256.
NACHMAN, R. I., FERRIS, L. B. and HIRSCH, J. C. (1971). *J. exp. Med.*, **133**:785.
NACHMAN, R. L., HUBBARD, A. and FERRIS, B. (1973). *J. biol. Chem.*, **248**:2928.
NATHENSON, S. G. and MURAMATSU, T. (1971). *Glycoproteins of Blood Cells and Plasma*, pp. 245–262. Ed. G. A. JAMIESON and T. J. GREENWALT. Philadelphia; J. B. Lippincott.
NEVILLE, D. M., JR (1960). *J. biophys. biochem. Cytol.*, **8**:413.
NEVILLE, D. M., JR. and GLOSSMANN, H. (1971). *J. biol. Chem.*, **246**:6335.
OSEROFF, A., ROBBINS, P. W. and BURGER, M. M. (1973). *A. Rev. Biochem.*, **42**:647.
PERDUE, J. F. (1973). *J. Cell Biol.*, **58**:265.
PERDUE, J. F., KLETZIEN, R. and MILLER, K. (1971a). *Biochim. biophys. Acta*, **249**:419.
PERDUE, J. F., KLETZIEN, R., MILLER, K., PRIDMORE, G. and WRAY, V. L. (1971b). *Biochim. biophys. Acta*, **249**:435.
PERDUE, J. F., KLETZIEN, R. and WRAY, V. L. (1972). *Biochim. biophys. Acta*, **266**:505.
PERDUE, J. F. and SNEIDER, J. (1970). *Biochim. biophys. Acta*, **196**:125.
PERDUE, J. F., WARNER, D. and MILLER, K. (1973). *Biochim. biophys. Acta*, **298**:817.
PHILLIPS, D. R. (1972). *Biochemistry*, **11**:4582.
PHILLIPS, D. R. and MORRISON, M. (1971). *Biochemistry*, **10**:1766.
PODUSLO, J. F., GREENBERG, C. S. and GLICK, M. C. (1972). *Biochemistry*, **11**:2616.
PONDER, E. (1952). *J. exp. Biol.*, **29**:605.
PRICER, W. E., JR. and ASHWELL, G. (1971). *J. biol. Chem.*, **246**:4825.
RAMIREZ, G., LEVITAN, I. B. and MUSHYNSKI, W. E. (1972). *J. biol. Chem.*, **247**:5382.
RAY, T. K. (1970). *Biochim. biophys. Acta*, **196**:1.
REIK, L., PETZOLD, G. L., HIGGINS, J. A., GREENGARD, P. and BARRNETT, R. J. (1970). *Science, N.Y.*, **168**:382.
RESCH, K., GELFAND, E. W., HANSEN, K. and FERBER, E. (1972). *Eur. J. Immun.*, **2**:598.
REYNOLDS, J. A. (1973). *Fedn Proc. Fedn Am. Socs exp. Biol.*, **32**:2034.
RODBELL, M. J. (1967). *J. biol. Chem.*, **242**:5744.
RODBELL, M. J. (1971). *Acta Endocr.*, **153**:337.
RODBELL, M. J. (1974). *J. biol. Chem.*, **249**:59.
RODBELL, M., KRANS, H. M. J., POHL, S. L. and BIRNBAUMER, L. (1971). *J. biol. Chem.*, **246**:1861.
ROTHFIELD, L. A. and FINKELSTEIN, A. (1968). *A. Rev. Biochem.*, **37**:675.
SCHER, I. and BARLAND, P. (1972). *Biochim. biophys. Acta*, **255**:580.
SCHIMMEL, S. D., KENT, C., BISCHOFF, R. and VAGELOS, P. R. (1973). *Proc. natn. Acad. Sci. U.S.A.*, **70**:3195.
SCHMIDT-ULLRICH, R., KNÜFERMANN, H. and WALLACH, D. F. H. (1973). *Biochim. biophys. Acta*, **307**:353.
SHARON, N. and LIS, H. (1972). *Science, N.Y.*, **177**:949.
SHIN, B. C. and CARRAWAY, K. L. (1973). *Biochim. biophys. Acta*, **330**:254.
SMITH, E. J. and CRITTENDEN, L. B. (1973). *Biochim. biophys. Acta*, **298**:608.
SOLYOM, A. and LAUTER, C. J. (1973). *Biochim. biophys. Acta*, **298**:743.
SOLYOM, A. and TRAMS, E. G. (1972). *Enzyme*, **13**:329.
STECK, T. L. (1972). *Membrane Molecular Biology*, pp. 76–114. Ed. C. F. FOX and A. KEITH. New York; Academic Press.
STECK, T. L., FAIRBANKS, G. and WALLACH, D. F. H. (1971). *Biochemistry*, **10**:2617.
STECK, T. L. and FOX, C. F. (1972). *Membrane Molecular Biology*, pp. 27–75. Ed. C. F. FOX and A. KEITH. New York; Academic Press.
STECK, T. L. and WALLACH, D. F. H. (1970). *Meth. Cancer Res.*, **5**:93.

STECK, T. L. and YU, J. (1973). *J. supramolec. Struct.*, **1**:220.
STEINER, S. and MELNICK, J. L. (1974). *Nature, Lond.*, **251**:717.
STONEHILL, E. H. and HUPPERT, J. (1968). *Biochim. biophys. Acta*, **155**:353.
SUTCLIFFE, H. S., MARTIN, T. J., EISMAN, J. A. and PILCZYK, R. (1973). *Biochem. J.*, **134**:913.
SUTHERLAND, E. W., ROBISON, G. A. and BUTCHER, R. W. (1968). *Circulation*, **37**:279.
TARONE, G., PRAT, M. and COMOGLIO, P. M. (1973). *Biochim. biophys. Acta*, **311**:214.
TAYLOR, J. E., NELSON, V. and SHIRACHI, D. Y. (1973). *Proc. west. pharmac. Soc.*, **16**:103.
THOMAS, L. and KINNE, R. (1972). *Biochim. biophys. Acta*, **255**:114.
TOUSTER, O., ARONSON, N. N., JR., DULANEY, J. T. and HENDRICKSON, H. (1970). *J. Cell Biol.*, **47**:604.
VAN BLITTERSWIJK, W. J., EMMELOT, P. and FELTKAMP, C. A. (1973). *Biochim. biophys. Acta*, **298**:577.
VITETTA, E., UHR, J. W. and BOYSE, E. A. (1972). *Cell Immun.*, **4**:187.
WALLACH, D. F. H. (1967). *The Specificity of Cell Surfaces*, pp. 129–163. Ed. D. B. DAVIS and L. WARREN. Englewood Cliffs, New Jersey; Prentice Hall.
WALLACH, D. F. H. (1972). *Biochim. biophys. Acta*, **265**:61.
WALLACH, D. F. H. and KAMAT, V. B. (1966). *Meth. Enzym.*, **8**:164.
WALLACH, D. F. H., KRANZ, B., FERBER, E. and FISCHER, H. (1972). *FEBS Lett.*, **21**:29.
WALLACH, D. F. H. and LIN, P. S. (1973). *Biochim. biophys. Acta*, **300**:211.
WARREN, L. and GLICK, M. C. (1969). *Fundamental Techniques in Virology*, pp. 66–71. Ed. K. HABEL and N. P. SALZMAN. New York; Academic Press.
WARREN, L. and GLICK, M. C. (1971). *Biomembranes I*, pp. 257–288. Ed. L. A. MANSON. New York; Plenum Press.
WARREN, L., GLICK, M. C. and NASS, M. K. (1966). *J. cell. Physiol.*, **68**:269.
WEINHOUSE, S. and ONO, T. (1973) (Eds). *Isozymes and Enzyme Regulation in Cancer*. Baltimore; University Park Press.
WEINSTEIN, D. B., MARSH, J. B., GLICK, M. C. and WARREN, L. (1969). *J. biol. Chem.*, **244**:4103.
WEINSTEIN, D. B., MARSH, J. B., GLICK, M. C. and WARREN, L. (1970). *J. biol. Chem.*, **245**:3928.
WEISER, M. M. (1973a). *J. biol. Chem.*, **248**:2536.
WEISER, M. M. (1973b). *J. biol. Chem.*, **248**:2542.
WERB, Z. and COHN, Z. A. (1972). *J. biol. Chem.*, **247**:2439.
WETZEL, M. G. and KORN, E. D. (1969). *J. Cell Biol.*, **43**:90.
WICKUS, G. G. and ROBBINS, P. W. (1973). *Nature, New Biol.*, **243**:65.
WILFONG, R. F. and NEVILLE, D. M., JR. (1970). *J. biol. Chem.*, **245**:6106.
WOLFF, J. and JONES, A. B. (1971). *J. biol. Chem.*, **246**:3939.
WRAY, V. P. and PERDUE, J. F. (1974). *J. biol. Chem.*, **249**:1189.
YU, J., FISCHMAN, D. A. and STECK, T. L. (1973). *J. supramolec. Struct.*, **1**:233.

ADDENDUM

Several volumes have been published concerning the isolation of surface membranes (Fleischer and Packer, 1974; Korn, 1974). Robbins and Nicolson (1975) discuss the use of the iodination technique in examining proteins from transformed cell surfaces. The use of lectins as membrane markers is reviewed (Nicolson, 1974) and additional specific markers have been reported, such as virus (Roesing, Toselli and Crowell, 1975) or α-bungarotoxin (Festoff and Engel, 1974).

REFERENCES TO ADDENDUM

FESTOFF, B. W. and ENGEL, W. K. (1974). *Proc. natn. Acad. Sci. U.S.A.*, **71**:2435.
FLEISCHER, S. and PACKER, L. (1974) (Eds). *Biomembranes, Methods in Enzymology*, Vols. 31 and 32. New York; Academic Press.
KORN, E. D. (1974) (Ed.). *Methods in Membrane Biology*, Vol. 2. New York; Plenum Press.
NICOLSON, G. L. (1974). *Int. Rev. Cytol.*, **39**:89.
ROBBINS, J. C. and NICOLSON, G. L. (1975). *Cancer: A Comprehensive Treatise*, Vol. 3. Ed. F. F. BECKER. New York; Plenum Press.
ROESING, T. G., TOSELLI, P. A. and CROWELL, R. L. (1975). *J. Virology*, **15**:654.

4

Physical studies of membranes

Y. K. Levine
Department of Physical Chemistry, University of Leeds

The molecular biology approach to studies of membrane structure is hampered by the very properties which endow the membrane with its unique functional capacity. Membrane components are organized in a highly mobile structure and the conventional physical techniques reflect structural properties, across the membrane thickness only, averaged over the entire plane of the membrane. Membranes cannot be crystallized for high-resolution X-ray diffraction studies and the combination of the inherent motions of their components together with the comparatively long recording time of diffraction patterns yields results corresponding to both a space- and time-averaged structure. At the same time these motions are not sufficiently fast and isotropic to allow the observation of high-resolution magnetic resonance (NMR) spectra from every part of the structure. Even if such spectra were available, their complexity would preclude definite assignments of resonance lines and it is therefore essential to label the membranes specifically for any meaningful studies of their dynamic properties. These difficulties have spurred the use of probe molecules in studies of the dynamic properties of membranes. In particular, electron spin resonance (ESR) studies of nitroxide spin-labelled analogues of membrane components have been most successful in probing the motions within the structure. These studies, however, utilize the insertion of a large extraneous molecule into the membrane and necessarily assume that its motion reflects accurately its dynamic microenvironment within the structure. It is still a moot point whether the presence of a probe molecule perturbs the local organization and behaviour of the membrane.

Nevertheless, despite the difficulties and reservations associated with the application of physical techniques, we now possess a cogent picture of the average structure of membranes. The skeletal organization is a phospholipid bilayer, which is penetrated in places by the protein. The extent of penetration is not yet gauged, but it appears that the lipid–protein interactions do not substantially alter the bilayer's properties. The membrane appears to be structurally uniform in its plane, with considerable freedom for lateral motion, but the structure is highly ordered if viewed in a direction perpendicular to its plane. This structural anisotropy is currently considered to determine many of the functional properties of membranes.

X-ray diffraction patterns from membrane preparations contain a prominent diffuse band centred at a spacing of 0.46 nm (Levine, 1973). Oriented arrays of membranes stacked naturally (nerve myelin and retinal rods) or artificially by centrifugation and partial dehydration exhibit a band of non-uniform intensity distribution, with a maximum on a line parallel to the planes of the membranes (myelin—Blaurock, 1967; retinal rods—Blaurock and Wilkins, 1969; erythrocytes—Knutton, 1970). The anisotropy in the intensity of the band indicates that it is due to diffraction by a substructure oriented preferentially in a direction perpendicular to the plane of the membrane. The degree of anisotropy, and hence the orientation of the substructure within the membrane, depends on the source of the membrane preparation; myelin and erythrocyte membranes exhibit a highly anisotropic band whereas retinal rod membranes show only weak anisotropy. Furthermore, the observation of a diffuse 0.46 nm band indicates that the substructure is not crystalline and thus the membrane appears to contain an anisotropic fluid environment within its structure.

Membranes of *Mycoplasma laidlawii* whose lipids are specifically enriched with palmitic acid (Engelman, 1971) or *Escherichia coli* membranes (Esfahani et al., 1971; Shechter, Gulik-Krzywicki and Kaback, 1972) exhibit a gradual thermal transition from a diffuse 0.46 nm diffraction band, to a sharp band at a spacing of 0.42 nm which corresponds to a crystalline substructure. On lowering the temperature the intensity of the 0.42 nm band increases, while that of the 0.46 nm band decreases, with both bands coexisting in the middle of the transition. The transition temperature varies with the supplemented fatty acid used in the growth medium. Palmitate-enriched *Mycoplasma* membranes exhibit a transition around 40 °C, while *E. coli* membranes enriched with oleate and elaidate exhibit transitions at $\sim$24 and $\sim$35 °C respectively. A transition temperature of $\sim$20 °C has been observed for unsupplemented *E. coli* membranes (Shechter, Gulik-Krzywicki and Kaback, 1972). It has been estimated (Engelman, 1970) that the crystalline regions in *Mycoplasma* membranes extend over regions $\sim$40 nm in width over the plane of the membrane. The ability of the anisotropic membrane substructure to exist in both the fluid and crystalline configurations can also be observed by measuring the absorption or emission of the latent heat of fusion associated with this change of state by differential scanning calorimetry techniques (DSC). Membranes and whole cells of *Mycoplasma* grown on unsupplemented media show a transition around 30–40 °C, but the transition temperature can be lowered by lowering the growth temperature of the culture (Melchior et al., 1970). It is interesting that the maximum heat absorption during the transition is always observed at a slightly higher temperature than the growth temperature of the culture. Reinert and Steim (1970) have shown that organisms harvested in the logarithmic phase continued to grow normally after simulated calorimetry experiments. The transition temperature is sensitive to the chemical nature of the fatty acid supplement to the growth medium and can be increased to 60 °C for stearate-enriched membranes or reduced to -20 °C if the medium is enriched with oleate instead (Steim *et al.*, 1969; Reinert and Steim, 1970). Reversible phase transitions, centred at 0 °C, have also been observed in rat liver mitochondrial and microsomal membranes by differential scanning calorimetry (DSC) (Blazyk and Steim, 1972).

The physical state of the membrane substructure also appears to determine the partition of the small nitroxide spin label TEMPO (2,2,6,6-tetramethylpiperidine-1-oxyl) between the membrane and its aqueous environment. Neural membranes, the excitable membrane of muscle and sarcoplasmic vesicles (Hubbell and McConnell, 1968), as well as *Acholeplasma* membranes (Metcalfe, Birdsall and Lee, 1972) bind TEMPO readily with a local concentration of less than 10^{-2} M. Erythrocyte membranes, on the other hand, show only a weak binding capacity (Hubbell and McConnell, 1968). The immediate environment of the label has a low viscosity and the label tumbles with a rotational correlation time of 10^{-9}–10^{-11} s. However, on traversing the thermal transition the label is extruded from *Acholeplasma* membranes. This change in partition is readily interpreted in terms of the decrease in the volume of fluid regions available to TEMPO within the membrane.

The X-ray diffraction bands observed from membranes with fluid and crystalline substructures bear a remarkable similarity to the band observed from liquid and solid paraffins, as well as from bilayers of their total lipid extract (Finean *et al.*, 1969). Indeed, it was this similarity that led the early workers (Schmitt, Bear and Clark, 1935) to believe that the band in myelin was due to diffraction from the hydrocarbon chains of the lipids. The lipid bilayer, with a hydrocarbon core sandwiched between two planes of hydrated polar headgroups, forms the basis of the Davson–Danielli model for membrane structure and it accounts for the 0.46 nm and 0.42 nm diffraction bands from membranes. Bilayers of total membrane lipids (beef heart mitochondria —Gulik-Krzywicki, Rivas and Luzzati, 1967; brain lipids—Luzzati and Husson, 1962; erythrocyte lipids—Rand and Luzzati, 1968; *E. coli*— Esfahani *et al.*, 1971) and egg lecithin (Levine and Wilkins, 1971) exhibit a diffuse 0.46 nm diffraction band corresponding to a fluid hydrocarbon interior. Bands from oriented multilayers of egg lecithin are anisotropic, with an intensity maximum on a line parallel to the planes of the bilayers. Furthermore, the anisotropy in the intensity distribution increases markedly if cholesterol is incorporated into the structure (Levine and Wilkins, 1971). At the same time the centre of the band shifts from a spacing of 0.46 nm for the pure egg lecithin bilayer to 0.475 nm for bilayers of an equimolar mixture of egg lecithin and cholesterol. Levine and Wilkins (1971) have shown that as the 0.46 nm band is due to diffraction from the lateral organization of the chains in the bilayer, the anisotropic intensity distribution was due to a preferential orientation of the chain axes in a direction perpendicular to the plane of the bilayer. The orientation of the long axis of a chain changes continuously throughout the thickness of the bilayer from the glycerol backbone to the centre (Levine, 1973). It should be emphasized, however, that the band is not understood in terms of the detailed lateral packing of the hydrocarbon chains. The average area per lipid molecule at the aqueous interface of the bilayer, calculated from the bilayer thickness (Levine, 1973) is 60–70×10^{-20} m^2. However, as the cross-sectional area of two fully extended fatty acid chains in the crystalline state is 40.6×10^{-20} m^2, this implies that on melting the chains shorten, but still maintain a marked preferential orientation of their long axis. Phillips, Williams and Chapman (1969) have estimated that the entropy change associated with the alkyl chain transition in lecithin bilayers (1.25 kcal deg^{-1} mol^{-1} per CH_2 group) is considerably smaller than the corresponding change found in the melting of n-alkanes (1.9 kcal deg^{-1} mol^{-1} per CH_2 group). Hence the configurational freedom of the alkyl chains in bilayers above the thermal transition is smaller than that of chains in the bulk liquid.

The transition temperature in lipid bilayers is highly sensitive to their chemical nature and the heterogeneity in their chemical composition. In general the transition temperature can be increased by lengthening the chains and by decreasing their unsaturation (Levine, 1972). Thus aqueous suspensions of dimyristoyllecithin exhibit a thermal transition at 23 °C, while the transition in bilayers of distearoyllecithin occurs at 60 °C (Chapman, Williams and Ladbrooke, 1967). The same effect is observed in aqueous suspensions of total *Mycoplasma* lipids, where the transition temperature can be reduced from 60 °C for stearate-enriched lipids to −20 °C for oleate-enriched lipids (Reinert and Steim, 1970). The thermal transition of natural membrane lipids is considerably broader, because of their heterogeneous chemical composition, and tends to occur at temperatures below 20 °C; the thermal transition in bilayers of total mitochondrial lipids is centred at about −5 °C (Gulik-Krzywicki, Rivas and Luzzati, 1967; Blazyk and Steim, 1972). It is interesting that binary mixtures of lecithins differing markedly in their chain lengths or unsaturation exhibit two separate transitions (Phillips, Williams and Chapman, 1970), suggesting a separation of molecular species within each bilayer. The transition temperature of the higher-melting component is broadened and shifted to lower temperatures, while that of the transition of the lower-melting species is unaffected. In contrast, bilayers of intramolecularly mixed chains exhibit a sharp transition at an intermediate temperature. It thus appears possible that crystalline and fluid domains coexist within the bilayers on traversing the thermal phase transition of the chains.

ESR studies of the partition of TEMPO between phospholipid bilayers and their aqueous environment indicate that above the thermal transition the label senses the same low-viscosity environment that exists in membranes (Hubbell and McConnell, 1968). The partition is reduced sharply on lowering the temperature below the thermal transition of the chains (Hubbell and McConnell, 1971), but can also be reduced, to a lesser extent, if cholesterol is introduced into the bilayer (Hoult, 1973).

Comparison of the thermal transitions in *Mycoplasma* and rat liver mitochondrial or microsomal membranes and bilayers of their total lipid extracts observed by DSC shows a strong correlation between the two systems. In addition, it is found that neither pronase treatment of the membrane nor the thermal denaturation of its proteins affects the thermal transition to any extent. However, X-ray diffraction studies of *E. coli* membranes and lipids indicate a poorer correlation, with the bilayers having substantially lower transition temperatures (Esfahani *et al.*, 1971). Finean *et al.* (1969) found that although the intensity contours of the diffuse diffraction bands from membranes and bilayers of their lipids at comparable hydration are similar, the centre of the membrane band was shifted to slightly lower spacings (0.445 nm and 0.45 nm for muscle microsomes and bilayers of their lipids respectively). This may well reflect changes in the lateral organization of the hydrocarbon chains in the membrane due to lipid–protein interactions, but contribution to the diffraction from membrane protein is also expected. Nevertheless, the anisotropy in the intensity distribution of the membrane band follows the chemical composition of the lipids in the same way that is observed in lipid bilayers. The anisotropy is poor in membranes with a highly unsaturated hydrocarbon composition and a low cholesterol content (retinal rods) but increases markedly in the cholesterol-rich myelin and erythrocyte membranes.

The total lipid bilayer content within the membrane structure can be estimated by comparing the partition of TEMPO into membranes and bilayers of their lipids through the thermal transition. In *Acheoplasma* membranes (Metcalfe, Birdsall and Lee, 1972) and muscle microsomes (Robinson *et al*., 1972), this method indicates that 75 percent of the membrane surface contains a lipid bilayer. A figure of 80–90 percent coverage is obtained from the ratio of the heats of transition in membranes and in bilayers of their lipids (Reinert and Steim, 1970; Blazyk and Steim, 1972). It is impossible to say from these experiments whether the bilayer structure is continuous throughout the membrane.

The evidence advanced so far demonstrates only the very close structural properties exhibited by the hydrocarbon regions of membranes and bilayers of their lipids. However, a direct demonstration of the existence of a lipid bilayer within the membrane is provided by X-ray diffraction studies of the electron density profile of the membrane in a direction perpendicular to its local surface. Stacked arrays of membranes lying with their planes parallel act as sets of parallel mirrors which reflect the X-radiation and give rise to Bragg interference patterns (Levine, 1973). The positions of the diffraction maxima in the pattern are related to the width, d, of the basic repeat unit of the periodic stack of membrane by Bragg's Law

$$2d \sin \theta = n\lambda \qquad (4.1)$$

where 2θ is the angle of scattering and n is an integer defining the order of the maximum. The diffraction maxima observed from membranes are all related to a single repeat spacing and are known as the lamellar diffraction orders. In the X-ray diffraction experiments the membranes appear to have a uniform structure in their planes and yield one-dimensional diffraction patterns only in a direction perpendicular to their planes. The repeat unit of the stack of membranes may consist of a membrane pair, as is the case in myelin and retinal rods, or of a single membrane, as in partially dehydrated pellets of erythrocyte ghosts. The main lamellar diffraction from intact retinal rod segments contains up to eleven Bragg orders of a lamellar spacing of 29–32 nm (Blaurock and Wilkins, 1969). The spacing can be increased further to 37 nm by immersing the rods in various concentrations of glucose-Ringer's solutions (Blaurock and Wilkins, 1972) or alternatively reduced if immersed in sucrose-Ringer's solutions (Corless, 1972). Myelin can be swollen to varying degrees in hypotonic Ringer's solutions, water or dilute salt and sucrose solutions from a normal spacing of $\sim$17 nm to 40 nm (Blaurock, 1971). It is interesting that the lamellar spacing of peripheral nerve myelin ($\sim$17.5 nm) is somewhat larger than that of CNS myelin

(~15.5 nm) (Blaurock and Worthington, 1969; Worthington and Blaurock, 1969). This has been ascribed to the difference in the fatty acid composition of the membranes (Caspar and Kirschner, 1971). Membrane pellets obtained by centrifugation do not exhibit a Bragg spectrum unless they are partially dehydrated to a membrane concentration of 70–80 percent by weight. At that stage the membranes stack and lie with their planes parallel. The array is characterized by a unique lamellar spacing: 11–12 nm for mitochondrial (Thompson, Coleman and Finean, 1968) and erythrocyte (Knutton et al., 1970) membranes, and 30 nm for the brush border membrane of guinea-pig intestinal epithelial cells (Limbrick and Finean, 1970). Electron micrographs reveal that the smaller spacings correspond to a repeat unit containing a single membrane and that the 30 nm spacing corresponds to a repeat unit containing a pair of membranes. Further dehydration of the pellet gives rise to additional diffraction which has been assigned to breakdown products of the membranes. Fully dried preparations exhibit complex and often non-lamellar diffraction patterns, which indicate a major structural reorganization of membrane components into chemically distinguishable phases. Unfortunately, there has been a tendency among workers to use fully dried membrane preparations to study membrane structure by infrared spectroscopic techniques (Levine, 1972). A lamellar diffraction pattern from erythrocyte membranes is obtained only if the ghosts are prepared in buffers of osmolarities less than 2–3 milliosmolar, so that the residual haemoglobin accounts for more than 30 percent of the total membrane protein. The complete removal of haemoglobin causes the membranes to become considerably more labile and they exhibit lamellar diffraction only from separate lipid and lipoprotein phases (Knutton et al., 1970). The purple membrane of *Halobacterium halobium*, in contrast to all other membrane preparations, retains its structural integrity at all stages of dehydration with a lamellar spacing of 4.9 nm (Blaurock and Stoeckenius, 1971).

The intensity variation in the lamellar Bragg diffraction orders contains the information about the electron density profile of the membrane structure. The profile is a measure of the local deviation of the electron density from the overall mean value of the structure. The electron density of unhydrated proteins and phospholipid headgroups is 0.45×10^{-30} e m^{-3}, while the densities of methylene chains and terminal methyl groups in the liquid state are 0.296×10^{-30} e m^{-3} and 0.165×10^{-30} e m^{-3}. We thus expect the proteins and lipid headgroups to appear as electron-dense peaks, while the hydrocarbon region of the membrane will be seen as a deep electron-deficient trough. The diffracted amplitude from the membrane structure in any direction is determined by the Fourier transform of the electron density profile (Levine, 1973). The lamellar Bragg orders sample the height of the transform at angular deviations corresponding to integral submultiples of the reciprocal lamellar spacing. Unfortunately, because only intensity is measured in the diffraction experiment, no information is available about the phases of the Bragg orders and in general it is difficult to reconstruct the electron density profile by the inverse Fourier operation. However, if we assume that the electron density profile of the membrane has a centre of symmetry, then the phases of the Bragg orders can only be 0° and 180°, corresponding to positive and negative amplitudes. The phase can change at those directions in which the diffracted amplitude, and hence the diffracted intensity, is identically zero. Hence the relative phases of the transform peaks can be determined by locating the transform zeros. This is done by swelling or shrinking the membrane array so as to change the lamellar spacing. In fact it is not essential to locate the actual zero, for the diffracted amplitude changes rapidly near a zero, but only slowly near a transform peak (Levine, 1973).

Useful information about the gross features of the electron density profile of the membrane can be obtained by calculating the Patterson map, using only the observed intensities and assuming that they all have the same phase (Levine, 1973). The Patterson map is the self-correlation map of the electron density profile. Effectively the map is the result of calculating the product of two electron density profiles over the region of overlap as the separation of their centres increases. However, because of the periodicity of the membrane array the map will have a centre of symmetry at half the lamellar spacing $d/2$. It will give rise to positive peaks when two regions of positive or negative densities overlap or correlate and will decrease to a negative trough for a correlation between a positive peak and negative region. It is apparent that the maximum structural information is obtained if the membranes occupy less than half the repeat unit, the rest being filled with a fluid of constant density.

It should be mentioned, however, that the experimentally determined diffracted intensities are not the true intensities because of distortions imposed by the geometry of the diffraction experiment (Levine, 1973). The measured intensities must be corrected by the relevant Lorentz factor, which is proportional to n^2 for Bragg reflections from a randomly oriented set of parallel planes. Unfortunately this was not widely appreciated in the past. In fact these geometrical distortions set the resolution limit of the diffraction experiment, for the attenuation of the diffracted intensity increases considerably with the order number of the Bragg reflection.

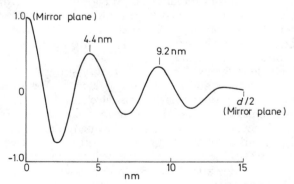

Figure 4.1 Patterson map for retinal rod membranes calculated from the data of Blaurock and Wilkins (1969). $d = 30$ nm

Patterson maps (*Figure 4.1*) of dark-adapted retinal rod membranes are easily interpreted in terms of the bilayer model for membrane structure. The maps show positive peaks near 4.4 nm and 9.2 nm with troughs around 2.0 nm and 6.0 nm. Blaurock and Wilkins (1969) assigned the 9.2 nm peak to the correlation between the two membranes, within the same repeat unit, whose centres are separated by that distance across the cytoplasmic space. The peak shape indicates further the lack of any gross asymmetries in the electron density profile of a single membrane. The 4.4 nm peak is then due to the correlation between the positive electron density peaks of a single membrane on either side of a hydrocarbon region of width ~2.0 nm. This identification is corroborated by optical measurements of the total solid contents of the rods by interference microscopy. The solids occupy 37 percent of the rod length, so that in a 30 nm wide repeat unit, the solids occupy an equivalent thickness of 5.5 nm per membrane. Blaurock and Wilkins (1969) concluded that the 4.4 nm peak could represent the centre-to-centre separation of the two membranes only if there were three times as much protein on the extracellular side of the membrane as on the cytoplasmic side, but such a high degree of asymmetry is inconsistent with the shape of the peak. Corless (1972) has confirmed this interpretation of the Patterson map and shown that the 9.2 nm peak shifts to smaller distances on shrinking the retinal rod membranes, while the 4.4 nm peak is unaffected. Patterson maps of the myelin membranes are not as informative as those for retinal rods because the membrane occupies considerably more than half the repeat unit. In peripheral nerve myelin the centre-to-centre separation of the two membranes within the same diffraction unit is 7.8 nm (Blaurock, 1971). Each membrane is located off-centre within its half of the repeat unit so that the cytoplasmic space is considerably narrower than the extracellular space between membranes in adjacent diffraction units (Worthington and Blaurock, 1969; Blaurock, 1971; Caspar and Kirschner, 1971). The arrangement of the membranes in CNS myelin is almost symmetrical, with the cytoplasmic space only slightly narrower than the extracellular space. In both forms of myelin the electron density profile of a single membrane appears to be largely symmetrical (Blaurock, 1971; Caspar and Kirschner, 1971).

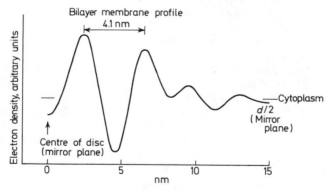

Figure 4.2 Electron density profile of a single retinal rod membrane calculated from the data of Blaurock and Wilkins (1969)

The Patterson map supports the assumption that the diffraction units of retinal rods and myelin possess a centre of symmetry, so that the diffracted amplitudes can be assigned phases of either 0° or 180°. It is found that the essential features of the electron density profiles are well defined at low resolution (<1.5 nm) by using four to six diffraction orders. The higher and weaker Bragg reflections add the necessary detail to resolve the small asymmetries in the electron densities. The profile of dark-adapted, intact, frog retinal rod membranes (*Figure 4.2*) (Blaurock and Wilkins, 1969, 1972; Corless, 1972) are characterized by a deep low-electron-density trough at the centre of the membrane, whose contours remain unaltered as the rods swell or shrink. This trough is bounded by two, fairly symmetric, electron-dense peaks, separated by 4.1 nm across the trough. Similar profiles are obtained for the myelin membrane (Blaurock, 1971; Caspar and Kirschner, 1971) at low resolution, but with a larger separation of the peaks across the

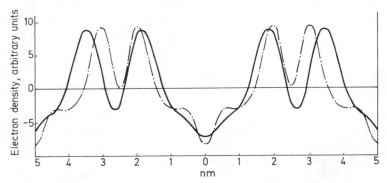

Figure 4.3 Electron density profile of egg lecithin bilayers at 14% water content (—·—·—·), d = 5.0 nm and at 21% water content (———), d = 5.15 nm. The central low-electron-density trough indicates the tendency of the terminal methyl groups to be localized near the centre of the hydrocarbon region. The smoother profile of the hydrocarbon region at 21% water content indicates that the terminal methyl groups are spread over a wider layer at the centre of the bilayer. Calculated from the data of Levine and Wilkins (1971)

trough, 4.6 nm. The profiles bear a marked resemblance to those found for phospholipid bilayers (*Figure 4.3*) with respect to both the shape of the low-electron-density trough and the overall dimensions of the structure: the separations of the electron-dense peaks across the hydrocarbon region, ∼3.7 nm in egg lecithin bilayers and ∼4.2 nm for bilayers of an equimolar mixture of egg lecithin and cholesterol (Levine, 1972) compare well with those found for the retinal rod and myelin membranes respectively. The constant contours of the trough upon swelling and shrinking the membranes strongly indicate that it corresponds to the hydrocarbon region of a lipid bilayer (Blaurock, 1971; Blaurock and Wilkins, 1972). This identification is confirmed in the case of myelin by the determination of the absolute electron density levels of the membrane profiles (Worthington and Blaurock, 1969; Blaurock, 1971). The hydrocarbon regions of phospholipid bilayers are not homogeneous in chemical composition and there is a marked tendency for the free terminal methyl groups of the lipid chains to be localized near the centre of the bilayers (Levine and Wilkins, 1971). This tendency is markedly reinforced if cholesterol is incorporated into the bilayer. The electron density profiles of myelin membranes indicate that this tendency exists in their hydrocarbon regions, with the terminal methyl groups localized in a layer about 1.5 nm wide at the centre of the region (Caspar and Kirschner, 1971; Blaurock, 1971). The high-resolution profile of the myelin membrane shows an asymmetric low-electron-density trough in the centre of the membrane. Caspar and Kirschner (1971) interpreted this asymmetry as being due to differences in the cholesterol contents in the two faces of the lipid bilayer: an approximately equimolar ratio of cholesterol and lipid on the extracellular side of the bilayer and a molar ratio of about 3:7 on the cytoplasmic side. This unequal distribution of cholesterol between the two sides is expected to cause the extracellular side of the bilayer to be somewhat wider than the cytoplasmic side, and indeed the centre of the hydrocarbon region is found to shift towards the cytoplasmic side. The contours of the trough could also be explained in terms of asymmetric penetration of proteins into the bilayer, but Caspar and Kirschner (1971) concluded that the low electron density of the profile rules out the presence of significant amounts of protein in the hydrocarbon region.

The comparable dimensions of electron density profiles of membranes and phospholipid bilayers indicate that the electron-dense peaks correspond to the polar headgroups of the lipids with fairly narrow layers of protein (Blaurock and Wilkins, 1969; Caspar and Kirschner, 1971). The marked lack of gross asymmetry in the profile of retinal rod membranes suggests that the rhodopsin molecules are distributed equally between the two sides of the disc. Corless (1972) attempted to locate the rhodopsin molecules in the structure by comparing the single membrane profiles of bleached and unbleached retinal rods. He concluded that the electron densities do not provide a reliable guide and that bleaching the retinae caused no major structural change within the membrane.

Aqueous suspensions of membranes and sonicated dispersions of bimolecular lipid vesicles yield remarkably similar diffraction patterns (Wilkins, Blaurock and Engelman, 1971). These non-periodic samples exhibit a continuous diffraction pattern consisting of a series of bands which are related directly to the Fourier transform of the single membrane electron density profile. No Bragg reflections are formed because of the lack of spatial order in the suspensions and hence the continuous intensity distribution is observed rather than simply sampled (Levine, 1973). Wilkins, Blaurock and Engelman (1971) showed that the continuous diffraction patterns from phospholipid vesicles corresponded to those predicted by simple models for the bilayer structure. The models indicate that the maxima of the intensity bands fall at spacings corresponding to integral fractions of the peak-to-peak separation of the phospholipid headgroups across the hydrocarbon region. The intensity variation of the bands was found to be determined by the homogeneity of the structure of the hydrocarbon region of the bilayer. Thus if the terminal methyl groups of the chains are localized at the centre of the hydrocarbon region, the odd-numbered bands are more intense than the even-numbered bands. The reverse is true if the hydrocarbon region is homogeneous, with a uniform distribution of the terminal methyl groups over its width. Sonicated dispersions of egg lecithin show an equivalent thickness of 3.6 nm, compared to $\sim$3.7 nm measured for the peak separation on electron density profiles calculated from the discrete Bragg spectrum at 20 percent water content.

If the lipid bilayer is sandwiched between two parallel sheets of *unhydrated* protein, then because of the similar electron densities of the protein and the polar headgroups of the lipids, the contribution of the protein layers to the continuous diffraction pattern would be limited to shifting the maximum of the first diffraction band to higher spacings (Wilkins, Blaurock and Engelman, 1971; Levine, 1973). However, if the protein were hydrated, and formed wide steps on either side of the bilayer, its contribution to the diffraction would superimpose a strong ripple on the first and possibly second bands of the continuous diffraction pattern (Engelman, 1971; Blaurock, 1973a). The diffraction from aqueous suspensions of erythrocyte, *Mycoplasma* and nerve ending membranes shows no contribution from protein layers external to the lipid bilayer and the patterns are dominated by the contribution from a bilayer membrane of equivalent thickness in the range 4.0–4.5 nm. *Mycoplasma* membranes provide the best evidence for this conclusion as the equivalent thickness of the palmitate-enriched membrane decreases (from 4.7 nm at 10 °C, to 4.1 nm at 52 °C) on passing through the thermal transition. The intensity distribution in the bands indicates that at 10 °C the terminal methyl groups of the lipids are well localized at the centre of the bilayer, as is to be expected in a crystalline hydrocarbon region. At 52 °C the band intensities indicate a fluid, though inhomogeneous, organization in the hydrocarbon region. Engelman (1971) has shown that if the protein were to form uniform layers $\sim$3.5 nm thick on each face of a lipid bilayer, its contribution to the diffraction pattern would be detected. Erythrocyte membranes, too, exhibit diffraction bands corresponding to a bilayer 4.5 nm thick with an inhomogeneous hydrocarbon region, similar to that observed in bilayers of its total lipid extract (Rand and Luzzati, 1968).

Protein contributions to the diffraction patterns have been recognized in rather specialized membranes such as the coat of the bacteriophage PM2 (Harrison *et al*., 1971) and cell envelopes of *Halobacterium halobium* (Blaurock, 1973a). Harrison *et al*. found a lipid bilayer 4.0 nm wide in the bacteriophage coat, with protein subunits extending from the lipid polar groups to the outer margin of the virus particle, a radial distance of 6.0 nm. The model system of phospholipid–cytochrome *c* dispersions (Blaurock, 1973b) also exhibits a bilayer diffraction pattern modulated by a ripple which is due to the contribution from external layers of the globular proteins.

The absence of any distinctive features in the diffraction from membranes which can be identified with the protein component leads to a model for membrane structure in which the protein is half-submerged in the lipid bilayer (Vanderkooi and Sundaralingam, 1970; Wilkins, Blaurock and Engelman, 1971; Blaurock, 1972). The protein penetration of the bilayer, however, appears to cause only small changes in the structural properties of the hydrocarbon region. Such penetration has been observed directly for the purple membrane of *Halobacterium halobium*, where there is an unusually high electron density at the centre of the hydrocarbon region of the membrane (Blaurock, 1972).

Support for this model of membrane structure is provided by studies of aqueous suspensions of lysozyme/cardiolipin mixtures (Gulik-Krzywicki *et al*., 1969) and serum albumin/lecithin–cardiolipin mixtures (Rand, 1971). The protein–lipid interaction causes a thinning of the lipid bilayer by up to 0.9 nm. Such thinning of the bilayer could be due either to protein penetration of the bilayer or to a marked increase in the area per molecule in a continuous lipid bilayer. This sort of thinning in the membrane could result in a structure for which the protein and lipid bilayer contributions to the diffraction overlap in such a way as to yield a bilayer-type diffraction pattern (Blaurock, 1973a).

Although the protein component of the membrane contributes little to the lamellar diffraction, all membranes exhibit a diffuse diffraction band at 1.0 nm Bragg spacing due to protein structural features in the plane of the membrane (Finean, 1969; Blaurock and Wilkins, 1969). Diffraction from the organization of protein molecules on the surface of the membrane has been recognized in retinal rod membranes (Blasie et al., 1965; Blasie and Worthington, 1969; Blaurock and Wilkins, 1969) and in the purple membrane of *Halobacterium halobium* (Blaurock and Stoeckenius, 1971). In both membranes the photopigment molecules have considerable long-range organization in the surface of the membrane. Blaurock and Stoeckenius (1971) have shown that in the purple membranes the molecules are organized in a quasi-crystalline manner in two well-defined planes. The centres of the molecules in the plane are 3.0 nm apart, with four molecules in a hexagonal cell unit: three centred in one plane and one on the opposite face of the membrane. The organization of rhodopsin molecules is less well defined and Blasie and Worthington (1969) analysed the diffraction bands in terms of a planar liquid arrangement of rhodopsin molecules, each molecule having three nearest neighbours at an average centre-to-centre separation of 5.6 nm.

Studies of the differential absorption of plane-polarized light (photo-dichroism) exhibited by the rhodopsin molecules in retinal rods indicate that the molecules are not fixed in the membrane surface, but possess considerable Brownian rotational motion (Hagins and Jennings, 1960; Brown, 1972; Cone, 1972). Dark-adapted rods are dichroic from the side, absorbing light which is plane-polarized perpendicularly to the long axes of the rods at least three times as effectively as light polarized in a direction parallel to the rod (Denton, 1954, 1959). This dichroism implies that the rhodopsin chromophore is oriented parallel to the plane of the disc membrane. The lack of photoinduced dichroism in the plane of the disc membrane indicates that the chromophores are randomly oriented, but the retina does not become dichroic even after partially bleaching the rhodopsin with plane-polarized light (Hagins and Jennings, 1960). Photodichroism in the plane of the membrane can be induced, however, if the retina is fixed with glutar-aldehyde prior to the partial bleaching with plane-polarized light (Brown, 1972). These observations are only consistent with the possibility that the rhodopsin molecule is free to undergo Brownian rotational motion within the plane of the disc membrane. This rotational motion is revealed by following the time course of the decay of the transient photodichroism induced in the retina by a flash bleach with polarized light. Cone (1972) measured a relaxation time of 20 μs for the Brownian rotational motion of the rhodopsin molecule, which corresponds to a viscosity of $\sim$2 P at the site of a molecule $\sim$2.5 nm in diameter. Similar values were obtained for the rotational motion of cytochrome oxidase in the inner mitochondrial membrane (Junge, 1972). The viscosity is comparable with those of light oils and suggests that the protein molecules are immersed in the lipid bilayer of the membrane.

Evidence that the surface layer of proteins is organized in a highly mobile state is furnished by studies of the translational diffusion of fluorescent antibodies attached to the surface antigens of tissue culture cells (Frye and

Edidin, 1970; Karnovsky and Unanue, 1973; Raff and de Petris, 1973). The observations of Frye and Edidin (1970) on the translational diffusion in cells of mouse and human origin indicate a diffusion coefficient of $\sim 10^{-10}$ cm^2 s^{-1}, corresponding to a viscosity in the region of 1–10 P at 20 °C.

Translational diffusion within the plane of the membrane is not confined to the protein layers. Nitroxide spin-labelled phospholipids incorporated into the lipid bilayer of the membrane of sarcoplasmic vesicles of rabbit skeletal muscle also diffuse laterally in the plane of the bilayer (Scandella, Devaux and McConnell, 1972). The diffusion coefficient of 6×10^{-8} cm^2 s^{-1} at 40 °C for the translational motion of the spin label is compatible with the values estimated for the larger protein molecules.

The membrane structure is thus in a considerable state of flux over its surface, but it exhibits a marked anisotropy in the organization of the hydrocarbon region in a direction perpendicular to its surface. This structural anisotropy is also sensed on a much shorter time scale ($\sim 10^{-6}$ s) by the motions of nitroxide spin labels buried in the interior of the membrane.

Fatty acid and phospholipid spin labels, with the paramagnetic NODO (N-oxyl-4′,4′-dimethyloxazolidine) ring attached rigidly to different carbon atoms along the hydrocarbon chain, can be incorporated into the membranes by exchange (Hubbell and McConnell, 1969), biosynthetically (Keith, Bulfield and Snipes, 1970; Tourtellotte, Branton and Keith, 1970) or by fusion with labelled phospholipid vesicles (Grant and McConnell, 1973). The degree of penetration of the NODO ring into the membrane structure may be inferred from the relative rates of reduction of different spin labels when the membrane is bathed in a solution of ionic reducing agents, e.g. ascorbic acid (Hubbell and McConnell, 1969). The penetration of the ring into membranes of sarcoplasmic vesicles can also be estimated by the differential broadening of the nuclear magnetic resonance signal of the choline N-methyl groups of the phospholipids (Eletr and Inesi, 1972). It is found that the further the ring is located from the carboxyl end of the chain, the deeper it penetrates into the hydrocarbon region of the membrane.

The ESR spectrum of the nitroxide ring is sensitive both to the polarity of its environment and to its orientation relative to the applied static magnetic field (Hamilton and McConnell, 1968; Jost, Waggoner and Griffith, 1971). If the ring is embedded in a crystalline matrix, its magnetic properties can be measured along the principal molecular axis (Libertini and Griffith, 1970; Hubbell and McConnell, 1971). The splitting of the nitroxide triplet varies from 31 G with the applied field oriented perpendicular to the plane of the ring (and parallel to the $2p\pi$ nitrogen orbital of the unpaired electron), to 6 G as the field is directed in the plane of the ring in directions both parallel and perpendicular to the N–O bond. These differences are averaged out if the ring tumbles isotropically with a rotational correlation time of 10^{-7}–10^{-12} s; the correlation time can be defined as the time it takes the molecule to rotate through an angle of 1 rad. The splitting of the averaged isotropic triplet—the isotropic coupling constant—changes from 15.2 G in hexane to 17.1 G in water, while simultaneously the centre line is shifted to higher field values. The rotational correlation time can be evaluated from the linewidths of the triplet (Hamilton and McConnell, 1968), and in general the slower the motion, the broader the lines.

At rates of motion less than 10^6 per second, there is an incomplete averaging of the magnetic parameters and a 'rigid glass' or 'polycrystalline' spectrum is observed from a randomly oriented ensemble of the magnetically axially-symmetric NODO rings (McConnell and McFarland, 1970). The spectra yield information about the coupling constants in directions parallel and perpendicular to the axis of symmetry of the ring. These spectra are observed not only in rigid solutions, but also if the NODO ring undergoes fast axial rotation, with a slow reorientation rate of the axis relative to the applied magnetic field. The axial motion effectively averages out the magnetic anisotropies in the plane perpendicular to the axis of rotation and the NODO ring behaves as though it were a static axially symmetric nitroxide label randomly oriented in space. As the reorientational rate of the axis increases the spectrum loses its anisotropic character and becomes a slow-moving isotropic spectrum. If the axis of rotation does not coincide with any of the principal molecular axes, it is possible to deduce the orientation of the NODO ring from the ESR spectrum. This is particularly easy for the fatty acid and phospholipid spin labels in which the $2p\pi$ orbital of the unpaired electron is parallel to the long axis of the hydrocarbon chain. Thus for a chain undergoing rapid axial rotation, the orientation of the $2p\pi$ orbital relative to the axis can be expressed in terms of an order parameter $S. S = \frac{1}{2} \langle 3 \cos^2 \alpha - 1 \rangle$, where α is the angle between the axis of rotation and the $2p\pi$ orbital (Hubbell and McConnell, 1971). If the NODO ring is attached at a position close to the carboxyl group, the motional freedom of the $2p\pi$ orbital is restricted by the motion of the entire chain within the membrane and it will experience a relatively polar environment. On the other hand, if the ring is located near the methyl end of the chain, its motion will be affected further by the flexibility of all carbon–carbon bonds between the carboxyl group and the ring. The overall motion then depends on the rigidity of the immediate environment of the NODO ring. Knowledge of the variation of the order parameter, S, with the position of the NODO ring along the fatty acid chain, provides a ready description of the order and anisotropy of the hydrocarbon region. It is readily seen that as the time-averaged angle α increases, the order parameter will decrease from a value of 1 towards 0.

Fatty acid spin labels incorporated into membranes execute fast axial rotation of the whole chain and the NODO ring undergoes a markedly anisotropic motion (Rottem et al., 1970; Hubbell and McConnell, 1971; Seelig and Hasselbach, 1971). It is found that in general the anisotropic character of the motion of the NODO ring decreases as it penetrates further into the hydrocarbon region of the membrane. In membranes of sarcoplasmic vesicles (Seelig and Hasselbach, 1971), the order parameter for a ring near the carboxyl end of the chain indicates a mean angle of deviation $\sim 30°$ and decreases slowly for the first seven carbon–carbon bonds, but with a pronounced increase in the flexibility of the chain thereafter. By the 10th bond the label tumbles almost isotropically. Similar behaviour is observed for the axonal membrane of the walking leg nerve of the Maine lobster *Homarus americanus* (Hubbell and McConnell, 1971) and in *Mycoplasma* membranes (Rottem et al., 1970). However, the hydrocarbon core of erythrocyte membranes is considerably more ordered, for even a ring attached to the 12th atom of the stearic acid chain exhibits a characteristically anisotropic spectrum (Hubbell and McConnell, 1969). Furthermore, the mean angle of deviation observed for the initial bonds of labels in erythrocyte membranes, $\sim 16°$, is roughly half that observed in membranes with low cholesterol content.

The fatty acid composition of the membrane appears to have a pronounced effect on the motional freedom of the label. *Mycoplasma* membranes labelled with the 4-stearic acid derivative exhibit a smaller-order parameter if enriched with oleic acid chains than if chains of the *trans* isomer, elaidic acid, predominate in the composition (Rottem *et al.*, 1970). Similar differences are observed in the apparent isotropic rotational correlation times of the 12-stearic acid derivative when incorporated *in vivo* into oleate-enriched and stearate-enriched *Mycoplasma* membranes (Tourtellotte, Branton and Keith, 1970). The spectra exhibited by the stearate-enriched membranes yielded larger correlation times, corresponding to slower motions, than those obtained from spectra of the oleate-enriched membranes. The spectra of labelled native *Mycoplasma* membranes indicate an even higher motional freedom of the NODO ring. It is interesting that neither glutaraldehyde fixation of the native *Mycoplasma* membranes nor changes in the membrane Mg^{2+} concentration affected the motions of the spin labels (Rottem *et al.*, 1970).

The anisotropic motion of the NODO ring within the hydrocarbon region can also be demonstrated by the preferential orientation of the $2p\pi$ orbital relative to the local membrane surface. In pellets of membranes of sarcoplasmic vesicles labelled with the 4-stearic acid derivative, the $2p\pi$ orbital orientation perpendicular to the membrane surface is at least 60 percent more probable than the parallel orientation (Eletr and Inesi, 1972), and a similar preference is observed for the 12-stearic acid label in erythrocyte membranes oriented by hydrodynamic shear (Hubbell and McConnell, 1969).

Comparison of the motional parameters of the fatty-acid spin labels in membranes and phospholipid bilayers indicates that the lipid–protein interactions tend to enhance the anisotropic character of the hydrocarbon region over the first seven or eight carbon–carbon bonds (Rottem *et al.*, 1970; Hubbell and McConnell, 1971; Seelig and Hasselbach, 1971). Although the absolute differences in the order parameters are only about 10 percent, the order parameters tend to decrease more rapidly across the hydrocarbon region of phospholipid bilayers than those of membranes. Hubbell and McConnell (1971) estimate that the anisotropy of the hydrocarbon region of the low-cholesterol axonal membrane of lobster is close to that observed for bilayers of 2:1 molar ratio mixture of egg lecithin–cholesterol. The enhancement of order in the hydrocarbon region of membranes is also reflected by the rotational correlation times of the 12-stearic acid label in *Mycoplasma* membranes and bilayers of its lipids (Tourtellotte, Branton and Keith, 1970). It was found that the differences in the rates of motion of the NODO ring between membranes and vesicles were greater for the oleate-enriched than the stearate-enriched systems.

A further, significant difference between membranes and phospholipid bilayers is the change in the polarity of the environment sensed by the label. In sarcoplasmic vesicles the polarity remains virtually constant for the first eight bonds and lies between that of water and ethanol. In contrast, the polarity in egg lecithin bilayers indicated by the same labels decreases continuously to a fairly hydrophobic value.

It must be emphasized, however, the ESR spin-label studies provide only a crude indication of the motions and organization of the hydrocarbon region of membranes (Hubbell and McConnell, 1971). It is encouraging, however, to note the consistency of these observations with those made by X-ray diffraction studies, particularly with regard to the lipid composition of the membrane.

Recent technical developments in nuclear magnetic resonance (NMR) have afforded a direct approach to the study of membrane dynamics by observing the resonance signals from protons (Davis and Inesi, 1971), from ^{13}C (Robinson et al., 1972) and ^{31}P (Davis and Inesi, 1972) nuclei of the phospholipid molecules in membranes of sarcoplasmic vesicles.

The proton spectrum indicates that the hydrocarbon region of the membrane contains two distinct populations of chains differing in their motions. About 80 percent of the hydrocarbon chains appear to be organized in a partially restricted and ordered structure, while the rest form a fluid environment and undergo fast, isotropic motions. It has been suggested that the fluid region contains a selected population of highly unsaturated hydrocarbon chains (Davis and Inesi, 1971). Robinson et al. (1972) found that there was little difference between the average motion of the carbon atoms of the chains of the lipids in bilayers and membranes. ^{13}C NMR studies of phospholipid bilayers indicate the existence of a fluidity gradient in the structure, with the motions of carbon atoms increasing from the glycerol backbone of the molecule both outwards to the aqueous headgroup region and inwards along the hydrocarbon chain (Levine et al., 1972). The motions of the phosphate groups of the lipids both in bilayers and in the membrane are approximately 10^2-10^3 times slower than those characteristic of the hydrocarbon chains and reflect the motional constraints of the glycerol backbone region.

The lipid–protein interactions in the sarcoplasmic membrane impose strong restrictions on the motions of the choline groups of the lipids, but these restrictions can be removed to a large extent by thermal denaturation or trypsin digestion of the protein (Davis and Inesi, 1971). In contrast, the structural alteration of the protein does not affect the motions of the phosphorus atoms. Surprisingly, the choline groups of nearly 60 percent of the lecithin molecules in the membrane exhibit a thermal transition from the rigid, motionally restricted state to a highly mobile state at around 40 °C. This transition correlates strongly with the temperature-dependent Ca^{2+} efflux. Agents which alter the protein structure and simultaneously increase Ca^{2+} efflux also increase the fraction of mobile choline methyl groups.

The investigations of membrane structure described above have revealed a membrane in which highly mobile protein molecules are submerged in a lipid bilayer. The hydrocarbon region of the bilayer has a non-uniform structure in a direction perpendicular to the membrane surface and is affected only to a small extent by the lipid–protein interactions. Little structural information about the protein is available as the spectroscopic techniques of circular dichroism and optical rotatory dispersion used in solution studies of the secondary structure of proteins are beset by large artefacts due to light scattering (for discussion see Levine, 1972; Gordon and Holzwarth, 1971). Nevertheless it appears that the average membrane protein has a considerable α-helical content (∼50%) with only a small proportion (0–10%) of the pleated sheet conformation.

REFERENCES

BLASIE, J. K. and WORTHINGTON, C. R. (1969). *J. molec. Biol.*, **39**:417.
BLASIE, J. K., DEWEY, M. M., BLAUROCK, A. E. and WORTHINGTON, C. R. (1965). *J. molec. Biol.*, **14**:143.
BLAUROCK, A. E. (1967). *Ph.D. Thesis*, University of Michigan, Ann Arbor.
BLAUROCK, A. E. (1971). *J. molec. Biol.*, **56**:35.
BLAUROCK, A. E. (1972). *Chem. Phys. Lipids*, **8**:285.
BLAUROCK, A. E. (1973a). *Biophys. J.*, **13**:281.
BLAUROCK, A. E. (1973b). *Biophys. J.*, **13**:290.
BLAUROCK, A. E. and STOECKENIUS, W. (1971). *Nature, New Biol.*, **233**:149.
BLAUROCK, A. E. and WILKINS, M. H. F. (1969). *Nature, Lond.*, **223**:906.
BLAUROCK, A. E. and WILKINS, M. H. F. (1972). *Nature, Lond.*, **236**:313.
BLAUROCK, A. E. and WORTHINGTON, C. R. (1969). *Biochim. biophys. Acta*, **173**:419.
BLAZYK, J. F. and STEIM, J. M. (1972). *Biochim. biophys. Acta*, **266**:737.
BROWN, P. K. (1972). *Nature, New Biol.*, **236**:35.
CASPAR, D. L. D. and KIRSCHNER, D. A. (1971). *Nature, New Biol.*, **23**:46.
CHAPMAN, D., WILLIAMS, R. M. and LADBROOKE, D. B. (1967). *Chem. Phys. Lipids*, **1**:445.
CONE, R. A. (1972). *Nature, New Biol.*, **236**:39.
CORLESS, J. M. (1972). *Nature, Lond.*, **237**:229.
DAVIS, D. G. and INESI, G. (1971). *Biochim. biophys. Acta*, **241**:1.
DAVIS, D. G. and INESI, G. (1972). *Biochim. biophys. Acta*, **282**:180.
DENTON, E. J. (1954). *J. Physiol.*, **124**:17.
DENTON, E. J. (1959). *Proc. R. Soc. Ser. B*, **158**:78.
ELETR, S. and INESI, G. (1972). *Biochim. biophys. Acta*, **282**:174.
ENGELMAN, D. M. (1970). *J. molec. Biol.*, **47**:115.
ENGELMAN, D. M. (1971). *J. molec. Biol.*, **58**:153.
ESFAHANI, M., LIMBRICK, A. R., KNUTTON, S., OKA, T. and WAKIL, S. J. (1971). *Proc. natn. Acad. Sci. U.S.A.*, **68**:3180.
FINEAN, J. B. (1969). *Q. Rev. Biophys.*, **2**:1.
FINEAN, J. B., KNUTTON, S., LIMBRICK, A. R. and COLEMAN, R. (1969). *Molec. Crystallogr.*, **7**:347.
FRYE, L. D. and EDIDIN, M. (1970). *J. Cell Sci.*, **7**:313.
GORDON, D. J. and HOLZWARTH, G. (1971). *Proc. natn. Acad. Sci. U.S.A.*, **68**:2365.
GRANT, C. W. M. and MCCONNELL, H. M. (1973). *Proc. natn. Acad. Sci. U.S.A.*, **70**:1238.
GULIK-KRZYWICKI, T., RIVAS, E. and LUZZATI, V. (1967). *J. molec. Biol.*, **27**:303.
GULIK-KRZYWICKI, T., SHECHTER, E., LUZZATI, V. and FAURE, M. (1969). *Nature, Lond.*, **223**:1116.
HAGINS, W. A. and JENNINGS, W. H. (1960). *Trans. Faraday Soc.*, **27**:180.
HAMILTON, C. L. and MCCONNELL, H. M. (1968). *Structural Chemistry and Molecular Biology*, p. 115. Ed. A. RICH and N. DAVIDSON. San Francisco; W. H. Freeman.
HARRISON, S. C., CASPAR, D. L. D., CAMERINI-OTERO, R. D. and FRANKLIN, R. M. (1971). *Nature, New Biol.*, **225**:543.
HOULT, J. R. S. (1973). *Ph.D. Thesis*, University of Cambridge.
HUBBELL, W. L. and MCCONNELL, H. M. (1968). *Proc. natn. Acad. Sci. U.S.A.*, **61**:12.
HUBBELL, W. L. and MCCONNELL, H. M. (1969). *Proc. natn. Acad. Sci. U.S.A.*, **63**:16.
HUBBELL, W. L. and MCCONNELL, H. M. (1971). *J. Am. chem. Soc.*, **93**:314.
JOST, P., WAGGONER, A. S. and GRIFFITH, O. H. (1971). *Structure and Function of Biological Membranes*, Chap. 3. New York; Academic Press.
JUNGE, W. (1972). *FEBS Lett.*, **25**:109.
KARNOVSKY, M. J. and UNANUE, E. R. (1973). *Fedn Proc. Fedn Am. Socs exp. Biol.*, **32**:55.
KEITH, A. D., BULFIELD, G. and SNIPES, W. (1970). *Biophys. J.*, **10**:618.
KNUTTON, S. (1970). *Ph.D. Thesis*, University of Birmingham.
KNUTTON, S., FINEAN, J. B., COLEMAN, R. and LIMBRICK, A. R. (1970). *J. Cell Sci.*, **7**:357.
LEVINE, Y. K. (1972). *Prog. Biophys. molec. Biol.*, **24**:1.
LEVINE, Y. K. (1973). *Prog. Surface Sci.*, **3**:279.

LEVINE, Y. K. and WILKINS, M. H. F. (1971). *Nature, New Biol.*, **230**:69.
LEVINE, Y. K., BIRDSALL, N. J. M., LEE, A. G. and METCALFE, J. C. (1972). *Biochemistry*, **11**:1416.
LIBERTINI, L. J. and GRIFFITH, O. H. (1970). *J. chem. Phys.*, **53**:1359.
LIMBRICK, A. R. and FINEAN, J. B. (1970). *J. Cell Sci.*, **7**:373.
LUZZATI, L. and HUSSON, F. (1962). *J. Cell Biol.*, **12**:207.
MCCONNELL, H. M. and MCFARLAND, B. G. (1970). *Q. Rev. Biophys.*, **3**:91.
MELCHIOR, D. L., MOROWITZ, H. J., STURTEVANT, J. M. and TSONG, T. Y. (1970). *Biochim. biophys. Acta*, **219**:114.
METCALFE, J. C., BIRDSALL, N. J. M. and LEE, A. G. (1972). *FEBS Lett.*, **21**:335.
PHILLIPS, M. C., WILLIAMS, R. M. and CHAPMAN, D. (1969). *Chem. Phys. Lipids*, **3**:342.
RAFF, M. C. and DE PETRIS, S. (1973). *Fedn Proc. Fedn Am. Socs exp. Biol.*, **32**:48.
RAND, R. P. (1971). *Biochim. biophys. Acta*, **241**:823.
RAND, R. P. and LUZZATI, V. (1968). *Biophys. J.*, **8**:125.
REINERT, J. C. and STEIM, J. M. (1970). *Science, N.Y.*, **168**:1580.
ROBINSON, J. D., BIRDSALL, N. J. M., LEE, A. G. and METCALFE, J. C. (1972). *Biochemistry*, **11**:2903.
ROTTEM, S., HUBBELL, W. L., HAYFLICK, L. and MCCONNELL, H. M. (1970). *Biochim. biophys. Acta*, **219**:104.
SCANDELLA, C. J., DEVAUX, P. and MCCONNELL, H. M. (1972). *Proc. natn. Acad. Sci. U.S.A.*, **69**:2056.
SCHMITT, F. O., BEAR, R. S. and CLARK, G. L. (1935). *Radiology*, **25**:131.
SEELIG, J. and HASSELBACH, W. (1971). *Eur. J. Biochem.*, **21**:17.
SHECHTER, E., GULIK-KRZYWICKI, T. and KABACK, H. R. (1972). *Biochim. biophys. Acta*. **274**:466.
STEIM, J. M., TOURTELLOTTE, M. E., REINERT, J. C., MCELHANEY, R. N. and RADER, R. L. (1969). *Proc. natn. Acad. Sci. U.S.A.*, **63**:104.
THOMPSON, J. E., COLEMAN, R. and FINEAN, J. B. (1968). *Biochim. biophys. Acta*, **150**:405.
TOURTELLOTTE, M. E., BRANTON, D. and KEITH, A. T. (1970). *Proc. natn. Acad. Sci. U.S.A.*, **66**:909.
VANDERKOOI, G. and SUNDARALINGAM, M. (1970). *Proc. natn. Acad. Sci. U.S.A.*, **67**:233.
WILKINS, M. H. F., BLAUROCK, A. E. and ENGELMAN, D. M. (1971). *Nature, New Biol.*, **230**:72.
WORTHINGTON, C. R. and BLAUROCK, A. E. (1969). *Biochim. biophys. Acta*. **173**:427.

ADDENDUM

Dupont, Harrison and Hasselbach (1973) have recently investigated the structure of the membranes of rabbit sarcoplasmic reticulum vesicles by X-ray diffraction techniques. The lamellar spacing was found to vary continuously from 27 nm for a pellet containing 50 percent by weight of water, to 17 nm at a water content of 25 percent. Further dehydration caused a reorganization of the membrane structure.

The Patterson map and electron density profile show an electron-dense peak in the extra vesicle space at a distance of 5–5.55 nm from the centre of the hydrocarbon region in addition to a largely symmetric bilayer structure. The separation between the electron-dense peaks (lipid headgroups and protein) across the hydrocarbon region was found to be 4 nm, comparable with other membranes. The asymmetry in the protein distribution about the phospholipid bilayer can be removed progressively by blocking the SH groups of the membranes (Dupont and Hasselbach, 1973). When more than four SH groups are blocked the asymmetry disappears, but with a halving of the lamellar spacing. This indicates that at this stage the diffraction unit consists of single symmetric bilayer membranes. The structural changes correlate well with the reduction in the ATPase activity of the membrane and this has been explained as being due to the progressive penetration of the protein projections into the phospholipid bilayers as the SH groups are blocked (Dupont and Hasselbach, 1973).

Blaurock (1972, 1973) has shown that the average thickness of solids calculated from the chemical composition of myelin, retinol rod and the purple membranes agrees well with the separations of the electron-dense peaks in the electron density profiles. This is also the case for sarcoplasmic reticulum membranes (Dupont, Harrison and Hasselbach, 1973). However, the average thickness of the lipid layer appears to be significantly smaller with the possible exception of myelin. The diffraction patterns of these membranes also show regular changes in proportion to their protein content, in a way that supports the idea of protein penetration (Blaurock, 1973).

These investigations argue in favour of a membrane structure in which the protein molecules are partially submerged in the lipid bilayer. These structures impose a chemical polarization on the protein molecules for only their hydrophobic portions should be in contact with the hydrocarbon region of the lipid bilayer.

REFERENCES TO ADDENDUM

BLAUROCK, A. E. (1972). *Nature, Lond.*, **240**:556.
BLAUROCK, A. E. (1973). *Nature, Lond.*, **244**:172.
DUPONT, Y. and HASSELBACH, W. (1973). *Nature, New Biol.*, **246**:41.
DUPONT, Y., HARRISON, S. C. and HASSELBACH, W. (1973). *Nature, Lond.*, **244**:555.

5

Physicochemical studies of cellular membranes

D. Chapman
*Department of Chemistry, University of Sheffield**

5.1 INTRODUCTION

In recent years considerable interest has grown in the study of the structure and function of cell membranes. Their permeability characteristics (to ions and organic molecules), their organizational aspects (of enzymes and pigments), their stability to drugs or disease processes and their associated triggering mechanisms are important and relevant to many areas of biology, pharmacology and physiology (Bittar, 1970). Cell recognition and some phenomena in immunology and cancer are also now becoming recognized as being related to membrane processes (Wallach, 1970).

Advances in our understanding of cell membrane structure have come from various directions. Techniques have been developed for the separation of cell membranes and subsequent analysis of their constituents, e.g. lipids, proteins or cholesterol. Electron microscope techniques have developed so that freeze-etch methods are now available and some of the concern about fixation methods and removal of water from the systems has been reduced. A whole range of physical techniques have been applied to membrane constituents and to cell membranes themselves, for example, X-ray methods, calorimetric methods, spectroscopic techniques (infrared, nuclear magnetic resonance, electron spin resonance, fluorescence and flash photolysis spectroscopy), and these have clarified our thinking and attitudes to the structure and dynamics of cell membrane systems (Chapman, 1968). We are now approaching an understanding of cell membrane structure and function at the molecular level in terms of constituent molecules.

In this chapter, we shall attempt to illustrate some of these developments.

* Present address: Department of Chemistry, Chelsea College, University of London.

5.2   THE CONSTITUENTS OF CELL MEMBRANES

The techniques required for the separation and isolation of cell membranes have been described in the preceding chapters. From the analysis of cell membranes (Guidotti, 1972) it is important to note that the ratio of protein to lipid varies from one membrane system to another.

Table 5.1   COMPOSITION OF CELL MEMBRANES

Membrane	Protein, %	Lipid, %	Carbohydrate, %	Weight fraction of protein	Ratio of protein to lipid
Myelin	18	79	3	0.18	0.23
Plasma membranes					
Blood platelets	33–42	58–51	7.5	0.4	0.7
Mouse liver cells	46	54	2–4	0.46	0.85
Human erythrocyte	49	43	8	0.49	1.1
Amoeba	54	42	4	0.54	1.3
Rat liver cells	58	42	(5–10)	0.58	1.4
Nuclear membrane of rat liver cells	59	35	2.9	0.59	1.6
Retinal rods, bovine	51	49	4	0.51	1.0
Mitochondrial outer membrane	52	48	(2–4)	0.52	1.1
Sarcoplasmic recticulum	67	33		0.67	2.0
Mitochondrial inner membrane	76	24	(1–2)	0.76	3.2
Gram-positive bacteria	75	25	(10)	0.75	3.0
Halobacterium purple membrane	75	25		0.75	3.0
Mycoplasma	58	37	1.5	0.58	1.6

This is indicated by the data shown in *Table 5.1*, where three general classes of membrane can be distinguished. The simplest is myelin, whose main function is thought to be as an insulator and whose major component is lipid.

Approximately one half of the mass of the plasma membranes of animal cells is made up of proteins. The main function of these proteins is probably in the enzymatic and transport functions of the membranes.

Another group of membranes, characterized by the plasma membranes of bacterial cells and by the inner membrane of the mitochondrion, contain a relatively large amount of protein (75 percent of the mass of the membrane). These membranes are associated with the functions of oxidative phosphorylation and synthesis of nucleic acids.

5.2.1   Lipids

When we look at a detailed analysis of the lipids and cholesterol present in myelin membranes of several vertebrate species (*Table 5.2*), we see a fairly consistent pattern, with a molar ratio of about 2:2:1 for cholesterol:total phospholipid:total glycolipid. The ratio of cholesterol to protein appears to be somewhat species-dependent (Dickerson, 1968).

The individual lipids present include various sphingolipids and phospholipids, such as cerebroside, sphingomyelin, lecithin (phosphatidylcholines),

Table 5.2 LIPID COMPOSITION OF MYELIN FROM THE CENTRAL NERVOUS SYSTEM AND PERIPHERAL NERVOUS SYSTEM OF DIFFERENT SPECIES (From Dickerson, 1968, courtesy of Blackwell Scientific Publications)

Values expressed as micromoles of lipid per μmole of cholesterol unless otherwise stated.

	Central nervous system								Peripheral nervous system	
	Man	Ox	Rat	Rabbit	Guinea-pig	Pigeon	Dogfish	Frog	Rat	Guinea-pig
Protein (mg/100 mg dry wt. myelin)	47.3	29.5	36	38		47.4	56	39		
Cholesterol (μmol/mg lipid)	0.700	0.622	0.670	0.665	0.640	0.758	0.558	0.660		
Cerebroside	0.354	0.454	0.331	0.315	0.50	0.388	0.214	0.203	0.4	1.8†
Sulphatide	0.073		0.066	0.075		0.059	0.131			
Total phospholipid	0.82	0.93	0.94	0.96	0.97	0.640	1.35	1.16	1.21	1.56
Phosphatidylinositol	0.013	0.029	0.042	0.024	0.028	0.016	0.034	0.026		
Serine phospholipid	0.154	0.174	0.146	0.146	0.123	0.069	1.13	0.088	0.23	0.53*
Cardiolipin, phosphatidic acid	0.004		0.025	0.074	0.019	0.057	0.088	0.021		
Sphingomyelin	0.160	0.143	0.080	0.159	0.12	0.051	0.199	0.107	0.24	
Lecithin	0.172	0.180	0.246	0.178	0.25	0.164	0.334	0.454	0.29	0.33
Ethanolamine phospholipid	0.425	0.330	0.421	0.346		0.288	0.600	0.462	0.45	0.7
Total plasmalogen	0.252	0.274	0.462	0.326	0.247	0.323	0.686			

* Value includes sphingomyelin.
† Authors say that this value was confirmed by five determinations on three preparations.

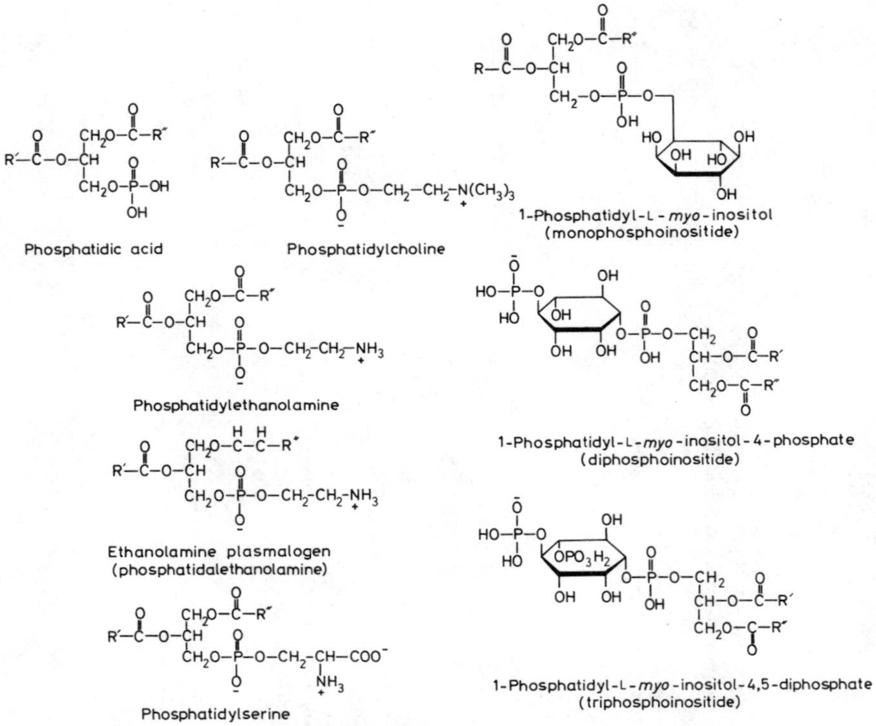

Figure 5.1 The structures of phospholipids found in cell membranes

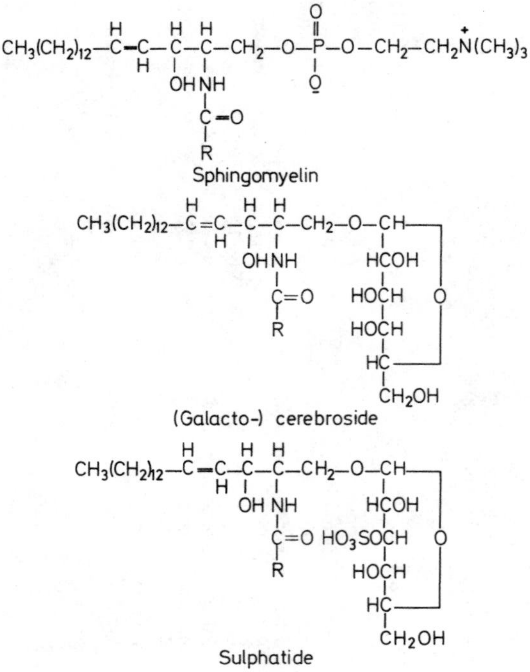

Figure 5.2 The structures of sphingolipids found in cell membranes

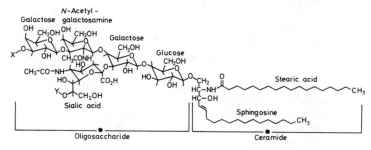

Figure 5.3 The structure of a ganglioside

phosphatidylethanolamines, plasmalogen and phosphatidylserines. The structures of these different lipid classes are shown in *Figures 5.1* and *5.2*. The cerebrosides and sulphate esters (*Figure 5.2*) appear to be typical constituents of myelin, whilst gangliosides (*Figure 5.3*) are primarily located in neurones, plasma membranes and synaptic membranes.

Another point to note is that there can be differences in the classes of lipids found for a particular membrane system. This is illustrated by studies

FATTY ACIDS

Common name	Systematic name	Formula
		SATURATED
Palmitic	n-Hexadecanoic	$CH_3(CH_2)_{14}COOH$
Stearic	n-Octadecanoic	$CH_3(CH_2)_{16}COOH$
Lignoceric	n-Tetracosanoic	$CH_3(CH_2)_{22}COOH$
Cerotic	n-Hexacosanoic	$CH_3(CH_2)_{24}COOH$
		UNSATURATED
Oleic	9-Octadecenoic	$CH_3(CH_2)_7CH=CH(CH_2)_7COOH$
Linoleic	9,12-Octadecadienoic	$CH_3(CH_2)_4CH=CHCH_2CH=CH(CH_2)_7COOH$
Linolenic	9,12,15-Octadecatrienoic	
Arachidonic	5,8,11,14-Eicosatetraenoic	$CH_3CH_2CH=CHCH_2CH=CHCH_2CH=CH(CH_2)_7COOH$
Nervonic	*cis*-15-Tetracosenoic	$CH_3(CH_2)_4(CH=CHCH_2)_4(CH_2)_2COOH$
		$CH_3(CH_2)_7CH=CH(CH_2)_{13}COOH$
		HYDROXY-ACIDS
Cerebronic	2-Hydroxytetracosanoic	
		$CH_3(CH_2)_{21}CHOH\ COOH$
Hydroxynervonic	2-Hydroxy-9-tetracosenoic	
		$CH_3(CH_2)_7CH=CH(CH_2)_{12}CHOH\ COOH$

STEROLS

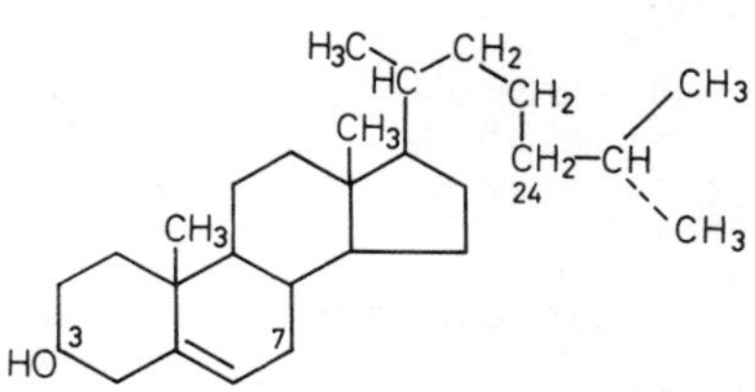

Cholesterol

*Figure 5.4 The structures of fatty acids and cholesterol found in cell membranes. Desmosterol differs from cholesterol only in having a double bond at C-24. Esters are formed by reaction of **OH** in position 3 with a fatty acid*

of the erythrocyte membrane. Thus the weight of total lipid is very similar in nearly all mammalian erythrocytes. The ratio of cholesterol to phospholipid is near 0.9 and the ratio of non-choline- to choline-containing phospholipid is similar in human, bovine and porcine species. However, the bovine red cell contains little lecithin but has a high content of sphingomyelin, while the porcine red cell has a particularly large amount of glycolipid (12–14%). Thus there are differences in the classes of lipids found in various erythrocytes (Hanahan, 1969). There are also differences within individual classes: porcine erythrocytes contain more phosphatidylserine than the other erythrocytes examined.

Associated with each of these different lipid classes is a range of fatty acids of various chain lengths and amounts of unsaturation. Their structures are shown in *Figure 5.4*. In the myelin membranes there are long fatty acids, 24 carbon atoms in length and of limited unsaturation, while in mitochondrial membranes there are large amounts of highly unsaturated C_{18} fatty acids (O'Brien, 1965).

It is interesting to note that phospholipids of brain and retina—both excitable tissues—are characterized by the presence of a long-chain, highly unsaturated fatty acid (22:6) as a significant proportion of the total phospholipid fatty acids. This is concentrated in the grey matter rather than white matter (Breckenridge, Gombos and Morgan, 1971).

The relative ease with which lipid components can be separated has thus allowed the lipid content of membrane systems to be carefully evaluated. A similar situation does not yet apply to membrane proteins, where a difficulty arises from the ability of the proteins to bind lipid tenaciously. Nevertheless, increasing attention is now being given to protein analysis.

Table 5.3 SOME TECHNIQUES USED FOR ISOLATION OF MEMBRANE PROTEINS (After Chavin, 1971)

Technique or reagent	Protein and source	Comments
Osmotic shock	High affinity binding proteins, gram-negative bacteria	Proteins released are probably pericytoplasmic or loosely bound to membrane rather than integral components of membrane
Distilled water	'Structural' protein from ghosts of human red cells	Membrane completely solubilized only if ghosts have been thoroughly deionized
Low ionic strength	ATPase, *Streptococcus faecalis*	Enzyme solubilized by repeated washes with 1 mM Tris-HCl buffer in absence of Mg^{2+}
High ionic strength	Protein from ghosts of human red cells	0.8 M NaCl solubilizes about 40 percent of membrane protein
Chelating agents	'Spectrin', ghosts of human red cells	5 mM EDTA, buffer of low ionic strength plus β-mercaptoethanol
Triton X-100	Acetylcholinesterase, electroplax electric organ	0.5% w/w detergent solution for 48 h at 4 °C. Sonication before detergent treatment. Enzyme partially inactivated
Deoxycholate	Acetylcholine receptor, electroplax electric organ	1% w/w detergent solution. Acetylcholinesterase also solubilized

5.2.2 Proteins

A number of techniques have been applied to the isolation of membrane proteins (Chavin, 1971) and some of these are shown in *Table 5.3*.

One approach to the solubilization of membrane proteins consists of the application of techniques used for aqueous systems. A variable, and sometimes large, amount of protein can be solubilized by fairly mild conditions such as high ionic strength (Rosenberg and Guidotti, 1969), chelating agents (Marchesi *et al.*, 1970) and even distilled water (Mazia and Ruby, 1968). Large amounts of protein can be solubilized from mitochondria of rat liver with either 0.8 M NaCl or 1 mM EDTA.

Many membrane proteins are difficult to solubilize under these conditions (Jones and Kennedy, 1969), while others tend to aggregate irreversibly (Rosenberg and Guidotti, 1969) and still others are unstable in aqueous solutions (Higashi, Siewert and Strominger, 1970). Furthermore, these difficulties may occur even in the presence of detergents and other denaturing reagents (Rosenberg and Guidotti, 1969). However, solubilization and fractionation may be considerably facilitated by using such reagents as ionic detergents, urea, guanidine and organic solvents.

Rosenberg and Guidotti (1968, 1969) have accomplished the most successful resolution so far of the proteins of erythrocyte membranes by combining several existing methods (Marchesi and Steers, 1968; Mitchell and Hanahan,

Table 5.4 COMPOSITION OF INTACT AND LIPID-EXTRACTED GHOSTS OF RED BLOOD CELLS

Component	Intact ghost, %	Lipid-extracted ghost, %
Protein	49.2	91.0
Lipid (total)	43.6	1.5
Phospholipid	32.5	1.5
Cholesterol	11.1	0
Carbohydrate (total)	7.2	7.5
Neutral sugars	4.0	2.5
Hexosamines	2.0	2.6
Sialic acids	1.2	2.4

1966). The composition of intact erythrocyte 'ghosts' is shown in *Table 5.4* and their amino acid composition is shown in *Table 5.5*. The procedure involves four steps: two aqueous extractions of the ghost, a lipid-extraction step, and solubilization in sodium dodecyl sulphate* followed by fractionation on Sephadex gels. Eight fractions (I–VIII) of membrane protein were observed after these procedures. The extraction of the membrane with EDTA has a significant effect on the subsequent extraction with NaCl. The distribution of proteins in the various fractions is shown in *Table 5.6*.

Subsequent analysis using Sephadex chromatography, gel electrophoresis and amino acid analysis showed the fractions to be both chemically and

---

* Recently, Dunn and Maddy (1973) have commented upon the molecular weight determinations of large polypeptides of erythrocyte membranes using sodium dodecyl sulphate solubilization. They point out that certain complexes can form, giving rise to anomalously high molecular weight.

physically distinct. In addition there was a non-uniform distribution of both neutral sugars and sialic acid among the subfractions. N-terminal analysis of each of the fractions indicated that there was a minimum of 12 different polypeptide chains in the membrane protein fractions. The average molecular weight of fractions IV–VIII ranged from 10 000 to 170 000, as determined by gel filtrations on Sephadex.

Table 5.5 AMINO ACID COMPOSITION OF LIPID-EXTRACTED MEMBRANE PROTEIN

Amino acid	Residues, %
Lysine	5.21
Histidine	2.44
Arginine	4.53
Aspartic acid	8.49
Threonine	5.86
Serine	6.26
Glutamic acid	12.15
Proline	4.26
Glycine	6.73
Alanine	8.15
Half-cystine	1.08
Valine	7.10
Methionine	2.02
Isoleucine	5.29
Leucine	11.34
Tyrosine	2.41
Phenylalanine	4.20
Tryptophan	2.49

One mole of NH_3 was formed for every three aspartic and glutamic acids.

Table 5.6 DISTRIBUTION OF PROTEINS IN THE VARIOUS FRACTIONS OF THE MEMBRANE PROTEINS

Fraction	Extraction procedure	Percentage of total membrane protein	
I	EDTA extraction	11	
II	NaCl extraction	41	
III	Lipid extraction	7	
IV		8	
V	Residue after lipid extraction	5	
VI	(gel-filtration in sodium	13	(Total 41%)
VII	dodecyl sulphate)	7	
VIII		8	
Total		100	

Marchesi *et al.* (1970) have reported the resolution of a specific protein from the erythrocyte membrane. The protein comprises approximately 20 percent of the ghost protein and is reported to be free of lipid, neutral sugar and sialic acid. The protein, known as spectrin, migrates as a single band on gel electrophoresis in the presence of sodium dodecyl sulphate, with an apparent molecular weight of 140 000. This value is corroborated

Table 5.7 MEMBRANE PROTEINS OF PLASMA MEMBRANES AND ENDOPLASMIC RETICULUM (After Guidotti, 1972)

Protein	Cell	Method of isolation	Molecular weight (native)	Chemical composition
1. Spectrin or tacktin A	Human erythrocyte	EDTA or salt-free H_2O	460 000	
'Actomyosin'	(Most animal cells)	EDTA	500 000	
Acetylcholinesterase	(Most animal cells)	Detergent		
2. Major glycoprotein	(Most animal cells)	Phenol extraction. detergent	10^5	65% carbohydrate
3. Minor glycoprotein	(Most animal cells)	Detergent	163 000	8% carbohydrate
Lipoprotein	(Most animal cells)	Sonication in 1-butanol	(lipid-free)	94% lipid
Glycoproteins	Human platelets	Trypsin	120 000	70% carbohydrate
			26 000	
Histocompatibility antigens	Mouse spleen, human lymphocytes	Papain	57 000	9% carbohydrate
Histocompatibility antigens	Human lymphocytes	Sonication or detergent	35 000	8% carbohydrate
			34 600	
4. Na^+,K^+-ATPase	Canine renal medulla, rabbit kidney, low brain	Detergent	Membrane-bound	30% lipid
5. Ca^{2+}-ATPase	Sarcoplasmic recticulum	Detergent	Membrane-bound	30% lipid

by results obtained from sedimentation equilibrium and viscosity measurements.

Recent work (Higashi, Siewert and Strominger, 1970) has indicated the possibility of extracting membrane proteins directly into organic solvents without denaturation. In principle this is a very sensible way of handling such proteins, although it may not always be successful. The enzyme, an ATP-dependent, C_{55} isoprenoid alcohol phosphokinase from the membrane of *Staphylococcus aureus*, has the following properties:

1. It is extractable in active form into acidic butanol at room temperature, from which it precipitates reversibly at $-15\,°C$.
2. It is stable in organic solvents such as acetone, heptane, ether and benzene, but not in chloroform–methanol (1:1).
3. It is insoluble in water and relatively unstable in aqueous buffers.
4. It undergoes a 50 percent increase in activity when heated in butanol at $100\,°C$ for 20 minutes.
5. Enzyme activity requires either phosphatidylglycerol or cardiolipin or both, but this requirement may be partially satisfied by certain detergents (Higashi, Siewert and Strominger, 1970; Higashi and Strominger, 1970).

A summary of the information available concerning the proteins found in various membranes is given in *Tables 5.7* and *5.8*.

5.3 PHYSICAL PROPERTIES OF THE MEMBRANE CONSTITUENTS

5.3.1 Lipid systems

The lipids from cell membranes all possess both a hydrophobic and a hydrophilic group within the molecule. They therefore orient themselves in aqueous surroundings so that the polar group is in the water and the hydrophobic group is away from the water. As we can see from *Figure 5.1*, the polar groups of the different classes may have a net neutral charge, as in lecithins, or a net negative charge, as in phosphatidylserines, at pH 7.5. The phosphoinositides can have varying degrees of negative charge corresponding to structures containing one, two or three phosphate groups.

We are still uncertain of the precise reasons for the existence of the different lipid polar groups within a membrane system. It may be that these different polar groups are important for different types of interaction with protein or various types of ion binding; for example, the phosphoinositides have been implicated in the flux of cations in nervous tissue.

The effects of variation in fatty acids can be illustrated by recent physical studies on phospholipids. Similar effects are also found with the sphingolipids. The capillary melting points of a number of pure phospholipids have been determined and are quite high; for example, the value for the diacyl phosphatidylethanolamines is about $200\,°C$, while for the phosphatidylcholines the melting points are at $230\,°C$. These values are independent of both the chain length and the degree of unsaturation of the fatty acid

Table 5.8 MEMBRANE PROTEINS OF MITOCHONDRIA (After Guidotti, 1972)

Protein	Cell	Method of isolation	Molecular weight (native)	Composition
Cytochrome c	Many	Acid extraction	12 000	1 haem
Cytochrome c_1	Beef heart	Deoxycholate extraction	300 000	6 haems
Cytochrome b	Beef heart	Deoxycholate, sodium dodecyl sulphate extraction		1 haem
Non-haem iron (complex III)	Beef heart	Deoxycholate, succinylation		2 iron atoms 2 sulphide groups
Cytochrome oxidase	Beef heart	Cholate extraction, emulsol dispersion	200 000 ($2 \times 100 000$)	14% lipid 2 haems 2 Cu^+
Succinate dehydrogenase	Beef heart	Detergent, chaotropic agents	100 000	7–8 iron atoms 7–8 sulphide groups 1 FAD
NADH–coenzyme A reductase	Beef heart	Detergent chaotropic agents		4 iron atoms 4 sulphide groups 1 FMN
ATPase	Beef liver	Mechanical disintegration, buffer extraction	280 000–360 000	
ATPase	Rat liver	Sonication	360 000–384 000	

residues associated with the phospholipid. The high values of these capillary melting points are, therefore, consistent with the occurrence of ionic linkages existing in the crystal associated with the polar groups of the phospholipid (Williams and Chapman, 1970).

In addition to the capillary melting point, other phase changes occur with phospholipids at lower temperatures; for example, when a pure phospholipid, dimyristoylphosphatidylethanolamine, containing two fully saturated chains, is heated from room temperature to the capillary melting point, a number of thermotropic phase changes (phase changes caused by the effect of heat) occur. This was first shown by IR spectroscopic techniques (Byrne and Chapman, 1964), then by thermal analysis, and has now been studied by a variety of physical techniques (Chapman, Byrne and Shipley, 1966). Above 120 °C, pressure on a coverglass with a needle causes the material to flow. When the temperature of the phospholipid reaches $\sim$120 °C the IR absorption spectrum undergoes a remarkable change*.

Above this temperature the spectrum loses all the fine structure and detail which was present at lower temperatures, becoming similar to that obtained with a phospholipid dissolved in a solvent such as chloroform. The IR spectra of this phospholipid at different temperatures are shown in *Figure 5.5*.

Differential thermal analysis (DTA) shows that a marked endothermic transition occurs at this temperature (Chapman and Collin, 1965). An additional heat change occurs at $\sim$135 °C and only a small heat change is involved near the capillary melting point of the lipid. This behaviour is similar to that which occurs with liquid crystals, such as *p*-azoxyanisole or cholesteryl acetate, which form nematic and cholesteric liquid-crystalline phases.

The X-ray long spacings show a dramatic reduction to some two-thirds of their original value at the first transition temperature with a further small reduction at the second transition temperature. The X-ray short spacings change at the first transition point from sharp diffraction lines to a diffuse spacing at 0.46 nm. Nuclear (proton) magnetic resonance (PMR) studies of these phospholipids show a gradual reduction in linewidth from about 15 G at liquid-nitrogen temperature until, at the first transition temperature, there is a sudden reduction in the linewidth to $\sim$0.09 G. This shows that molecular motion increases gradually as the temperature increases until, at the transition temperature, a considerable increase in the molecular motion takes place (Chapman and Salsbury, 1966).

The main conclusions from these various studies are as follows. (a) Even with the fully saturated phospholipid at room temperature, some molecular motion occurs in the solid. This is evident from the PMR spectra (Chapman and Salsbury, 1966), and from the IR spectra taken at liquid nitrogen and at room temperatures. (Note the difference between the IR spectra at -186 °C and room temperature.) (b) When the phospholipid is heated to a higher temperature, it reaches a transition point at which a marked endothermic change occurs and the hydrocarbon chains in the lipid 'melt', exhibiting a very high degree of molecular motion. This is evident both in the appear-

---

* These studies of phospholipids followed an earlier study of simple soap systems. Using IR spectroscopy it was shown that a marked change in the spectrum of sodium stearate occurred from a crystalline-type spectrum to a liquid-type spectrum some 200 °C below the final melting point of the soap (Chapman, 1958) and associated with the appearance of isomers.

ance of the IR spectrum and also in the narrow NMR linewidth. On the one hand, the broad diffuse appearance of the IR spectrum is consistent with the flexing and twisting of the chains and with a 'break-up' of their all-planar *trans* configuration.

When phospholipids contain shorter chains, or unsaturated bonds, the marked endothermic phase transitions occur at lower temperatures. The temperatures at which these transitions occur parallel the behaviour of the melting points of the related fatty acids. The transition temperatures are

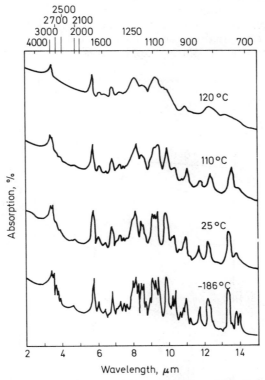

Figure 5.5 The IR spectrum of 1,2-dilauroyl-D,L-phosphatidylethanolamine at different temperatures. (From Chapman, Byrne and Shipley, 1966, courtesy of The Royal Society)

high for the fully saturated long-chain phospholipids, but are lower when there is a *trans* double bond present in one of the chains and lower still when there is a *cis* double bond present. This variation of transition temperature also confirms that the phase transition is primarily associated with a 'melting' of the hydrocarbon chains of the phospholipid, while this in turn is a reflection of the dispersion forces between the chains.

Only one main 'melting of the chains' occurs even when there are two different types of chain present in the phospholipid molecule. The transition temperatures for different phospholipid classes vary even though they contain exactly the same fatty acid residues.

Biological membranes exist in an aqueous environment and it is therefore important to study the effects of water on the lipid systems. Small amounts of water can have unusual effects upon the mesomorphic behaviour. Thus, the lecithins exhibit additional liquid crystalline forms between the first transition temperature and the capillary melting point. The intermediate liquid crystalline form exhibits X-ray spacings consistent with a cubic-phase organization. On the other hand, if all the water is removed from the phospholipid, the lipid will no longer exhibit this cubic phase (Chapman, Williams and Ladbrooke, 1967).

When phospholipids are examined in increasing amounts of water, the various physical techniques, such as microscopy, NMR spectroscopy or differential thermal analysis (DTA), show that the endothermic transition temperature for a given phospholipid decreases as the amount of water

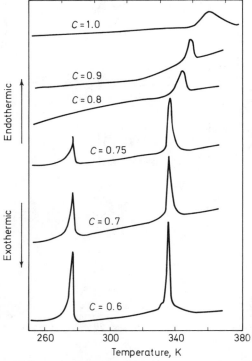

Figure 5.6 Differential scanning calorimetric (DSC) heating curves for 1,2-distearoyllecithin in increasing amounts of water

increases, reaching a limiting value independent of the water concentration (*Figure 5.6*). We can understand this if we regard the effect of water as leading first to a 'loosening' of the ionic structure of the phospholipid crystals which, in turn, affects the whole crystal structure and leads to a reduction, up to a certain limit, of the dispersion forces between the hydrocarbon chains. Large amounts of energy are still required to counteract the dispersion forces between the chains so that quite high temperatures are still required to cause melting. These limiting transition temperatures parallel the melting point behaviour of the analogous fatty acids, becoming lower with increasing unsaturation.

Values of the heat involved in the endothermic transition of some lecithins are given in *Table 5.9*. These data have been analysed to provide information about the degree of disorder in the lipid chains in the liquid crystalline phase (Phillips, Williams and Chapman, 1969).

Table 5.9 THERMAL DATA FOR PHOSPHOLIPIDS (1,2-diacyl phosphatidylcholines)

Acyl chain length	Solid (β crystals) → liquid crystalline		Solid (α form) → liquid crystalline		Gel → liquid crystalline	
	T, °C	ΔH, kcal mol^{-1}	T, °C	ΔH, kcal mol^{-1}	T, °C	ΔH, kcal mol^{-1}
C_{22} (behenoyl)	120	16.7	105	7.1	75	14.9
C_{18} (stearoyl)	115	13.8	97	6.2	58	10.7
C_{16} (palmitoyl)			93	4.3	41	8.7
C_{14} (myristoyl)			89	3.4	23	6.7
C_{12} (lauroyl)					~0	
C_{18-2} (oleyl)					−22	7.6

Some of the water added to the phospholipid appears to be 'bound' to the lipid; for example, 1,2-dipalmitoylphosphatidylcholine binds about 20 percent water (Chapman, Keough and Urbina, 1974). This does not freeze at 0 °C and calorimetric studies made with lipid–water mixtures show (Chapman, Williams and Ladbrooke, 1967) that only after 20 percent water has been added to the lipid is a peak at 0 °C observed. Some differential scanning calorimetry (DSC) traces for lipid plus water mixtures of different concentrations are shown in *Figure 5.6*. The 'bound' water may have considerable relevance to interactions of anaesthetics, drugs and ions with biological membranes.

There are a number of important features associated with the transition temperature for the lipid when it is in the presence of water. The first is that the ease of dispersing the lipid in water increases markedly above the transition temperature. Only those phospholipids in water which have transition temperatures below or near to room temperature, spontaneously form myelin figures. Fully saturated phospholipids which have high transition temperatures do not form myelin figures at room temperature. However, if the temperature is raised to the transition point, these phospholipids, such as 1,2-dipalmitoylphosphatidylcholine, spontaneously form myelin figures. The phase diagram for this lipid is shown in *Figure 5.7*.

The ability to form myelin figures with saturated phospholipids has been confirmed by light microscope and electron microscope investigations. It has been used to attempt to provide information about the situation of the osmium used after osmium tetroxide fixation procedures (Chapman and Fluck, 1966). NMR and ESR spectroscopic studies (Veksli, Salsbury and Chapman, 1969; Hubbell and McConnell, 1969) have been made of these lamellar lecithin systems, and the molecular motion of the $N(CH_3)_3$ group and of different methylene groups along the chain has been studied. The molecular motion is greatest at the methyl end of the hydrocarbon chains.

A number of other studies of thermotropic phase transitions of phospholipids have been made. The endothermic transition is accompanied by a decrease in the thickness of the lipid bilayer, a change in bilayer volume, a marked decrease in PMR linewidth, a decrease in the order parameter of ESR probes and a decrease in the fluorescence polarization of various probes.

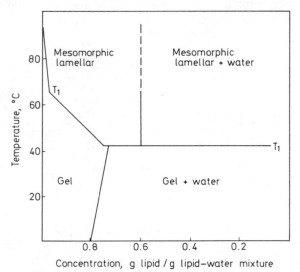

Figure 5.7 *Phase diagram of the 1,2-dipalmitoyllecithin–water system. (From Chapman, Williams and Ladbrooke, 1967, courtesy of Associated Scientific Publishers)*

Diffusion of phospholipids along the length of the bilayer has been studied (Kornberg and McConnell, 1971; Devaux and McConnell, 1972; Sackmann and Trauble, 1972) using spin-label methods, and diffusion coefficients have been determined to be $D = 1.8 \times 10^{-8}$ cm^2 s^{-1}. A direct measurement using NMR spin-echo methods which does not involve strong perturbation effects has recently been applied to lipid diffusion studies (G. Lindblom, D. Chapman and W. Derbyshire, unpublished work, 1973). The value of the diffusion coefficient obtained using this method is $D = 0.5 \times 10^{-7}$ cm^2 s^{-1} for dimyristoyllecithin above the phase transition temperature.

Another interesting feature of the lipid phase transition is shown by freeze-etch electron microscopy (*Figure 5.8*). When a phospholipid is above the phase transition temperature before quenching to low temperature then smooth faces are observed. If, however, the lipid is in the gel condition before quenching, then a rippled appearance results (Fluck, Henson and Chapman, 1969).

An important aspect of the physical properties of soaps and phospholipids is that the long-range organization (the type of phase which exists at a particular temperature) and the short-range organization (the lattice packing of the hydrocarbon chains) are determined by the type of polar group and its interaction. For example, the transition of stearic acid salts from crystal to liquid crystal, as revealed by NMR and X-ray diffraction, occurs at 114 °C for the sodium salt and at 170 °C for the potassium salt. The transition temperature of phospholipids with the same type of hydrocarbon chains also depends upon the nature of the polar group. Thus, the lipid dimyristoyl-phosphatidylethanolamine in excess water shows a transition temperature almost 30 °C higher than the corresponding lecithin. The most straightforward explanation of this appears to be that the bulkier trimethyl-ammonium group and the associated water structure of lecithin prevent the lipid chains from packing as effectively as can occur with phosphatidyl-ethanolamine. This thermotropic behaviour has its parallel in the monolayer

Figure 5.8 Freeze-etch micrograph of a 2% dispersion of dipalmitoyllecithin in water. A, end-on view of corrugations; B, large, irregular water spaces; C, one lamella with corrugations similar to those above it. (From Fluck, Henson and Chapman, 1969, courtesy of Academic Press)

behaviour of these compounds. Thus monolayers of phosphatidylethanol-amine molecules are much more condensed than are monolayers of corresponding phosphatidylcholines with the same hydrocarbon chains. The condensation of monolayers of long-chain acids with an increase in the pH of the subphase is another example of this effect of polar group packing affecting lipid chain packing.

Pure lipids, particularly saturated ones, in excess water give relatively sharp transitions over only a few degrees, corresponding to a highly cooperative transition. This is the case even with a lipid containing different fatty acid chains, e.g. stearoyl and elaidoyl chains (Phillips, Ladbrooke and Chapman, 1970). However, mixtures of lipids containing a range of unsaturated fatty acid chains (e.g. egg yolk lecithin) show a wider range of transition (Ladbrooke and Chapman, 1969). Mixtures of 1,2-dimyristoyl-lecithin and 1,2-dipalmitoyllecithin, examined in water, co-crystallize below the transition temperature (T_c) to give a series of solid solutions (Figure 5.9). This is similar to the behaviour of mixtures of 1,2-dipalmitoyl- and 1,2-distearoyllecithin (Chapman, Keough and Urbina, 1974). In both cases ideal mixing of the two lipids appears to be occurring and monolayer studies are consistent with this. When the difference of chain length between lecithins is increased to four methylene groups, monotectic behaviour without solid solution becomes apparent (Chapman, Keough and Urbina, 1974). Thus, the lattice arrangement can accommodate small chain-length differences so that co-crystallization can occur but with greater differences; migration of lecithin molecules within a given bilayer occurs as the system

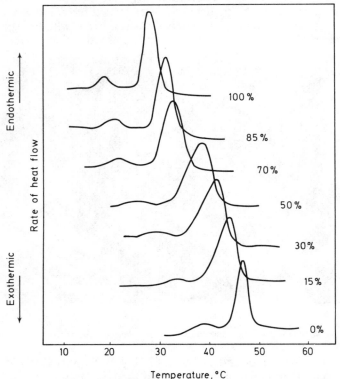

Figure 5.9 DSC *heating curves for various mixtures of 1,2-dimyristoyl-lecithin with 1,2-dipalmitoyllecithin. (From Chapman, Keough and Urbina, 1974, courtesy of the American Society of Biological Chemists)*

is cooled to give regions corresponding to the two separate components. Mixtures, in excess water, of dioleoyllecithin with a number of long-chain saturated lecithins also exhibit monotectic behaviour with two well-defined transitions corresponding to the phase changes of the individual components. These properties are important for the interpretation of the behaviour of cell membrane systems where two different fatty acids are incorporated biosynthetically.

Mixtures of the two different lipid classes of the same chain length, for example 1,2-dimyristoyllecithin and 1,2-dimyristoylphosphatidylethanol-amine, show very different melting behaviour (*Figure 5.10*) compared with simple lecithin mixtures (Ladbrooke and Chapman, 1969). The first point to note is that the presence of only 5 percent dimyristoylphosphatidyl-ethanolamine within the lecithin bilayer causes the pretransition endotherm of the lecithin to be substantially broadened, and to disappear entirely at amounts of about 10 percent. This shows that the mixture of relatively small amounts of 'foreign material' (even of related lipids) into the headgroup region can affect the packing and hence the long-range order of the organization of the lipid. Similar effects are observed when cholesterol is included in lecithin bilayer systems, and is also observed with gramicidin A. Such behaviour may be of biological significance, and in some cases may underline the way in which important signals can be transmitted from one part of the membrane surface to another.

The asymmetry and broadening in the DSC heating curves which occurs at higher concentrations of phosphatidylethanolamine (*see Figure 5.10*) is

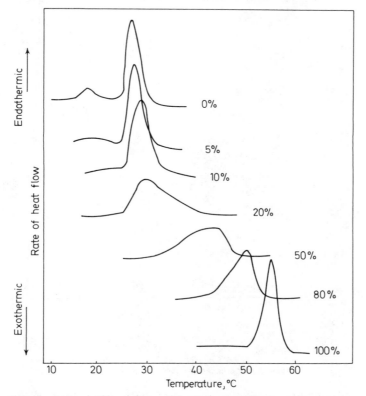

Figure 5.10 DSC *heating curves for various aqueous mixtures of 1,2-dimyristoyllecithin with 1,2-dimyristoylphosphatidylethanolamine. (From Chapman, Keough and Urbina, 1974, courtesy of the American Society of Biological Chemists)*

particularly interesting. It shows that clusters of lipids in a gel and in liquid crystalline forms can coexist within this transition range. The transition is of lower cooperativity than occurs with the individual components. Some clustering of the different lipid classes may be occurring.

A second feature of lipid–water systems is their monolayer behaviour. Monolayer studies of phospholipids have been carried out for a considerable number of years, usually with natural phospholipid mixtures and, in most cases, with egg yolk phosphatidylcholine, but in recent years studies have been made with pure synthetic phospholipids (Chapman, Owens and Walker, 1966b). These studies show that the fully saturated phospholipids exhibit, at room temperature, monolayers which are more condensed than are the unsaturated phospholipids containing *cis* hydrocarbon chains, i.e. the saturated lipids occupy less area at low surface pressures than do the unsaturated derivatives (*Figure 5.11*).

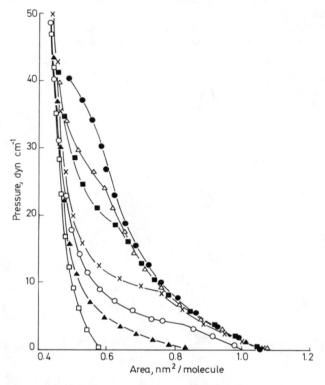

Figure 5.11 Monolayer curves of 1,2-dipalmitoyllecithin at different temperatures. (●) 34.6 °C, (△) 29.5 °C, (■) 26.0 °C, (×) 21.1 °C, (○) 16.8 °C, (▲) 12.4 °C, (□) 6.2 °C. (From Phillips, Ladbrooke and Chapman, 1970, courtesy of Associated Scientific Publishers)

Monolayers obtained with phosphatidylcholines are observed to be much more expanded than are the corresponding phosphatidylethanolamines containing the same acyl chains. These results can be compared with the DTA results discussed earlier. A high transition temperature for liquid crystal formation is, in general, correlated with a condensed-type monolayer and a low transition temperature with an expanded film. (The DTA transition temperatures are higher for the phosphatidylethanolamines than with the corresponding phosphatidylcholines.) Phospholipids containing *trans*-elaidoyl unsaturated chains have higher transition temperatures than those containing *cis*-oleoyl chains. Phospholipids containing one fully saturated chain and one *trans*-unsaturated chain give condensed monolayers similar to those observed with completely saturated phospholipids.

Electrostatic interactions of the polar head of phospholipids can also affect their thermotropic mesomorphism (Chapman, Keough and Urbina, 1974). The effect of different divalent cations on the endothermic phase transition of dipalmitoylphosphatidylcholine (DPL) and ox-brain phosphatidylserine (PS) has been studied. From monolayer and liposome experiments, uranyl ions (UO_2^{2+}) are known to interact stoichiometrically with the polar groups of phosphatidylcholine (PC) and to bind in the membrane surface of different biological membranes, presumably to phosphate groups. We have confirmed by IR spectroscopy that the strong binding of this ion is specifically to the phosphate groups. It is also known that PS forms a stoichiometric complex with Ca^{2+} and other divalent ions.

These interactions have very strong effects on the thermotropic mesomorphism of both lipids. The original transition temperature is 41.4 °C, which is finally shifted in the 1:1 complex to 46 °C. With PS, which contains a net negative charge, the binding of Ca at 1:1 proportion shifts the transition from 16.9 °C, observed with free PS, to 21.9 °C.

5.3.2 Lipid–cholesterol systems

An interaction which is of considerable biological importance is that of cholesterol with phospholipids. Cholesterol occurs in many membranes, such as the myelin sheath and red blood cell membranes. The interaction of egg yolk lecithin and cholesterol has been studied in bulk systems in the

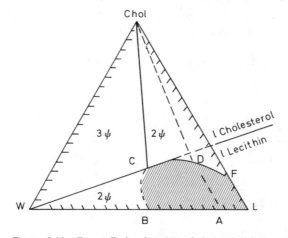

Figure 5.12 Egg yolk lecithin (L)–cholesterol (chol)–water (W) system at 25 °C. The number of phases is indicated. (From Small and Bourgès, 1966, courtesy of Gordon & Breach)

presence of water using microscopic and X-ray techniques (Small and Bourgès, 1966). The three-component phase diagram of this system is shown in *Figure 5.12*. Along the water–cholesterol side of the triangle the cholesterol is totally insoluble in water at 25 °C. However, along the lecithin–water side, from about 12–45 percent water (A–B in *Figure 5.12*), lecithin forms with water a lamellar paracrystalline phase. Mixtures containing more than 45 percent water separate into this lamellar mesomorphic phase which floats in the excess of water in the form of myelin figures and anisotropic droplets. All mixtures containing the three components appear as points inside the triangular diagram.

From the phase diagram it appears that cholesterol can be added to lecithin up to a molecular ratio of 1 : 1 to give the same lamellar paracrystalline phase already given by lecithin and water. Any excess above this proportion separates as cholesterol crystals. X-ray analysis shows that, in general, the addition of cholesterol to lecithin at a constant water concentration tends to make the lipid layer slightly thicker.

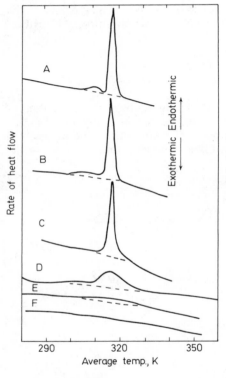

Figure 5.13 DSC *heating curves of aqueous 1,2-dipalmitoyllecithin–cholesterol mixtures. A, 0.0 mol %; B, 5.0 mol %; C, 12.5 mol %; D, 20.0 mol %; E, 32.0 mol %, and F, 50.0 mol % cholesterol. (From Ladbrooke, Williams and Chapman, 1968a, courtesy of Associated Scientific Publishers)*

Recent studies of lecithin–cholesterol–water interactions have been made by DSC (Ladbrooke, Williams and Chapman, 1968a). This work has shown that addition of cholesterol to dipalmitoyllecithin in water lowers the transition temperature between the gel and liquid crystalline phases and decreases the heat absorbed at the transition. No transition is observed with an equimolar ratio of the lecithin with cholesterol. Unsaturated lecithins and the lipid extract of human erythrocyte ghosts exhibit similar behaviour.

The DSC curves between 6.9 and 86.9 °C for a series of varying ratios of 1,2-dipalmitoyllecithin and cholesterol, each containing 50 percent by weight of water, are shown in *Figure 5.13*. The effect of cholesterol is to disrupt the ordered array of the hydrocarbon chains of the lipid in the gel phase and, when cholesterol and lecithin molecules are present in equimolar proportions, all the chains are in a fluid condition. The presence of equimolar amounts of cholesterol and lipid causes the phospholipid–cholesterol mixtures to be dispersible in water over a much wider temperature range than occurs with the individual phospholipid.

Studies using NMR spectroscopy (Chapman and Penkett, 1966; Oldfield, Chapman and Derbyshire, 1971; Darke *et al.*, 1972) and ESR (Barratt, Green and Chapman, 1969; Hubbel and McConnell, 1971) have shown that the presence of cholesterol in the liquid crystalline phase inhibits some of the hydrocarbon chain motion. These various results seem to show that, at a particular temperature, the presence of cholesterol causes the hydrocarbon chains of differing phospholipid molecules to be in an 'intermediate fluid' condition. Those lipids which would normally be above their limiting transition temperature may show a certain amount of inhibition of chain motion, while the hydrocarbon chains of those lipids which would normally be in a gel condition are given greater fluidity.

5.3.3 Lyotropic mesomorphism

Phospholipids in the presence of water can also exhibit various types of lyotropic mesomorphism; for example, they can exhibit different types of liquid crystalline organization (Luzzati and Husson, 1962; Luzzati, 1968). Some phospholipids give both lamellar- and hexagonal-type structures (Luzzati *et al.*, 1966). Lecithin appears to exhibit only a lamellar-type

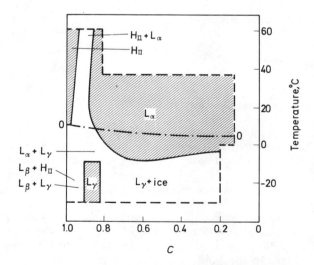

Figure 5.14 The phase diagram of mitochondrial lipid–water systems. Four phases are observed—one hexagonal (H_{II}), three lamellar (L_α, L_β, L_γ). (From Gulik-Krzywicki, Rivas and Luzzati, 1967, courtesy of Academic Press)

structure over a large range of water concentration. Other phospholipids, such as phosphatidylethanolamines and samples of brain lipid, exhibit hexagonal and lamellar organization, depending upon the concentration in water. Gulik-Krzywicki, Rivas and Luzzati (1967) have shown that four phases can be distinguished in a mitochondrial lipid extract. The composition of the extract from beef heart mitochondria was 34 percent lecithin, 29 percent phosphatidylethanolamine, 10 percent phosphatidylinositol, 20 percent cardiolipin, 2 percent cholesterol, and 5 percent neutral lipids. The phases observed were hexagonal and lamellar (*Figure 5.14*).

5.3.4   Model membranes

The spontaneous formation by phospholipids, in water and above their transition temperature, of globules containing bilayers of lipid separated by aqueous layers has led to their use as model membranes (Bangham, Standish and Watkins, 1965). The argument is that if real membranes contain bilayers of lipid, then here is a natural system whose constitution can be chosen to mimic the lipid characteristic of a real membrane. Ions can then be trapped in the aqueous compartments and the permeability of these ions across the lipid bilayer studied. Admittedly, there is no protein associated with the system but this may be included in later studies. Using this model membrane system various effects which occur with real membranes can be mimicked, for example anaesthetic and drug effects.

A second model membrane system also uses the liquid crystalline behaviour of phospholipids (Mueller *et al.*, 1962; Mueller and Rudin, 1967). In this system a phospholipid above its transition temperature is brushed across an orifice separating two aqueous compartments. It spontaneously thins down and forms a single bilayer. This model membrane can be used to study ion permeability, water flow, electrical resistance and excitability characteristics. The model membrane is fairly stable and some of this stability is probably associated with the flow characteristics of the phospholipid in its liquid crystalline phase.

5.3.5   Lipid–polypeptide–water systems

Only a few studies have been made of lipid–polypeptide systems. Although they might be thought to be useful model systems for studying electrostatic or hydrophobic interactions, the studies that have been made have usually been associated with ion-transport properties. Thus much work has been done on polypeptides and related molecules such as the antibiotics valinomycin, nonactin, gramicidin A and alamethicin.

In some cases, such as gramicidin A and alamethicin, it has been postulated that these molecules actually form channels bridging the width of the bilayer (Urry, 1971; Hladky and Haydon, 1972). In agreement with this idea is the finding that freezing the lipid below its transition temperature does not affect the ability of gramicidin A to facilitate ion transport across lipid bilayers (Krasne, Eisenman and Szabo, 1972). In any case, there is good evidence that some of these molecules readily penetrate monolayers and bilayers and can modify the lipid. Thus alamethicin molecules are thought to be situated at a lipid–water interface, with the polar side chains in the aqueous phase and the non-polar groups among the hydrocarbon chains. (It is of interest that the amino acid composition of alamethicin is particularly hydrophobic.) The evidence for this orientation at the lipid–water interface comes from the high surface activity of alamethicin observed in monolayer studies and from microelectrophoresis measurements which show that lecithin particles have a net negative charge (Hauser, Finer and Chapman,

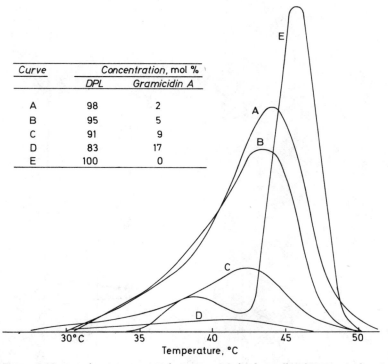

Figure 5.15 DSC heating curves of aqueous 1,2-dipalmitoyllecithin–gramicidin A mixtures. (From Chapman, Keough and Urbina, 1974, courtesy of the American Society of Biological Chemists)

1970). Studies using calorimetry heating curves show that as the concentration of gramicidin A added to lipid increases, the heat associated with the endothermic phase transition of dipalmitoyllecithin decreases (*Figure 5.15*). This is consistent with the polypeptide being interdigitated among the lipid chains (Chapman, Keough and Urbina, 1974).

The interaction of gramicidin S with lecithin is interesting in that it shows that the polypeptide solubilizes the lipid even when it is at a temperature corresponding to the gel phase of the lipid (Pache, Chapman and Hillaby, 1972).

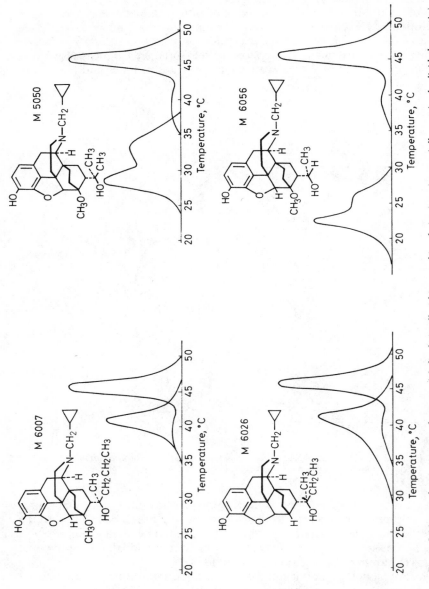

Figure 5.16 DSC heating curves of aqueous 1,2-dipalmitoyllecithin–morphine-derivative mixtures *(all at equimolar lipid:drug ratio)*. *(From Cater et al., 1974, courtesy of Associated Scientific Publishers)*

5.3.6 Lipid–drug interactions

Recent studies (Cater et al., 1974) have sought to discover how various drugs (for example, morphine derivatives and antidepressants) interact with lipid–water systems. They have shown that the interactions depend critically upon the specific molecular structure. Some drugs cause large shifts of transition temperature whilst others cause very little shift. These effects may perhaps be related to their pharmacological mode of action. Some effects are shown in *Figure 5.16*.

5.3.7 Lipid–protein–water systems

A number of model systems incorporating protein into lipid–water systems have been studied by physical techniques such as X-ray diffraction, NMR and ESR spectroscopy. Shipley, Leslie and Chapman (1969) used X-ray diffraction to study the interaction of phospholipids with cytochrome c in the aqueous phase. The lipid was a mixture of lecithin and phosphatidylserine, with the ratio of phospholipid to cytochrome c being about 300:1. Cytochrome c modifies the swelling characteristics of the lipid dispersion and two lamellar lipoprotein systems (I and II) were produced which, in

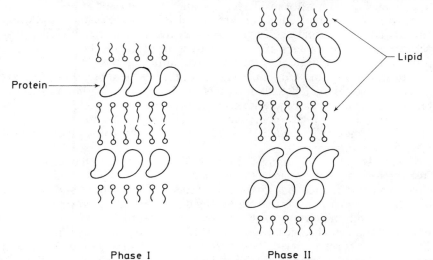

Figure 5.17 Schematic diagrams illustrating the way in which certain protein and lipid systems, e.g. cytochrome c and phosphatidylserine–water systems, are organized. (From Shipley, Leslie and Chapman, 1969.) A similar arrangement occurs with cytochrome c and phosphatidylinositol. (From Gulik-Krzywicki et al., 1969, courtesy of Macmillans.) (Two lamellar phases, I and II, were observed with structures as shown. The main interaction is considered to be electrostatic)

excess water, gave dimensions of 8.7 nm and 11.6 nm. These dimensions were consistent with the incorporation of one or two cytochrome c molecules (diameter 3.1 nm) between the lipid bilayers (*Figure 5.17*). The presence of water was necessary for the structural integrity of these phases. When the overall ratio of protein to lipid was increased, yet another lamellar phase

(III) was observed. The dimensions of these model complexes were observed to be somewhat similar to those of myelin, having a prominent second-order reflection at 8.6 nm. Similarly, erythrocyte and mitochondrial membranes, on centrifugation to a pellet, give diffraction spacings of 11.5–12.0 nm, similar to the repeat pattern of lamellar system III.

Gulik-Krzywicki et al. (1969) used X-ray techniques to investigate complexes of ferricytochrome c with phosphatidylinositol, cardiolipins and mitochondrial lipids. They also studied mixtures of lysozyme with cardiolipin and with phosphatidylinositol. A variety of phases was observed. The ferricytochrome c complexes were lamellar and their structures were interpreted as indicating layers of protein/water between the lipid lamellae, some containing one protein layer and others two. However, other systems, such as those containing lysozyme, gave a variety of phases, including hexagonal lattices, depending on the nature of the lipid. This work showed that the polymorphism of protein–lipid–water systems may be quite complex. Rand (1971) has recently studied lecithin–cardiolipin mixtures with bovine serum albumin. He concludes that electrostatic interaction brings the components together so that shorter-range polar and hydrophobic interactions can occur, causing subsequent gross conformational changes in the protein.

Bilayers have also been used to study protein–lipid complexes. Thus Steinemann and Lauger (1970) studied the interaction of cytochrome c with phosphatidylinositol in bilayer systems. At low ionic strength, about 10^{13} cytochrome c molecules/cm^2 are bound to the lipid surface, and these authors concluded that the interaction was mainly electrostatic. The fast desorption of the protein after a rise in ionic strength showed that the protein did not penetrate into the lipid layer appreciably.

Clowes, Cherry and Chapman (1972) studied the interaction of tetanus toxin with lecithin ganglioside mixtures. In this case the toxin binds to the bilayer. The thickness of the lipid–protein complex increased from 7.9 ± 0.2 nm to 11.9 nm, which is consistent with the toxin maintaining its structure and not forming an extended layer over the surface, as was observed with erythrocyte protein.

Cherry, Berger and Chapman (1971) studied the interaction of sialic-acid-free erythrocyte apoprotein with erythrocyte lipids. They showed that the conductance increased by up to three orders of magnitude and that the bilayer became electrically and physically unstable, tending to rupture.

Studies with lipid vesicles have shown that soluble proteins such as lysozyme and cytochrome c cause a marked increase in permeability to ions. This has been correlated with the penetration of protein into the hydrocarbon region of the lipid membranes (Kimbelberg and Papahadjopoulos, 1968, 1971). A membrane protein, spectrin, isolated from erythrocyte ghosts has been studied in the vesicles. When lipid vesicles are added to phosphatidylserine vesicles the ^{23}Na$^+$ diffusion rate increases. Ca^{2+} and protein have a synergistic effect on ^{23}Na$^+$ diffusion rate with phosphatidylserine and also phosphatidylserine–lecithin vesicles. In contrast to protein alone, the effect of calcium plus protein is seen at both neutral and acid pH. However, the nature of this interplay between protein and calcium is unclear (Juliano, Kimelberg and Papahadjopoulos, 1971).

Thus, the importance of both ionic and hydrophobic associations has been pointed out in a variety of model systems. This has also been discussed by Green and Fleischer (1963) for mitochondrial membrane components, by Zwaal and van Deenen (1970) for the recombination of red blood cell lipids and apoproteins, and by Sweet and Zull (1969) for the interaction of albumin and lipid vesicles.

5.3.8 Permeability studies in model systems

Permeability properties are affected by the degree of unsaturation of the fatty acids present in the lipids. Finkelstein and Cass (1968) report that an increase in permeability to water is observed in model membrane systems with increased amounts of unsaturated lipids.

It has also been shown with liposome systems (de Gier, Mandersloot and van Deenen, 1968) that there is a pronounced increase in the rate of penetration of glycerol with an increase in unsaturation of the lipid. This is also the case for glycol and erythritol (de Gier *et al.*, 1971). The liposomes containing saturated lipids exhibit abnormal temperature effects, but this effect is removed by the presence of cholesterol. It is concluded that these alcohols cross the lipid barrier as single molecules without water of hydration, and the activation energies are consistent with this concept.

Experiments with liposome systems containing lecithins from rats deficient in essential fatty acids and from normal rats (Chen, Lund and Richardson, 1971) show that glucose permeability is less with the lecithin from the rats which are deficient in essential fatty acids. The passive diffusion rate of glucose is 50 percent of that of lecithin from normal rats. This is clearly related to the greater degree of unsaturation of the fatty acids in the lecithin from the control rats.

With this background of information on lipid–water and lipid–cholesterol water systems, let us turn to the membranes themselves.

5.4 PHYSICOCHEMICAL PROPERTIES OF CELL MEMBRANES

While specific membrane types are dealt with in considerable detail elsewhere in this series, it is convenient to summarize physical and chemical data here for certain selected membrane types.

5.4.1 Chemical properties

5.4.1.1 MYELIN MEMBRANES

There is evidence that only three classes of polypeptide chains exist in myelin membranes (Eng *et al.*, 1968; Gonzalez-Sastre, 1970; Brostou and Eylar, 1971). One is very well characterized and has been called the basic protein, because it can be extracted with weak acids from myelin and because a large fraction of its constituent amino acids are basic. The molecular weight of this component is approximately 18 000 and the amino acid sequence of the protein from bovine (Eylar *et al.*, 1971) and human (Carnegie, 1971) myelin has been determined. It appears to be a random coil with no ordered secondary or tertiary structure, but residues (–Pro–Arg–Thr–Pro–Pro–Pro) near the centre of the linear sequence of the molecule are thought to cause the peptide chain to make a U bend. The protein is easily degraded by proteolytic enzymes during isolation.

The basic protein component has the ability to induce antibodies which can cause allergic encephalomyelitis in guinea-pigs and rabbits. At least two peptides which are antigenic and responsible for the allergic encephalomyelitis can be derived from the basic protein. Both peptides are encephalitogenic in rabbits (Brostou and Eylar, 1971). The fact that antibody to this protein may cause an allergic inflammation in an intact animal indicates that the antigen (protein) must be accessible to the antibody or must be exposed on the surface of the myelin membrane.

The other major protein component of myelin has been given the name of proteolipid (Folch and Lees, 1951). The two remarkable characteristics of this group of polypeptide chains are that they are soluble in chloroform–methanol (2:1) and that they have a unique amino acid composition in which the non-polar amino acids account for approximately 50 percent or more of all the amino acids on a molar basis (Tenenbaum and Folch-Pi, 1966). The molecular weight of the polypeptide chain of the proteolipid protein (Green and Reynolds, 1971) is approximately 25 000.

The remaining 20 percent of the protein present in myelin is called the Wolfgram protein and is characteristically soluble in acidified chloroform–methanol (Wolfgram, 1966); very little else is known about this component.

5.4.1.2   ERYTHROCYTE MEMBRANES

Chemical studies of the erythrocyte membrane show that there are two proteins on the exterior surface (Bretscher, 1971a). One has a molecular weight of 31 000–55 000 daltons (Bretscher, 1971b) and is the major sialoglycoprotein of the membrane. Two-thirds of the mass of the molecule is carbohydrate, and it contains most of the sialic acid and half of the hexose and hexosamine of the human erythrocyte. It has A, B and MN blood group activities and carries the receptors for influenza virus and phytohaemagglutinin. The molecule, by virtue of its high content of carbohydrate, travels anomalously on sodium dodecyl sulphate gel electrophoresis and does not stain with the dye Coomassie Blue, used to analyse the gel for protein. Thus it appears only on those gels stained for carbohydrate. The sugar chains are attached to the protein both through O-glycosidic and N-glycosidic linkages between the anomeric carbon of N-acetylhexosamine residues and the hydroxyl groups of serine and threonine and the amide group of asparagine. At least two types of oligosaccharide are attached to the protein (Winzler, 1969). The concept that at least a portion of this glycoprotein is located on the outer surface of the erythrocyte membrane derived from the observation that most of the bound sialic acid could be removed from intact erythrocytes by neuraminidase and that with intact cells trypsin would hydrolyse sialoglycopeptides containing 35–40 percent of the total sialic acid of the erythrocyte membrane (Winzler et al., 1967).

The other surface protein is the molecule with a mass of approximately

100 000 daltons which reacts in intact cells with labelling reagents that cannot pass through the membrane (Steck, Fairbanks and Wallach, 1971). It is hydrolysed in intact cells by pronase into a polypeptide with a mass of 70 000 daltons and some small peptides. This component contains a small amount of bound carbohydrate (5–8 percent by weight) and represents a considerable fraction of the erythrocyte membrane (25–30 percent of the protein) (*Table 5.1*). It has an unusual amino acid composition in that approximately 50 percent of the residues are non-polar.

Bretscher (1971a) has suggested that both the main sialoglycoprotein and the 100 000 dalton protein penetrate and traverse the lipid bilayer so that parts of these proteins are exposed on both the outer and inner surfaces of the membrane. This conclusion is deduced from the fact that formylmethionyl sulphone methyl phosphate, a labelling reagent that cannot pass through the membrane, reacts with some parts of these molecules in intact cells (when only the outer surface can react) and with other parts of these proteins in ghosts (when both sides of the membrane are exposed to the reagent). The conclusion is correct if the arrangement of proteins and lipids is the same in cells as in isolated membranes, a hypothesis for which there is insufficient support. A similar arrangement in the membrane for the sialoglycoprotein and the 100 000 dalton protein was deduced by enzymatic digestion of the membrane (Steck, Fairbanks and Wallach, 1971).

A possible arrangement of the various major protein and glycoprotein components of the erythrocyte membranes is shown in *Figure 5.18*.

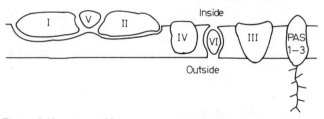

Figure 5.18 *A possible arrangement of the various proteins and glycoprotein components. Components I, II and V have been pictured as being rather superficial in location and physically associated* in situ *while component VI has been assigned an indeterminate position. PAS 1–3 represents the sialoglycoprotein components. (From Steck, Fairbanks and Wallach, 1971, courtesy of the American Chemical Society)*

Recently, chemical studies (Bretscher, 1972) have also suggested that there is an asymmetric distribution of the lipids of the erythrocyte membrane with the outer half of the bilayer constituted of lecithin and sphingomyelin whilst the inner half of the bilayer contains the lipids phosphatidylethanolamine and phosphatidylserine. Enzyme studies appear to support this concept (Low, Limbrick and Finean, 1973).

5.4.1.3   PLASMA MEMBRANES

The plasma membranes of mouse liver cells, rat liver cells, kidney cells, hamster kidney fibroblasts and platelets are all composed of a large collection (10–20) of polypeptide chains, some of which are glycoproteins (Guidotti,

1972). The same heterogeneity occurs in the membranes of the endoplasmic reticulum and the surface membranes of bacterial cells (Inouye and Guthrie, 1969). These analyses give a minimum estimate of the number of protein in each membrane because each size class can be heterogeneous with respect to amino acid sequence of the chains and because many polypeptides present in small amounts will not be observed. Specialized membranes like retinal rods of frogs and cows and purple membrane of *Halobacterium halobium* are made up almost completely of one distinct protein (Guidotti, 1972).

5.4.2 Physical properties

The study of membrane structures is now very active and a variety of techniques are being used. The myelin membrane has received particular attention.

5.4.2.1 MYELIN MEMBRANE

The molecular order in the nerve myelin sheath was characterized by Schmitt, Bear and Palmer (1941) from the results of their X-ray diffraction and optical studies. By use of the electron microscope, Geren (1954) demonstrated that the periodic lamellar structure of the myelin sheath is formed from a spirally wrapped cell membrane. Analyses of X-ray patterns of myelin have been based on diffraction data to about 3 nm spacing. Recently, electron-density profiles of three types of myelin membrane have been constructed by comparing intensity measurements on their diffraction patterns with 1 nm spacing. These studies have been made with rabbit and frog sciatic nerves and rabbit optic nerve (Caspar and Kirschner, 1971).

The conclusion from the many techniques applied to myelin is that the myelin membrane is essentially a bilayer of lipid, although there may be a small amount of protein extending across it, and the structure is arranged so that a pair of membranes lie with their cytoplasmic surfaces opposed. The electron density profile does not provide a chemical identification, so that any particular arrangement of molecules in the myelin membrane is made on the basis of data from other methods.

Particular features of the interpretation of the data are that protein and water are distributed in the spaces between the membrane bilayers but the concentration of protein may be higher near the lipid surface. The cholesterol and polar lipid are present in equimolar ratio in the external side of the bilayer. It is suggested that there is about half as much cholesterol in the inner side of the hydrocarbon region. The asymmetrical distribution of cholesterol (Caspar and Kirschner, 1971) is presumed to result from specific interactions with protein; however, the protein does not seem to modify the local packing arrangement of the lipid significantly. It is suggested from the electron-density profile that the methyl ends of the lipid chain are more disordered than that part of the chain near to the polar group, as observed in model lipid–cholesterol systems using NMR and ESR spectroscopy. The hydrocarbon thickness of the myelin membranes is estimated at 3.8 nm for rabbit myelin and 3.5 nm for frog myelin (*Figure 5.19*).

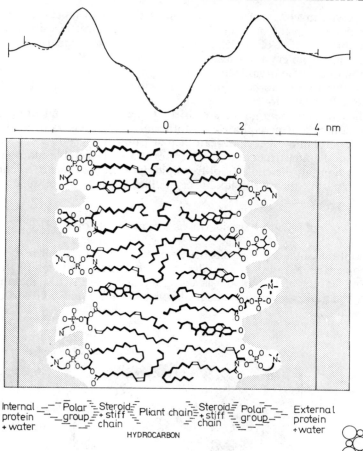

Figure 5.19 Schematic illustration of the myelin membrane structure. The electron density profiles of myelin from rabbit optic nerve (dotted) and sciatic nerve are shown above. The boundaries of the membrane unit of sciatic and optic nerve myelin correspond to the wider and narrower margin respectively. The distinctive portions of the structure which are identified with features of the density profile are shown below. For clarity, the lipid molecules are represented as all lying in the same plane. The composition illustrated is—6 cholesterols; 5 glycerolipids (2 lecithins, 2 ethanolamine plasmalogens, 1 serine phosphoglyceride; 4 sphingolipids (2 sphingomyelins, 2 cerebrosides). A section of the sodium chloride lattice is represented at the same scale (bottom right). (From Casper and Kirschner, 1971, courtesy of Macmillans)

Myelin membranes have also been studied by calorimetry (Ladbrooke et al., 1968b). Cholesterol removes the endothermic transition observed with lipids in model lipid–cholesterol systems at equimolar ratios. Since myelin membranes contain considerable quantities of cholesterol, it is interesting to study myelin isolated from white matter of ox brain in a similar manner. The results of such experiments show that:

1. With wet myelin, thermal transitions are not detectable. In this case the cholesterol and other lipids appear to be organized in a single phase. The organization of cholesterol in the membrane appears to prevent the lipids from crystallizing.
2. To maintain the organization of the lipid in myelin a critical amount of water appears to be required. This water fails to freeze at 0 °C and may correspond to 'bound' water.
3. When myelin membranes are dried, the cholesterol and other lipid crystallizes and precipitates. Endothermic transitions associated with cholesterol and other lipids can then be observed.
4. The total lipid extract in water does not show a detectable endothermic transition but the cholesterol-free lipid does. In the absence of cholesterol, part of the myelin lipid is crystalline at body temperature.

5.4.2.2 OUTER SEGMENT MEMBRANE OF THE RETINAL ROD

A model of the outer segment has been developed, using chemical, electron microscopic and X-ray data (Vanderkooi and Sundaralingam, 1970). An important feature of this model is the globular nature of the membrane proteins, the molecular arrangement of which is shown in *Figure 5.20*. The

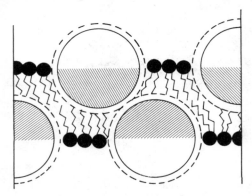

Figure 5.20 A suggested arrangement for the membrane structure of the retinal rod. The filled circles are the polar heads of the lipids, the large circles are the proteins. (The solid line represents a 4.05 nm sphere and the dashed line a 4.6 nm sphere. The cross-hatched region is predominantly nonpolar). (From Vanderkooi and Sundaralingam, 1970)

authors suggest that there are protein–protein, protein–lipid and lipid–lipid interactions present and that these will be constantly changing because of thermal motion. They argue that the lipids are hydrophobically bonded to the proteins, as shown by the difficulty in obtaining rhodopsin free of lipid in aqueous media.

^{13}C NMR spectra of the lipid in intact bovine retinal-rod membranes have also been reported (Millet, Hargrave and Raftery, 1973). Cone (1972), using flash illumination, studied the transient dichroism of the photo products of rhodopsin in rod outer segment membranes. He exploited the rapid absorption changes which can result from the illumination of the rhodopsin molecule and concluded that it was rotating rapidly with a relaxation time of about 20 s.

5.4.2.3 ERYTHROCYTE MEMBRANE

Electron microscope studies of freeze-etched material have shown that the fracture faces of most biological membranes (not myelin) appear as smooth sheets interrupted by numerous particles (in erythrocyte membranes about 8.5 nm in diameter, uncorrected for shadow thickness). When it became clear that the fracture face represents a plane through the interior of the membrane, it could be concluded that the particles represent localized regions where the bilayer membrane continuum is interrupted. In this context, it seemed legitimate to consider the membrane particles as the morphological representation of entities which are intercalated (but not necessarily sequestered) in the hydrophobic matrix of the bilayer membrane (Branton, 1971). It is important to note that the particles observed in the erythrocyte membrane account for only about 30 percent of the total protein of the membrane (Singer, 1971).

Freeze-etching has been used to follow the course of proteolytic digestion in red blood cell ghosts (Engstrom, 1970). If the particles represent proteins within the membrane matrix, they should disappear when protein is removed from the ghosts before freeze-etching. Red blood cells were incubated with the proteolytic enzyme pronase for varying times. The course of protein hydrolysis was monitored by simultaneously measuring protein loss and particle loss. Preliminary experiments indicated that pronase attacked all the cells in the preparation with equal vigour. Loss of membrane protein initially caused extensive aggregation of the particles with little decrease in number. The initial resistance of the particles to digestion by pronase is consistent with the view that the particles are buried with the hydrophobic region of the membrane matrix into which pronase diffusion would be difficult. After more prolonged treatment with pronase, most of the particles were lost. This eventual removal of particles from extensively digested ghosts is consistent with the hypothesis that the particles are protein. These particles are considered to traverse the hydrophobic space of the bilayer and apparently can move when subjected to media of different pH. It is suggested that the plasma membrane is a planar fluid domain formed by a bilayer which is interrupted by localized yet mobile intercalations.

NMR studies of erythrocyte membranes (Chapman et al., 1968; Kamat and Chapman, 1968) have been compared with similar studies on serum lipoproteins (Chapman et al., 1969). The spectra of the serum lipoproteins show signals from the aromatic amino acids, and signals from the various molecular segments of the lipids which are present, for example methylene groups, cholesterol ring methyl groups, double bonds, etc. The spectra of membranes in either the unsonicated or sonicated condition do not show the aromatic amino acid signals unless treated with detergent, while lipid signals become more apparent only after treatment with sodium deoxycholate.

The general conclusion is that the lipid and protein structures occurring in the membrane are tighter and more compact than those of serum lipoproteins.

Further conclusions were that the polar groups of the lipids, for example the trimethylammonium group, are in an environment similar to that observed with sonicated lecithin in D_2O and have similar freedom of molecular motion.

1. The protons of the sugar residues are situated in a local environment that allows considerable freedom.
2. The lipid hydrocarbon chains are restricted in their molecular freedom arising from lipid–cholesterol and/or lipid–protein interactions.
3. Studies of erythrocyte membranes (Metcalfe et al., 1971; Keough et al., 1973) using ^{13}C NMR spectroscopy are consistent with these conclusions; in this case the membranes are not sonicated.

The circular dichroism spectra of erythrocyte ghosts in the region of 190–230 nm, as well as certain other membrane systems, show some of the features characteristic of proteins in a partially helical conformation with certain anomalous features such as low values of Θ near 190 nm and a red shift above 220 nm. The red shift has been attributed to a number of different structural features of the membrane, including interactions between protein helices and an apolar environment for the protein helices. There is now convincing evidence that these anomalies are due to optical artifacts arising from the particulate, turbid nature of the samples (Urry, 1972). Urry (1972) pointed out methods of correcting these distortions and suggested that about 50 percent or more of the red blood cell membrane protein is in an α-helical conformation. This has been interpreted to indicate that proteins of the membrane are predominantly globular rather than spread out over the membrane surface. Models consistent with globular proteins being intercalated in the membrane alternating with lipid bilayer structure have been suggested (Lenard and Singer, 1966; Wallach and Zahler, 1966).

5.4.2.4 SARCOPLASMIC RETICULUM MEMBRANES

Sarcoplasmic membranes have been isolated from muscle tissue in the form of vesicular fragments which sustain Ca^{2+} uptake and have been studied by NMR spectroscopy (Davis and Inesi, 1971). A reversible, temperature-dependent structural transition was observed involving choline methyl groups of lecithin. The variation of the mobility of the methyl groups as a function of temperature correlated well with the temperature-dependent efflux of Ca^{2+}. Other agents which alter protein structure and simultaneously increase Ca^{2+} efflux also increase the fraction of mobile choline methyl groups. Irreversible heat denaturation of the membrane protein also alters the temperature dependence of the transition. Thus, Davis and Inesi concluded that the observed thermal transition involves protein as well as lipid.

5.4.2.5 *Acholeplasma laidawii* MEMBRANES

Thermal (Steim *et al.*, 1969) and X-ray studies (Engelman, 1971) have shown that thermal phase transitions occur with these mycoplasma membranes. The temperature of the transition varies depending on the nature of the fatty acids used to supplement the growth medium. These results are consistent with the idea that appreciable regions of the membrane contain lipid in a bilayer form. This particular plasma membrane contains some 50 percent of protein in the membrane mass. The X-ray evidence (Engelman, 1971) is interpreted to show that most of the protein is not in layers at the surface of the bilayer. This has been discussed on the basis of calorimetric data (Chapman and Urbina, 1971). Studies with deuterated fatty acids (Oldfield, Chapman and Derbyshire, 1972) incorporated into the membranes using deuteron magnetic resonance spectroscopy lead to the conclusion that appreciable portions of the membranes could have 'rigid' lipid chains, i.e. heterogeneity of lipid packing occurring with both solids and fluids present.

A possible model for the structure of the mycoplasma membrane which is consistent with the evidence suggests that the proteins are inserted almost entirely into the lipid bilayer, some perhaps penetrating it completely, some extending halfway through. The hydrophilic groups on the proteins would be in contact with the aqueous phase and the hydrophobic groups would be localized in the lipid hydrocarbon region. Engelman (1971) suggests that the central feature of membrane, compared with non-membrane, proteins would then be that they can be substituted for groups of lipid molecules in the bilayer without greatly perturbing it.

Tourtellotte, Branton and Keith (1970) have studied this membrane using a variety of techniques. The freeze-etch results show smooth areas and particulate components. The smooth component occupies about 80 percent of the membrane area leaving the particles to occupy about 20 percent.

5.4.2.6 THE PURPLE MEMBRANE OF *Halobacterium halobium*

Quite recently crystalline packing of visual-pigment-like molecules has been observed in the 'purple membrane' fragment of the cell membrane of *Halobacterium halobium* (Blaurock and Stoeckenius, 1971). The diffraction pattern of orientated sheets of purple membrane includes a number of sharp reflections centred on the axis (called the equator) in the plane of the membranes. The equatorial reflections are all accounted for by a regular, close packing of large numbers of hexagons in a plane; the centres of the hexagons are 6.3 nm apart. The scale of the array points to a crystalline arrangement of the protein molecules. The extracted lipids, dispersed in water, show no sign of the equatorial reflections. The sharp reflections appear as lines extended at right angles to the equator, indicating that the protein molecules are arranged with their centres in planes parallel to the membrane surfaces. The variation of intensity along the line suggests that the protein molecules are located in two planes, one on each side of the centre plane of the membrane. The extensive equatorial pattern means that the protein molecules must be located in the membrane with unusual precision.

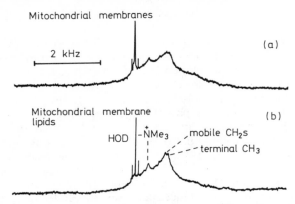

Figure 5.21 270 MHz proton NMR spectra at 18°C in D$_2$O of (a) mitochondrial membranes and (b) mitochondrial membrane lipids. (From Keough et al., 1973, courtesy of Associated Scientific Publishers)

Flash photolysis studies (Naqvi *et al.*, 1973) of these membranes have enabled a rotational correlation time to be determined for the single protein present. The lifetime was found to be longer than 20 ms, indicating that the protein is *not* freely rotating in the membrane. Recently, studies have been made of the flexibility of the lipids in *Halobacterium cutirubrium* cell envelopes (Esser and Lanyi, 1973).

Thus, the membrane proteins cause all but a narrow centre portion of the lipid bilayer to become particularly ordered. It is suggested that the lipids in this system are especially rigid.

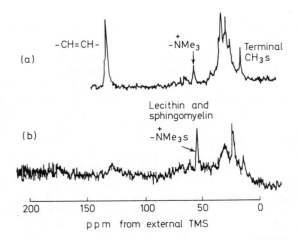

Figure 5.22 25.2 MHz ^{13}C NMR spectra at 37°C of (a) mitochondrial membrane, and (b) erythrocyte ghosts. (From Keough et al., 1973, courtesy of Associated Scientific Publishers)

5.4.2.7 MITOCHONDRIAL MEMBRANES

PMR and ^{13}C NMR spectroscopy have been applied to the study of mitochondrial membranes (Keough et al., 1973). The spectra show that unsaturated lipid chains and trimethylammonium groups have considerable mobility. The peptide backbones probably have considerably less rotational freedom than the lipid molecules. The proteins are apparently more 'rigid' in this membrane than the lipids (Figures 5.21 and 5.22).

Calorimetric studies made with mitochondrial membrane systems (Blazyk and Steim, 1972) also show that all the lipid of the membrane is fluid at body temperature.

5.4.2.8 Escherichia coli MEMBRANES

Changes in the slopes of Arrhenius plots for transport in *Escherichia coli* membranes have been observed with cells that have a relatively simple fatty acid composition in the membrane lipids (Overath, Schairer and Stoffel, 1970). These characteristic temperatures have been shown to correlate with those temperatures that define changes of phase of the membrane phospholipids. The highest characteristic temperature corresponds to the point at which the formation of 'rigid' patches of membrane lipids occurs (Linden et al., 1973).

Such lipid phase transitions (similar to those obtained with simple lipid/water systems) have been observed using spin labels (Linden et al., 1973), fluorescent probes, light scattering and dilatometry (Overath and Trauble, 1973).

REFERENCES

BANGHAM, A. D., STANDISH, M. M. and WATKINS, J. C. (1965). *J. molec. Biol.*, **13**:238.
BARRATT, M. D., GREEN, D. and CHAPMAN, D. (1969). *Chem. Phys. Lipids*, **3**:140.
BITTAR, E. (Ed.) (1970). *Membranes and Ion Transport*. New York; Wiley–Interscience.
BLAUROCK, A. E. and STOECKENIUS, W. (1971). *Nature, New Biol.*, **233**:152.
BLAZYK, J. F. and STEIM, J. M. (1972). *Biochim. biophys. Acta*, **266**:737.
BRANTON, D. (1971). *Phil. Trans. R. Soc., Ser. B*, **261**:133.
BRECKENRIDGE, W. C., GOMBOS, G. and MORGAN, I. G. (1971). *Brain Res.*, **33**:581.
BRETSCHER, M. S. (1971a). *J. molec. Biol.*, **58**:775.
BRETSCHER, M. S. (1971b). *Nature, New Biol.*, **231**:229.
BRETSCHER, M. S. (1972). *J. molec. Biol.*, **71**:523.
BROSTOU, S. and EYLAR, E. H. (1971). *Proc. natn. Acad. Sci. U.S.A.*, **68**:765.
BYRNE, P. and CHAPMAN, D. (1964). *Nature, Lond.*, **202**:987.
CARNEGIE, P. R. (1971). *Nature, Lond.*, **229**:25.
CASPAR, D. L. D. and KIRSCHNER, D. A. (1971). *Nature, New Biol.*, **231**:46.
CATER, B., CHAPMAN, D., HAWES, S. and SAVILLE, J. (1974). *Biochim. biophys. Acta*, **363**, 54.
CHAPMAN, D. (1958). *J. chem. Soc.*, 784.
CHAPMAN, D. (1968). *Biological Membranes*, Vol. 1, p. 125. Ed. D. CHAPMAN. London; Academic Press.
CHAPMAN. D., BYRNE, P. and SHIPLEY, G. G. (1966). *Proc. R. Soc., Ser. A*, **290**:115.
CHAPMAN, D. and COLLIN, D. T. (1965). *Nature, Lond.*, **296**:189.
CHAPMAN, D. and FLUCK, D. J. (1966). *J. biophys. Cytol.*, **30**:1.
CHAPMAN, D., KEOUGH, K. M. and URBINA, J. (1974a). *J. biol. Chem.*, **249**:2512.
CHAPMAN, D., OWENS, N. F. and WALKER, D. A. (1966). *Biochim. biophys. Acta*, **120**:148.

CHAPMAN, D. and PENKETT, S. A. (1966). *Nature, Lond.*, **211**:1304.
CHAPMAN, D. and SALSBURY, N. J. (1966). *Trans. Faraday Soc.*, **62**:2607.
CHAPMAN, D. and URBINA, J. (1971). *FEBS Lett.*, **12**:169.
CHAPMAN, D., WILLIAMS, R. M. and LADBROOKE, B. D. (1967). *Chem. Phys. Lipids*, **1**:445.
CHAPMAN, D., KAMAT, V. B., DE GIER, J. and PENKETT, S. A. (1968). *J. molec. Biol.*, **31**:101.
CHAPMAN, D., LESLIE, R. B., HIRZ, R. and SCANN, A. (1969). *Biochim. biophys. Acta*, **176**:524.
CHAVIN, S. I. (1971). *FEBS Lett.*, **14**:269.
CHEN, L. F., LUND, D. B. and RICHARDSON, T. (1971). *Biochim. biophys. Acta*, **225**:108.
CHERRY, R. J., BERGER, J. U. and CHAPMAN, D. (1971). *Biochem. biophys. Res. Commun.*, **44**:644.
CLOWES, A., CHERRY, R. J. and CHAPMAN, D. (1972). *J. molec. Biol.*, **67**:49.
CONE, R. A. (1972). *Nature, Lond.*, **236**:39.
DARKE, A., FINER, E. G., FLOOK, A. G. and PHILLIPS, M. C. (1972). *J. molec. Biol.*, **63**:265.
DAVIS, D. G. and INESI, G. (1971). *Biochim. biophys. Acta*, **241**:1.
DE GIER, J., MANDERSLOOT, J. G. and VAN DEENEN, L. L. M. (1968). *Biochim. biophys. Acta*, **150**: 666.
DE GIER, J., MANDERSLOOT, J. G., HUPKES, J. V., MCELHANEY, R. N. and VAN BECK, W. P. (1971). *Biochim. biophys. Acta*, **122**:610.
DEVAUX, P. and MCCONNELL, H. M. (1972). *J. Am. chem. Soc.*, **94**:4475.
DICKERSON, J. W. T. (1968). *Applied Neurochemistry*, p. 48. Ed. A. M. DAVISON and J. DOBBING. Oxford; Blackwell Scientific Publications.
DUNN, M. J. and MADDY, A. H. (1973). *FEBS Lett.*, **36**:79.
ENG, L. F., CHAO, F. C., GERSTL, B., PRATT, D. R. and TAVASTSJERNA, M. G. (1968). *Biochemistry*, **7**:4455.
ENGELMAN, D. M. (1971). *J. molec. Biol.*, **58**:153.
ENGSTROM, L. H. (1970). *Ph.D. Dissertation*. University of California, Berkeley.
ESSER, A. F. and LANYI, J. K. (1973). *Biochemistry*, **12**:1933.
EYLAR, E. H., BROSTOFF, S., HASHIM, G., CAECAM, J. and BURNETT, P. (1971). *J. biol. Chem.*, **246**:5770.
FINKELSTEIN, A. and CASS, A. (1968). *J. gen. Physiol.*, **52**:1455.
FLUCK, D. J., HENSON, A. F. and CHAPMAN, D. (1969). *J. Ultrastruct. Res.*, **29**:416.
FOLCH, J. and LEES, M. (1951). *J. biol. Chem.*, **191**:807.
GEREN, B. B. (1954). *Expl Cell Res.*, **7**:558.
GONZALEZ-SASTRE, F. (1970). *J. Neurochem.*, **17**:1049.
GREEN, D. E. and FLEISCHER, S. (1963). *Biochim. biophys. Acta*, **70**:554.
GREEN, H. O. and REYNOLDS, J. A. (1971). *Fedn Proc. Fedn Am. Socs exp. Biol.*, **30**:1065.
GUIDOTTI, G. (1972). *A. Rev. Biochem.*, **41**:731.
GULIK-KRZYWICKI, T., RIVAS, E. and LUZZATI, V. (1967). *J. molec. Biol.*, **27**:303.
GULIK-KRZYWICKI, T., SCHECTER, E., LUZZATI, V. and FAURE, M. (1969). *Nature, Lond.*, **223**:1116.
HANAHAN, D. J. (1969). *Red Cell Membrane—Structure and Function*, p. 83. Ed. G. A. JAMIESON and T. J. GREENWALT. Philadelphia; J. B. Lippincott.
HAUSER, H., FINER, E. G. and CHAPMAN, D. (1970). *J. molec. Biol.*, **53**:413.
HIGASHI, Y., SIEWERT, G. and STROMINGER, J. L. (1970). *J. biol. Chem.*, **245**:3683.
HIGASHI, Y. and STROMINGER, J. L. (1970). *J. biol. Chem.*, **245**:3691.
HLADKY, S. B. and HAYDON, D. A. (1972). *Biochim. biophys. Acta*, **274**:294.
HUBBELL, W. L. and MCCONNELL, H. M. (1969). *Proc. natn. Acad. Sci. U.S.A.*, **64**:20.
HUBBELL, W. L. and MCCONNELL, H. M. (1971). *J. Am. chem. Soc.*, **93**:314.
INOUYE, M. and GUTHRIE, J. P. (1969). *Proc. natn. Acad. Sci. U.S.A.*, **64**:957.
JONES, T. H. D. and KENNEDY, E. P. (1969). *J. biol. Chem.*, **244**:5981.
JULIANO, R., KIMELBERG, H. K. and PAPAHADJOPOULOS, D. (1971). *Biochim. biophys. Acta*, **241**:894.
KAMAT, B. B. and CHAPMAN, D. (1968). *Biochim. biophys. Acta*, **163**:411.
KEOUGH, K. M., OLDFIELD, E., CHAPMAN, D. and BEYNON, P. (1973). *Chem. Phys. Lipids*, **10**:37.
KIMELBERG, H. and PAPAHADJOPOULOS, D. (1968). *Biochim. biophys. Acta*, **150**:333.
KIMELBERG, H. and PAPAHADJOPOULOS, D. (1971). *J. biol. Chem.*, **246**:1142.
KORNBERG, R. D. and MCCONNELL, H. M. (1971). *Proc. natn. Acad. Sci. U.S.A.*, **68**:2564.
KRASNE, S., EISENMAN, G. and SZABO, G. (1972). *Science, N.Y.*, **174**:412.
LADBROOKE, B. D. and CHAPMAN, D. (1969). *Chem. Phys. Lipids*, **3**:304.
LADBROOKE, B. D., WILLIAMS, R. M. and CHAPMAN, D. (1968a). *Biochim. biophys. Acta*, **150**:333.
LADBROOKE, B. D., JENKINSON, T. J., KAMAT, V. B. and CHAPMAN, D. (1968b). *Biochim. biophys. Acta*, **164**:101.
LENARD, J. and SINGER, S. J. (1966). *Proc. natn. Acad. Sci. U.S.A.*, **56**:1828.

LINDEN, C. D., WRIGHT, K. L., MCCONNELL, H. M. and FOX, C. F. (1973). *Proc. natn. Acad. Sci. U.S.A.*, **70**:2271.
LOW, M. G., LIMBRICK, A. R. and FINEAN, J. B. (1973). *FEBS Lett.*, **34**:1.
LUZZATI, V. (1968). *Biological Membranes*, Vol. 1, p. 103. Ed. D. CHAPMAN. London; Academic Press.
LUZZATI, V. and HUSSON, F. (1962). *J. Cell Biol.*, **12**:207.
LUZZATI, V., REISS-HUSSON, F., RIVAS, E. and GULIK-KRZYWICKI, T. (1966). *Ann. N.Y. Acad. Sci.*, **137**:409.
MARCHESI, S. L., STEERS, E., MARCHESI, V. T. and TILLACK, T. W. (1970). *Biochemistry*, **2**:50.
MARCHESI, V. T. and STEERS, E. (1968). *Science, N.Y.*, **159**:203.
MAZIA, D. and RUBY, A. (1968). *Proc. natn. Acad. Sci. U.S.A.*, **61**:1005.
METCALFE, J. C., BIRDSALL, N. J. M., FEENEY, J., LEE, A. G., LEVINE, Y. K. and PARTINGTON, P. (1971). *Nature. Lond.*, **233**:199.
MILLET, F., HARGRAVE, P. A. and RAFTERY, M. A. (1973). *Biochemistry*, **12**:3591.
MITCHELL, C. D. and HANAHAN, D. J. (1966). *Biochemistry*, **5**:51.
MUELLER, P. and RUDIN, D. O. (1967). *Biochem. biophys. Res. Commun.*, **26**:398.
MUELLER, P., RUDIN, D. O., TIEN, H. T. and WESTCOTT, W. C. (1962). *Nature, Lond.*, **194**:979.
NAQVI, R. K., GONZALEZ-RODRIGUEZ, J., CHERRY, R. J. and CHAPMAN, D. (1973). *Nature, New Biol.*, **245**:249.
O'BRIEN, J. S. (1965). *Science, N.Y.*, **147**:1099.
OLDFIELD, E., CHAPMAN, D. and DERBYSHIRE, W. (1971). *FEBS Lett.*, **16**:102.
OLDFIELD, E., CHAPMAN, D. and DERBYSHIRE, W. (1972). *Chem. Phys. Lipids*, **9**:69.
OVERATH, P., SCHAIRER, H. V. and STOFFEL, W. (1970). *Proc. natn. Acad. Sci. U.S.A.*, **67**:606.
OVERATH, P. and TRAUBLE, H. (1973). *Biochemistry*, **12**:2625.
PAPAHADJOPOULOS, D., JACOBSON, K., NIR, S. and ISAC, T. (1973). *Biochim. biophys. Acta*, **311**:330.
PACHE, W., CHAPMAN, D. and HILLABY, R. (1972). *Biochim. biophys. Acta*, **255**:358.
PHILLIPS, M. C., LADBROOKE, B. D. and CHAPMAN, D. (1970). *Biochim. biophys. Acta*, **196**:35.
PHILLIPS, M. C., WILLIAMS, R. M. and CHAPMAN, D. (1969). *Chem. Phys. Lipids*, **3**:234.
RAND, R. P. (1971). *Biochim. biophys. Acta*, **241**:823.
ROSENBERG, S. and GUIDOTTI, G. (1968). *J. biol. Chem.*, **243**:1985.
ROSENBERG, S. and GUIDOTTI, G. (1969). *J. biol. Chem.*, **244**:5118.
SACKMANN, E. and TRAUBLE, H. (1972). *J. Am. chem. Soc.*, **94**:4492.
SCHMITT, F. O., BEAR, R. S. and PALMER, K. J. (1941). *J. cell. comp. Physiol.*, **18**:31.
SHIPLEY, G. G., LESLIE, R. B. and CHAPMAN, D. (1969). *Nature, Lond.*, **222**:561.
SINGER, S. J. (1971). *Structure and Function of Biological Membranes*, p. 145. Ed. L. ROTHFIELD. New York; Academic Press.
SMALL, D. M. and BOURGÈS, M. (1966). *Molec. Crystals*, **1**:541.
STECK, T. L., FAIRBANKS, G. and WALLACH, D. F. H. (1971). *Biochemistry*, **10**:2617.
STEIM, J. M., TOURTELLOTTE, M. E., REINERT, J. C., MCELHANEY, R. N. and RADER. R. L. (1969). *Proc. natn. Acad. Sci. U.S.A.*, **63**:104.
STEINEMANN, A. and LAUGER, P. (1970). *J. Membrane Biol.*, **4**:74.
SWEET, C. and ZULL, J. E. (1969). *Biochim. biophys. Acta*, **173**:94.
TENENBAUM, D. and FOLCH-PI, J. (1966). *Biochim. biophys. Acta*, **115**:141.
TOURTELLOTTE, M. E., BRANTON, D. and KEITH, A. D. (1970. *Proc. natn. Acad. Sci. U.S.A.*, **66**:909.
URRY, D. W. (1971). *Proc. natn. Acad. Sci, U.S.A.*, **68**:672.
URRY, D. W. (1972). *Rev. Biomembranes.* **265**:115.
VANDERKOOI, G. and SUNDARALINGAM, M. (1970). *Proc. natn. Acad. Sci. U.S.A.*, **67**:233.
VEKSLI, Z., SALSBURY, N. J. and CHAPMAN, D. (1969). *Biochim. biophys. Acta*, **183**:434.
WALLACH, D. H. F. (1970). *Biological Membranes*, Vol. 1, p. 145. Ed. D. CHAPMAN and D. H. F. WALLACH. New York; Academic Press.
WALLACH, D. H. F. and ZAHLER, P. H. (1966). *Proc. natn. Acad. Sci. U.S.A.*, **56**:1552.
WILLIAMS, R. M. and CHAPMAN, D. (1970). *Prog. Chem. Fats*, **11**:1.
WINZLER, R. J. (1969). *Red Cell Membranes—Structure and Function*, p. 157. Ed. G. A. JAMIESON and T. J. GREENWALT. Philadelphia; J. B. Lippincott.
WINZLER, R. J., HARRIS, E. D., PEKAS, D. J., JOHNSON, C. A. and WEBER, P. (1967). *Biochemistry*, **6**:2195.
WOLFGRAM, F. (1966). *J. Neurochem.*, **13**:461.
ZWAAL, R. F. and VAN DEENEN, L. L. M. (1970). *Chem. Phys. Lipids*, **4**:311.

6

Mechanical properties of cellular membranes

Peter B. Canham
Department of Biophysics, Health Sciences Centre,
The University of Western Ontario, London, Ontario

6.1 INTRODUCTION

Cell membranes are a living, dynamic part of the cell playing a role far beyond that of a sack or partition. The recent acceptance of this fact has stimulated numerous investigators to examine the physical properties of membranes with the goal of improving comprehension of cellular function. The revelation of the mechanical properties of cell membranes is a challenging undertaking; it requires the application of mathematical and physical principles to membrane biology, a task necessitating a conceptual leap from the traditional materials of engineering to the living cell membrane, a delicate and complex structure which is easily damaged during experimentation. Very delicate techniques must be evolved before conclusive results can be obtained. Experimentation in this area has mushroomed in the past few years and a considerable number of data have been published. However, there are many unresolved conflicts in the varied reports, as could well be expected whenever any new field comes under scientific investigation. In this chapter a determined effort has been made to include all significant developments in this research but emphasis is given to those theories which best encompass the experimental observations and which are mutually corroborating.

Cell membranes can be the external covering of the cell, such as the plasma membrane, or any internal membranous structure contained by the cell, such as the endoplasmic reticulum and the mitochrondrial membranes. Since the mechanics of membranes is the study of the motion of membranes and the action of forces on them, experimental work can best be done on membranes which are physically accessible. As a result, most of the research has been carried out on the membrane of the mammalian red blood cell, i.e. the plasma membrane of one particular kind of cell. It is acknowledged that the mechanical properties determined for the red cell membrane may not apply to the membranes of other cells. Even different types of membranes within a particular cell may have only a few common properties. Yet an exhaustive study of the red cell membrane could readily lead to the determination of the general properties of membranes and the relationship of structure to physical behavior. However, the following discussion in combination with accumulated data from membrane chemistry and ultramicroscopy may make possible the prediction of mechanical properties for those membranes which do not lend themselves to physical investigation.

6.2 MECHANICAL PROPERTIES OF THE RED CELL MEMBRANE

6.2.1 Behavior of the red cell

Observation of red cells moving in the microcirculation or through micropipettes or other glass tubes leads one to the realization that the red cell membrane is a structure which behaves more as a liquid layer than a solid. The films discussed by Brånemark (1971) and Brånemark and Lindström (1973) show the cell, sometimes singly, often in small rouleaux, moving fluidly past corners at bifurcations, adopting umbrella shapes or sausage shapes reversibly, and demonstrating almost no elastic stiffness or permanent deformation upon returning to a biconcave shape. A film borrowed from A. W. L. Jay in Calgary showed the movement of cells in the microcirculation of the spleen in a squashed spleen preparation. The red cells, in contrast to the rigid white cells, would move like droplets of immiscible fluid in the plasma, adopting the biconcave shape whenever the moving plasma slowed or stopped. In contrast with the rapid movement of the red cells, the white cells appeared virtually immovable, although an occasional white cell was seen to dislocate and travel with the flow. When struck by red cells singly or associated into rouleaux, the white cell remained stationary and the red cells would slide and slip off its surface. The smaller white cells may have a plasma membrane of mechanical properties similar to those of the erythrocyte membrane, yet they appear rigid. This rigidity is due to two conditions: first, the presence of cytoplasmic gel and organelles in the cell, and secondly, the spherical shape of the white cell. A sphere can have no flexibility unless the surface is able to expand. An essential but not sufficient condition for the deformability of the red cell is its excess membrane area, i.e. its area in excess of that required to contain the cell volume in a spherical shape. The larger white cells, for example the polymorphonuclear leukocytes, are not spherical and consequently are more flexible than the small white cells. They are still much less deformable than the red cell because of the cellular interior. The flexibility of the membrane rather than the deformability of the whole cell is an important distinction to make (Braasch, 1969; LaCelle, 1972).

6.2.2 Membrane thickness

The thickness of the membrane may be described in several ways. In terms of diffusion, it may be considered as the perpendicular distance between the plane on the outer side of the membrane at which the extracellular mobility of an approaching molecule is altered by a certain amount, and the parallel plane on the inner side where the same molecule would undergo a similar limitation of its intracellular mobility as it approached the membrane from inside the cell. What will present a barrier to a diffusing molecule depends on the properties of that molecule. Another value for membrane thickness may be derived through measurements from X-ray data of repeating membrane layers. Or the distance between two dense lines in a photomicrograph from an electron microscopic image of a thin section may be measured (Stein, 1967, p. 4).

The mechanical thickness would require a different definition. It could be the perpendicular distance between two points directly opposite each other across the membrane, the points being the places at which an inert, uncharged solid object (large relative to the membrane thickness) encounters a prescribed amount of increased resistance on slowly approaching the membrane from one side and then the other of the membrane. This assumes that the cell is free and in a suspending material of low viscosity which is relatively free of structure, and that the contents of the cell are also relatively free of structure and of substance unlike that of the membrane, at least in areas immediately internal to the membrane. Consequently, this concept of mechanical thickness is more applicable to the red cell than to nucleated cells with cytoplasmic organelles and gels which might be partly continuous with the inner membrane surface. Even the hemoglobin in the red cell is thought to be structurally associated with the inner surface (Ponder, 1961). Another definition in terms of mechanics would be to define the thickness in terms of its strength by comparing the strength of the thin sheet of membrane with the strength of membrane material in bulk. However, the plasma membrane is not a homogeneous solid or semisolid and the different components of the membrane contribute different mechanical properties to the membrane. Perhaps of even greater importance to the mechanical properties is the interaction of the chemically discrete components of the membrane.

Depending then on the various interpretations of what membrane thickness constitutes, the value arrived at could vary substantially. Reported values for the red cell have ranged from 0.007 to 0.1 µm (Ponder, 1961). The generally accepted value, arrived at through X-ray studies and electron microscopy, is 0.01 µm (Stoekenius and Engelman, 1969). One is encouraged from thin-section electron micrographs to believe that the erythrocyte membrane is of uniform thickness, as shown in work by J. D. Robertson (1964) and H. Swift (1962) (cited in Stein, 1967, pp. 4 and 6). However, such evidence does not preclude the possibility that the membrane has different physical properties in different regions.

6.2.3 Structure of the red cell membrane

The interpretation of mechanical experiments on cells requires an understanding of the basic structure of the membrane. (For details of the membrane architecture the reader is referred to Chapters 1 and 10 of this volume.) There are two general classes of models of the erythrocyte membrane, the micellar arrangement and the Davson–Danielli bilayer (Stein, 1967; Stoeckenius and Engelman, 1969; Singer and Nicolson, 1972; Nystrom, 1973). Because it is more stable physicochemically and is thought to be representative of most membrane structure, the Davson–Danielli model is assumed correct for this discussion. The lipid bilayer, made up of phospholipid and cholesterol in a 3:1 weight ratio, is coated with structural protein, making a protein:lipid weight ratio for human red cells of 1:1 (Weed, Reed and Berg, 1963; Dodge, Mitchell and Hanaham, 1963). In his review (1969), Korn pointed out that the lipid ratio is constant and that the protein:lipid ratio is species-specific. The pattern of the protein on or in the surfaces of the membrane is not yet established; however, the distribution of some of the surface and membrane proteins has been reported (Blaurock, 1973; Marchesi et al., 1973). A preference for the micellar model has been indicated by permeability studies and by some work on model lipid membranes (reviewed by Korn, 1969), but what is considered as the strongest contradiction of the Davson–Danielli model is the abundant evidence of particles and pits in the membrane illustrated by freeze fracture techniques (Weinstein and Koo, 1968; Weinstein, 1969; Lessin, 1972; Seeman and Iles, 1972). Nonetheless, the membrane may still be interpreted mechanically in terms of a continuum provided that the probes used to investigate the mechanical properties of the membrane are large relative to the size of the irregularities and their repeating distances. This condition is easily met when micropipettes, microprobes, or visible light microscopy are used. The high-resolution dimensional studies of Evans and Fung (1972), obtaining image reconstruction from interference microscopy, are insensitive to the irregularities because of the averaging techniques. Hence, with regard to the sameness of the membrane of the adult human red cell, it is proposed that the membrane is structurally the same from one patch, of area 0.01 μm^2, to the next.

The following mechanical description of the erythrocyte membrane is then possible and, indeed, is gaining wide acceptance. The membrane is an isotropic continuum in the plane of the membrane, with surfaces which are viscoelastic, not resisting (or resisting weakly) shear at equilibrium, but capable of strong elastic resistance to tensile two-dimensional surface stress. That is, the membrane resists one-directional strain primarily as a viscous liquid, and two-dimensional strain like an elastic solid. The surface of the membrane is coated with a fibrous matrix which is thought to contribute to the membrane the property of a weakly elastic two-dimensional surface. This description is the result of: (a) direct experimental results, for example the pipette experiments of Rand and Burton (1964), Leblond (1972), and Jay (1973), the sphering experiments of Ponder (1948), Rand and Burton (1963), Canham and Parkinson (1970), Evans and Fung (1972), Evans and Leblond (1972, 1973), the one-directional deformation of the red cell in experiments by Hochmuth and Mohandas (1972) and Hochmuth,

Mohandas and Blackshear (1973), and the X-ray crystallographic studies of Rand and Luzzati (1968); (b) the interpretation of these experiments by their authors and by others (Ponder, 1961; Katchalsky *et al.*, 1960; Stoeckenius and Engelman, 1969; Rand, 1968; Burton, 1970), and (c) in particular, the mechanical considerations of the red cell membrane by Fung and Tong (1968), Evans (1973a, b) and Skalak (1973).

6.2.4 The area of the red cell membrane

The area of the red cell membrane is a critical parameter in the discussion of membrane mechanics, but is not a parameter which can be measured easily. The long-established method of using the theorem of Pappus, probably first done for the red cell by Ponder (1948), involves tracing around the image of the membrane. For this the cell must be photographed in the on-edge position, and then treated as a surface of revolution; this was done by Rand and Burton (1963), and Jay (1975), and has been used by the present writer to obtain the areas and volumes of several thousand cells under different experimental conditions (standard measurements in Canham and Burton, 1968); for example, young and old cells (Canham, 1969a), swelling cells (Canham and Parkinson, 1970). A completely different method, developed by Evans and Fung (1972), allows the reconstruction of the geometric parameters and obtains the overall cell thickness, the cell thinness at the center and the major diameter to an accuracy of approximately 1 percent. In their technique the red cell is assumed to be an object of cylindrical symmetry containing a uniform index of refraction. A holographic image of the cell is obtained using an interference microscope. Subsequently the phase image of the cell is transformed into a mathematical cell profile fitting the empirical thickness equation

$$D(r) = [1-(r/R_0)^2]^{\frac{1}{2}} [C_0 + C_2(r/R_0)^2 + C_4(r/R_0)^4] \qquad (6.1)$$

where $D(r)$ is the ordinate and r the abscissa on a Cartesian plot. The C_i values define each cell. From this equation the surface area and volume are computed with a reported accuracy of approximately 1 percent. The technique is elegant and very precise and it is reassuring that the results are in close agreement with the more direct method of Canham and Burton (1968), except for the mean cellular volume which was 106 μm³ ±18 compared with 94 μm³ ±14 (Evans and Fung, 1972). A discussion comparing the two methods is inappropriate; however, one of the conclusions of Evans and Fung should be clarified. These authors reported that the membrane area is constant during all but the latter stages of swelling, at which a 7.5 percent increase in area occurred. It is expected that the entire 7.5 percent might be accounted for when the experimental results of Danon (1961) and the theoretical results of Canham (1969b) are considered. Danon's cinemicrographic sequence of gradually swelling red cells showed that small cells hemolyze first. This is explained by the more spherical shape of the smaller cells (Canham and Burton, 1968). It is estimated that between 50 and 90 percent of the cells would be hemolyzed in a medium of osmolarity 131 milliosmolar, the tonicity of the most dilute suspending medium in the experiments of Evans and Fung (1972). This would leave cells with a predicted mean area between 6 and 15 percent higher than the mean. Earlier theoretical evidence for increased surface area of 7 percent during osmotic hemolysis (Canham, 1969b) should also be reduced in view of the more recent results of others. Seeman *et al.* (1969) demonstrated a prelytic potassium loss (10–20 percent) from the red cell, when the membrane was under stress. The more rapid the hemolysis, the less the potassium lost from the cells. It is well established that cells hemolyze in media of high tonicity if the medium is diluted rapidly, and the prelytic potassium loss has been interpreted as a passive protective mechanism against lysis (Canham and Parkinson, 1970).

Unpublished experiments by A. W. L. Jay and the author, in which several human erythrocytes were gradually drawn into a 2 µm cylindrical glass pipette with negative pressure, add further evidence to the concept of an inextensible membrane. Initially the captured red cell adopted a florence-flask shape and then it gradually entered the pipette; rather than suffer a perceptibly stretched membrane the volume decreased and the cell moved into the pipette unhemolyzed. The decreasing volume for one cell entering a narrow pipette is shown in *Figure 6.1*. Characteristically for the eight cells

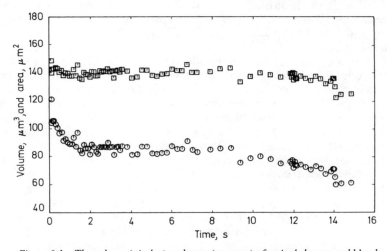

Figure 6.1 The volume (circles) and area (squares) of a single human red blood cell drawn into a 2.1 µm pipette by a linearly increasing negative pipette pressure to 80 mmHg. Of particular importance is the constancy of the area at 140 µm² despite the increasing tension, which rose to a value of 35 dyn cm^{-1}. The measured geometry of the cell indicates that the cell enters the pipette by losing volume. (Unpublished results of A. W. L. Jay and P. B. Canham)

studied, this cell did not demonstrate any membrane stretch. The initial rapid volume loss is interpreted as an artefact associated with transient distortions in the membrane surface during the first 0.5 s, or an effect caused by the nearness of the cell to glass. Subsequent volume loss was interpreted to be due to (a) a hydraulic loss in water, resulting in an osmotic imbalance, associated with the high intracellular pressure, and (b) a loss of cell water associated with a solute loss, characteristic of red cell membranes under tensile stress (Seeman *et al.*, 1969). The pipette studies, unlike the previous sphering studies (Canham and Parkinson, 1970), allowed a measurement of the area proportional to one linear dimension (length of the sausage-shaped cell) rather than proportional to the square of the spherical diameter. It is acknowledged that whereas the accuracy in an absolute single measurement is low in the direct photographic method compared with the technique of holographic reconstruction, the accuracy is greatly improved for measuring trends in geometry in response to gradually altering circumstances.

An effort has been made to support the concept that the membrane of the red cell cannot be stretched without hemolysis. The studies of Burton (1970) concerning the behavior of holes in model membranes in response to the stretching of the membrane indicate how the holes, or pores, in a cell membrane could be increased in an area proportionately far more than the membrane itself. The expansion ratio (change in hole diameter:change in surface diameter) for a thin isotropic elastic membrane is $(2-2\sigma)/(1-2\sigma)$, where σ is the Poisson ratio. As Burton made clear, the expression for the expansion ratio is probably not directly applicable; however, it emphasizes that for biological membranes the expansion ratio can be extremely high (σ is thought to be close to 0.5 for biological membranes).

The multitude of shapes which the red blood cell will adopt (Bessis, 1972), in addition to the multitude of external agents which will cause the same shape change (Weed and Chailley, 1972), constitutes a substantial challenge to any unifying theory of membrane behavior. The isovolumetric transformation from a biconcave cell (discocyte) to crenated sphere (spheroechinocyte), which was described by Ponder (1948), has recently been further described by Evans and Leblond (1973), who showed the repeatability of the phenomenon, and by Brecher and Bessis (1972), who described the morphology. The isovolumetric transformation is of particular interest in terms of the area of the membrane because some shapes, in particular the spheroechinocyte type II and the spherostomatocyte type II (Bessis, 1972, Figures 6 and 10), seem to have greatly reduced surface area. Neither the few small spicules in the spheroechinocyte nor the irregularities in the hilum of the spherostomatocyte (Bessis, 1972, Table 2) seem capable of accounting for the apparent reduction of the membrane area of the cell when in its normal biconcave shape. The 40 short spicules for the spheroechinocyte type I (Bessis, 1972, Figure 5) which appear to be approximately 1.5 μm high and between 0.25 and 0.5 μm in diameter can account for about one-half to two-thirds of the difference in area; that there is no physical disappearance of the membrane is shown by the reversibility (Evans and Leblond, 1972). An explanation is proposed involving the 'Gorter and Grendel game' discussed by Rand and Hoffman (Rand, 1968). The average surface area of a population of normal human red cells is 135 μm^2 and the volume is approximately 100 μm^3. The isovolumetric smooth sphere (spheroechinocyte II) has an area of approximately 103 μm^2, which is only 75 percent of the original membrane area. Rand reported that 20–30 percent more phospholipid and cholesterol (depending on their interaction) than had actually been determined experimentally would be needed in order to make a complete lipid bilayer around the human red cell. Possibly the membrane of the spheroechinocyte is a condensed form of phospholipid and cholesterol in a continuous two-layered lipid envelope around the cell, whereas the larger membrane area of the biconcave cell suggests the more open structure of lipid interacting with some of the protein. Sirs and Stolinsky (1971) concluded from transmission electron microscopy and freeze fracture techniques that the spheroechinocyte had a membrane eight times as thick as the biconcave cell. It is unclear how their model of disc-like membranous subunits could be interpreted in terms of a phospholipid–cholesterol matrix.

6.2.5 Viscoelastic properties

Viscoelasticity is the property of a material which exhibits some of the properties of elasticity in combination with some properties of viscosity. The term *viscoelastic* spans such a broad spectrum of behaviour that one must be specific in how the term is to be applied, and if possible be quantitative. Synovial fluid, a joint lubricant, and the blood vessel wall, both of which are viscoelastic, have very different properties. The red cell membrane, possibly characteristic mechanically of many plasma membranes, exhibits the quality of viscoelasticity when subjected to microprobes and pipettes (Jay and Burton, 1969). A slowly advanced 0.1 µm pipette did not penetrate into the cell but caused the membrane to deform reversibly around the pipette, whereas a rapidly speared cell membrane was penetrated, sealing around the pipette and then occasionally resealing the hole when the pipette was withdrawn. It would seem that the fabric or loose weave of the protein surface layer was distorted in the manner of a pinned woolen necktie, and the permeability barrier of lipid deformed as a liquid film would around the pipette. It is also possible that the surface protein coats the outer surface of the pipette, which upon withdrawal leaves a discontinuity in the lipid continuum as suggested by the high incidence of red cell lysis in these experiments on cellular potentials (Jay and Burton, 1969). The ability to penetrate almost every kind of cell more easily than the red cell (e.g. in the cellular communication studies of Loewenstein, 1968) suggests that the latter are more resilient. However, the gel-like interior of cells with intracellular organelles, and their ability to fabricate membranes, suggests that comparisons must be made with caution.

The membrane of the mammalian red blood cell is viscoelastic in that it strongly resists two-dimensional stretch elastically, as shown in the experiments of cells drawn into small glass pipettes (Rand, 1964; Rand and Burton, 1964; Jay, 1973) but will resist one-dimensional deformation primarily as a viscous material with a weak elastic constant (Hochmuth and Mohandas, 1972). The red cell drawn into a pipette is not only elastically tough but has been shown to behave in a manner dependent on the size of the pipette (*Figure 6.2*). Another form of viscous cellular deformation was described by Bull (1972), in which individual red cells were induced by a fluid surface shear to exhibit a 'tank tread' motion about a fixed point of attachment. Accepting that we have some appreciation of what is meant by viscoelasticity as applied to the red cell membrane, let us examine some of the quantitative mechanical properties.

Rand (1968) combined the results of X-ray crystallographic studies on red cell lipids in water with results from studies on the behavior of red cells drawn into glass micropipettes. The results of these experiments where swollen red cells were drawn into different pipettes are shown in *Figure 6.3*. Initially an individual cell was captured and held with 0.5 mmHg sucking

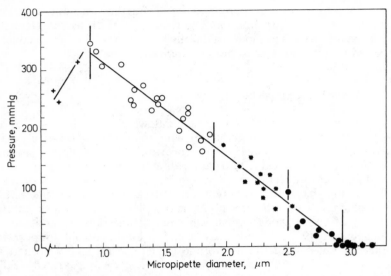

Figure 6.2 The pressure required to draw human erythrocytes rapidly into glass cylindrical micropipettes plotted against the inside diameter of the pipette. (●) Reversible, (∗) crenated, (○) hemolyzed, (+) stretched. 'Reversible' as used in the work of Jay (1973) meant that a cell ejected from the pipette was not crenated (with spicules). The term 'stretched' implied an increase in area which was not measured. In view of other studies, in particular those of Hochmuth et al. (1973), on the long thread tethers (discussed in the text), it is probable that there was only extensive strain. (Reproduced from Jay, 1973, courtesy of the Rockefeller University Press. The author was pleased to have had this material prior to its being published)

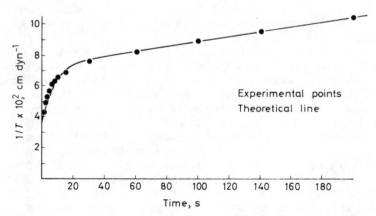

Figure 6.3 The inverse of membrane tension plotted against the time required to draw hypotonically swollen human red cells into micropipettes. The time was measured from the application of a step increase in sucking pressure after the cells had been trapped individually at the entrance of the pipette by 0.5 mmHg pressure. The numbers on the ordinate axis have been multiplied by 100, as indicated. These results led to the viscoelastic model of the red cell membrane, shown in Figure 6.4. (Reproduced from Rand, 1964, courtesy of the author and the Rockefeller University Press)

pressure; then (at time zero) a step increase in sucking pressure was applied and the time required to suck the cell completely into the pipette was recorded. The cell hemolyzed in the process. The data and model (*Figure 6.4*) suggest that the membrane will always yield and become permeable to solutes and hemoglobin even if the applied tension (negative pipette pressure) is very low. Rand (1964) found that, at a tension of 10 dyn cm^{-1}, the cells took an average of 200 s to enter the pipette whereas at 15 dyn cm^{-1} the average cell took only 6 s. The implication that the membranes will always yield,

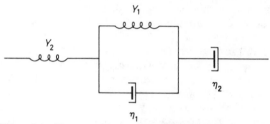

Figure 6.4 *The viscoelastic model proposed by Rand, which has the same stress–strain relation as that exhibited by human erythrocytes drawn into pipettes. Y_1 and Y_2 are elastic stiffness parameters with units of* dyn cm^{-2} *and η_1 and η_2 are viscous parameters with units of poise. The elements in this model cannot be directly equated with the physical elements in the red cell membrane, as pointed out by Rand. (Reproduced from Rand, 1964, courtesy of the author and the Rockefeller University Press)*

implied by the use of η_2 in series (*Figure 6.4*), is probably not true but at the present time the author is unaware of any other time studies, despite the tremendous recent enthusiasm for sucking cells into pipettes. Rand determined an elastic modulus of 7.3×10^6 to 3×10^8 dyn cm^{-2}. The larger value is more acceptable, for in Rand's analysis his critical strain was estimated to be a 16 percent area increase and more recent data suggest the cell is intolerant of more than 4 percent (Canham and Parkinson, 1970). In truth we do not know the critical strain and are forced to accept that the elastic modulus is in excess of 10^8 dyn cm^{-2}. In addition, Rand made an estimate of the viscous component of the membrane by fitting his data to the model shown in *Figure 6.4*. The same two critical parameters, viz. the critical strain and the thickness of the membrane, were needed for obtaining both the viscous and the elastic elements. And, again, the higher values of $\eta_2 \simeq 37 \times 10^{10}$ and $\eta \simeq 9 \times 10^{10}$ P (critical strain 16 percent area increase, thickness 0.01 μm) seem more appropriate. Rand (1964) pointed out that the viscoelastic model was a behavioral simulation model and not a model directly relevant for the structure of the membrane. In his later presentation (Rand, 1968) on the comparison between the behavior of lipid models and the behavior of red cells entering pipettes, he cautioned that the liquid crystalline bilayer of model membranes is a mechanically fragile, viscous liquid and cannot account for the tough, viscoelastic behavior of the intact red cell membrane. He proposed that the necessary strong bonding in the plane of the membrane might come from the surface protein. The possible role of the protein of the membrane has been examined very recently by Evans (1973a, b) and will be discussed later.

It seems evident from the ever-increasing literature discussing red cells and micropipettes that the insight of Rand and Burton (1963, 1964) launched an enormous series of experiments which can supply considerable information about the mechanics of the membrane. However, the inclusion of a series viscous element in Rand's viscoelastic model is probably compatible with survival of only the circulating red cell, which is undergoing continual motion and deformation and is rarely exposed to a constant stress for more than a few seconds. Many other cells, for example endothelial cells, which are only partly restrained in their environment but under constant stress, could not survive with the membrane which is modeled in *Figure 6.4*.

Recent unpublished evidence of A. W. L. Jay and the author suggests an explanation for the apparent yielding of the red cell membrane under stress. We propose that the membrane does not yield but, in fact, maintains its normal surface area by accommodating extreme shape changes (as are required when the cell enters a micropipette) through a loss of volume which occurs in two stages. As sucking pressure is first applied to the red cell, it changes shape in response to a very low negative pressure. However, once the deformation has progressed to the degree at which the membrane is taut, the intracellular pressure rises and forces water out of the cell hydraulically (*Figure 6.1*). The cell will then remain static (only slightly inside the micropipette) until the negative pressure is increased, causing the tension in the membrane to approach the high prelytic stress value (Seeman *et al.*, 1969). Then an additional volume loss, this time of water and solutes, will occur owing to an increased membrane permeability to solutes. In his study of the stretching of pores in model membranes, Burton (1970) demonstrated how a red cell might become permeable to solutes or hemoglobin without a detectable increase in the overall area. Since the volume of the red cell is kept constant owing to the membrane's impermeability to solutes (Stein, 1967, p. 242), it follows that once the membrane does become permeable to solutes and hemoglobin, the volume can adjust to any requirement. Of course, the cell can sustain only a limited tensile stress in the membrane before hemolysis is inevitable.

The two-stage volume loss is supported by our work and that of Rand (1964). In Jay and Canham's unpublished work, the red blood cell is described as entering a pipette smaller than the minimum cylindrical diameter* by becoming smaller as the tension of the membrane increased. A sucking pressure of less than 0.5 mmHg is sufficient to cause a biconcave cell (diameter 8.0 μm, volume 96 μm^3, area 136 μm^2) to begin to enter a 2.1 μm pipette by adopting a florence-flask shape; the initial contact of the pipette is made at the dimple of the cell (schematically shown in *Figure 6.5*). Two circumferential bands of membrane are shown, each 0.25 μm wide, the one an equatorial band on the right, and the second, approximately 0.5μm to the left, destined to enter the pipette as shown at the bottom of the figure. This shape transformation illustrates the amount of deformation which is required and is accommodated by the red cell without a detectable change in

---

* The minimum cylindrical diameter, D, is a parameter of a red blood cell and is the diameter of the thinnest long tube through which an individual cell can pass without alteration of its area or volume. It is calculated (Canham and Burton, 1968) by solving the following equation in D.

$$\text{Vol.} = \text{Area} \times 0.25D - \frac{\pi D^3}{12}$$

the total area or volume. To complete the entry of the cell into the pipette involves less total deformation than the initial stage, and yet a pressure in excess of 50 mmHg is required. This is due to the necessity of reducing the volume of the cell hydraulically and possibly of rendering the membrane permeable to solutes through heightened membrane tension.

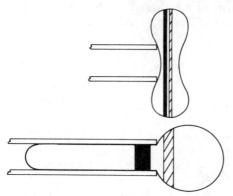

Figure 6.5 Schematic drawing of a biconcave cell drawn into a 2 μm pipette with a small negative pressure (~ 1 mmHg). The property of possible large surface strains within the framework of constant area permits the calculation of the destiny of each membrane element. This is shown in two bands, each 0.25 μm wide. The right band (hatched) is an equatorial band which is deformed into a thicker band outside the pipette entrance, whereas a closely adjacent band has an extension ratio of approximately 0.25, entering the pipette completely

Rand's experiments (1964) were slightly different in that the cells were pre-swollen osmotically, and they hemolyzed before completely entering the pipettes. He reported that a low sucking pressure of 0.5 mmHg would hold the red cell on the end of the pipette with a small tongue already in the pipette. Then upon applying the step increase in sucking pressure, he observed that the cell resisted momentarily (a time of a few seconds up to three minutes, which is probably the period of water loss) and then hemolyzed and disappeared into the pipette. It might have been better to have determined parameters of viscosity from the period during which the swollen cell was caught and held with a pressure of 0.5 mmHg.

6.2.6 Second elastic constant from tethering experiments

The red cell will attach to glass or plastic surfaces. If the suspending medium flows over an attached cell the cell will deform with the flow and sometimes keep its point of attachment. It takes on the appearance of a flattened tear drop (dacryocyte) and returns to its circular symmetry when the flow stops. This was demonstrated by Hochmuth and Mohandas (1972), who analyzed the deformation as a function of the fluid shear stress. They said 'with certainty the low strain modulus of elasticity of human red cells in uniaxial loading is on the order of 10^4 dyn cm^{-2}'. This deformation modulus of elasticity is a measure of the erythrocyte's tendency to return to the biconcave shape and was based on the incompressibility of the membrane. Incompressibility is defined as occurring when the product of the extension ratios, λ_x and λ_y, in the plane of the membrane, is unity, or when a rectangular element of membrane can be formed into any shape provided the area is constant (*Figure 6.6*). Although the analysis of Hochmuth and

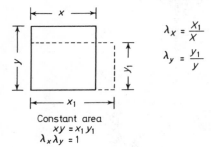

Figure 6.6 *The deformation of a rectangular element of membrane under the condition of constant area. The lipid core of the membrane reorganizes as a viscous fluid, and is thought not to retain any deformation energy. The extension ratios λ_x and λ_y are expressions of the degree of deformation*

Mohandas (1972) and Hochmuth, Mohandas and Blackshear (1973) utilized methods for materials which can be strain-hardened, they emphasized that the discocyte–dacryocyte transformation was repeatable at least six times. This is important in that, despite extension ratios in excess of 200 percent, the membrane does not 'remember' previous stress.

Evans (1973b), using a more formal approach to the shear of the red cell surface, obtained the same value for the shear modulus of 10^4 dyn cm^{-2} from the data of Hochmuth, Mohandas and Blackshear (1973).

6.2.7 The membrane: A two-dimensional elastomer

The possibility of two elastic moduli, the one in excess of 10^8 dyn cm^{-2} (Rand, 1964), and the second 10^4 dyn cm^{-2} (Hochmuth and Mohandas, 1972; Evans, 1973b) is not in conflict with the existing knowledge of the membrane. Evans (1973a) made an important contribution to understanding the behavior of the red cell membrane when he assembled and formulated ideas which, although informally a part of the thinking of several membrane biologists, brought together the concept of surface indistensibility and the concept of surface deformability. He endorsed the concept of the membrane as a lipid sandwich covered with a protein layer and assumed, on the basis of increasing evidence, that the membrane is incompressible, that its thickness is constant, and that the surface area was therefore constant. Like Hochmuth and Mohandas (1972), he ascribed the tendency to return to the equilibrium biconcave shape to a weak elastic property of the membrane, calling the membrane an incompressible two-dimensional elastomer. Evans suggested that the elastic deformability of the membrane was due to a 'two-dimensional matrix of randomly kinked cross-linked chains which are elongated and oriented upon stretching', a property ascribed to its protein component, and the two-dimensional inextensibility was ascribed to the liquid bilayer of phospholipid (which he preferred to

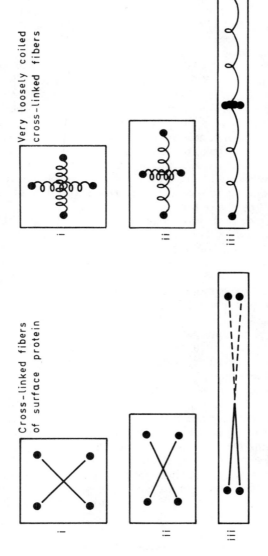

Figure 6.7 A schematic interpretation of cross-linked surface protein fibers. The left-hand set of figures shows the limit of extension without breaking the links. The right-hand scheme is much more deformable because the links themselves are elastically extensible. (i) Square element of membrane on the rim of a biconcave cell deformed into (ii) a rectangular element on the same cell osmotically sphered, or into (iii) a rectangular element on the same cell deformed into a sausage-shaped cell. Sphering a cell osmotically causes the most surface shear at the rim (Canham, 1972), but sphering requires much less surface shear than the isovolumetric transformation into sausage or tether formations (Hochmuth, Mohandas and Blackshear, 1973)

call a liquid crystal bilayer of phospholipid and cholesterol, reflecting his concern for ordered structure). Subsequently (Evans, 1973b), using continuum mechanics, he interpreted and found compatible the original pipette experiments of Rand and Burton (1963, 1964) which yielded an elastic modulus for two-dimensional extension in excess of 10^8 dyn cm^{-2}, the elegant and precise pipette experiments of Leblond (1972), and the fluid-shear experiments of Hochmuth and Mohandas (1972) and Hochmuth, Mohandas and Blackshear (1973). *Figure 6.7* represents an interpretation of the deformation process for an equatorially positioned element of membrane on the discocyte surface (a) when the cell is osmotically sphered, and (b) so that the equatorial extension ratio is approximately 0.8 (8.0:6.6 μm), or when the cell is drawn gradually from the dimple into a pipette of diameter 3.0 μm in which the equatorial extension ratio is 0.38 (8.0:3.0 μm). *Figure 6.7* (left-hand side) illustrates the irreversible alteration imposed on the cell if the cross-linked matrix is made up of inextensible links. Osmotic sphering can be accommodated but not the passage into a slim pipette. The surface proposed in the left-hand side of *Figure 6.7* would yield if deformed into a sausage shape. Tethering experiments imposed even more distortion (Hochmuth and Mohandas, 1972) and probably would overextend the loose bonding proposed in *Figure 6.7* (right-hand side). The graphic macromodel in *Figure 6.7* (right-hand side) is exampled in certain types of rubber in which reversible elastic strains of several hundred percent are possible (Treloar, 1967).

Evans made the assumption that the biconcave profile represents the unstressed, or least stressed, state of the surface of the red cell membrane, that is to say the red cell is preformed in some way. He has not yet discussed the consequences of the assumption in his published work but it is my interpretation from an exchange of ideas shared with him early in 1973 that the immature red cell might be thought of as having a flaccid, irregular shape. Shaping forces, perhaps associated with the uniform charge distribution in the membrane (Lew, 1970, 1972), might become important during the maturing process of the adult erythrocyte. The surface matrix of randomly coiled, or kinked, fibers might gel, or bond and provide the precast backbone of the adult red cell membrane. In conjunction with the concept of a preformed adult red cell, one might think of the surface matrix of randomly coiled fibers as if the matrix were being cold-worked by the shearing forces imposed by the environment in the circulation and that the 'average' shape of the cell would become the shape of minimum stress.

The elastomer concept of the red cell membrane is particularly attractive at the present time because it has brought together in a unified theory the property of two-dimensional non-distensibility with one-directional deformability.

6.2.8 An alternative to the two-dimensional elastomer concept

The membrane concept proposed by Evans (1973a) is not unique in explaining the reported experimental results on the red cell membrane. An investigation into the normal equilibrium shapes of the red cell and the shapes obtained during osmotic swelling was made by Canham (1970), who proposed that the red cell membrane might resist bending deformations, i.e. curvature, elastically. In developing the concept of stiffness associated with curvature, Canham made the assumptions that the membrane was inextensible, that it had a constant thickness, that it was labile in the plane of the membrane so that it resisted uniaxial stress as a viscous surface (i.e. not storing any elastic energy), and that it was mechanically identical over the entire surface of the membrane. The assumptions conform with the concept of the membrane proposed by Evans (1973a) except for the important difference that Canham assumed that there was no elastic resistance to uniaxial strain and that the membrane did resist curvature. He showed that, for a wide variety of normal and swollen biconcave cells, the shape was predicted by minimizing the function

$$F = 0.5D\int_A (1/R_1)^2 + (1/R_2)^2 \, dA \qquad (6.2)$$

where R_1, R_2 are the principal radii of curvature. D is the bending rigidity of the membrane: $D = h^3 E/[12(1-v^2)]$, where h is the thickness, v is Poisson's ratio, A is the surface area of the membrane, and E is Young's modulus. The concept that the membrane could resist curvature mechanically has been severely criticized (Fung, 1966; Fung and Tong, 1968) because the membrane is very thin (0.01 µm) relative to the radius of curvature ($\geqslant 1.0$ µm). Other workers (Lew, 1970, 1972; Lopez, Duck and Hunt, 1968; Adams, 1973a, b) have discussed the equilibrium shapes of the red blood cell in terms of the membrane potential (associated with the relatively fixed charges in the membrane) and in this way support the concept of a resistance to curvature but caused by electrical forces. The profiles of cells with a wide range of cellular areas and volumes, including osmotically swollen cells, were predicted by minimizing the above function F. For the average cell (area 137 µm^2, vol. 107 µm^3), F_R ranged from 2.4 to 3.3, where $F_R = F/F_s$ and F_s is the equivalent F for a sphere and equal to $4\pi D$. Despite the success of the hypothesis, the implication is that the inner and outer layers are not free to move relative to each other in the plane of the membrane without elastic resistance. Although there are reported glycoprotein molecular 'pegs' across the membrane (Marchesi et al., 1973), it is unlikely that there would be any significant resistance to side slip between the inner and outer layers. Altering the hypothesis of resistance to curvature to allow for unrestrained slip between the inner and outer lipid layers possibly leads to a clearer understanding of the preference for the biconcave shape over three allowable geometries—the flat-sided, the sausage shape, and the tear-drop shape—and makes more straightforward the explanation of the echinocyte II (a flat red cell with smoothly capped spicules). The alteration means that a planar wavy surface, in which the total outer surface is unstretched relative to the total inner surface, stores no elastic deformation. The potential difficulty with the bending energy hypothesis as proposed earlier (Canham, 1970) is illustrated by considering the sausage-shaped profile, for which $F_R \simeq 2.7$ (sausage diameter is 3.3 µm, length of sausage 13 µm) and that the energy available to return the cell to the biconcave state is ΔF, the $F_{\text{sausage}} - F_{\text{disc}}$, which is 2.7−2.4. It is acknowledged that there is the possibility of finding a combination of pipette diameter, cellular area and volume, that would predict, on the curvature hypothesis of Canham (1970), that the sausage shape would be more stable than the biconcave shape. For this reason a further interpretation in mechanical terms of the resistance to curvature was needed.

To change the shape of the red cell without increasing the volume requires stretching the outer layer and compressing the inner layer such that no extension of area in the midplane is necessary. If the resistance to increasing the membrane area is a property of the core layers in the midplane of the membrane (i.e. of those layers of the membrane which have a thickness of rather less than a nanometer), then there would be no resistance to curving the membrane. However, if the layers of the membrane resisting stretch exist at a distance from the central plane, then their unstretchability would contribute to a resistance to curvature. Using a sphere to illustrate this, the inner layers in the membrane would be compressed and the outer layers extended. The difference in surface area, ΔA, between the inner and outer layer of a membrane around a sphere of radius r is $\Delta A = 8\pi r \Delta r$, where Δr is the separation between the inner and outer layers. For a sausage-shaped cell, $\Delta A_{sa} = (8\pi r + 2\pi l)\Delta r$, where r and l are the sausage radius and length of the cylindrical section respectively. For a biconcave cell, $\Delta A_{bi} = 62\Delta r$, from the empirical expression $A = 62.0r - 112.0$ μm^2, where r is the major cellular radius (Canham and Burton, 1968). For a 13 μm sausage with a 3.0 μm diameter, ΔA is approximately $110\Delta r$, which is equal to 1.1 μm^2 for a spacing of 0.01 μm, representing 0.8 percent for $\Delta A/A$. Consequently there is a force returning the sausage shape to the biconcave ($\Delta A_{sa} - \Delta A_{bi}$) provided the membrane is so structured that the unstretchable layers (having an elastic modulus in excess of 10^8 dyn cm^{-2}) are separated by a significant part of the thickness of the membrane. The pancake profile was used before to illustrate the increased stability of biconcavity. Its outer surface is stretched relative to its inner surface by an amount $(2\pi^2 h + 4\pi w^2) \times \Delta w$, where w is the radius of the rim and h is the radius of the flat portion of the pancake (Canham and Burton, 1968). Consequently the three profiles, viz. the biconcave, the flat-sided and the sausage, have the outer membrane layer stretched relative to the inner layer by amounts $62\Delta r$, $96\Delta r$ and $110\Delta r$ respectively, for the average cell with an area and volume of 135 μm^2 and 100 μm^3.

The experiments of Leblond (1972) provided evidence in apparent conflict with the results of Rand and Burton (1964), who reported that there was no significant difference mechanically between the rim region and the central region of the cell. It is probable that the technique of Leblond was sensitive to the deformation of the cell and that of Rand and Burton was not. Leblond measured the length of cylindrical tongue of red cell membrane that could be pulled into a 1.5 μm pipette using a negative pressure of 3 mmH$_2$O. The results (*Figure 6.8*) showed that approximately twice as much tongue was drawn from the rim as from the dimple region. There are two interpretations, in addition to that of membrane folding, proposed by Evans and Leblond (Leblond, 1972, see open discussion), which assume that there is no mechanical difference in the properties of the membrane between the rim region and the dimple region, and that the tongue of membrane is not folded. Neither has been tested mathematically. The first explanation, based on the two-dimensional elastomer concept of Evans, is that the tongue taken at the rim would involve less surface strain (or surface shear) than a tongue taken from the relatively uncurved dimple region. Similarly it is anticipated that the modified hypothesis of the resistance to curvature, as stated above, could explain the difference in that the total differential stretching, ΔA, of the outer membrane layer relative to the inner layer would be anticipated to be larger for the total cell when the tongue is formed from the dimple region rather than the rim region.

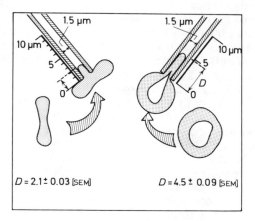

Figure 6.8 Schematic presentation of an experiment by Leblond (1972) in which the length of tongue, D, pulled into a 1.5 μm pipette, with constant pressure of 3 mmH$_2$O, is compared when taken from the region of the dimple of the cell (left) and when taken from the region of the rim (right). (SEM is 'standard error of the mean'.) Care was taken to ensure that the part of the cell outside the pipette was not spherical. (Reproduced from Leblond (1972) with kind permission of the author and the editors of Nouvelle Revue française d'Hématologie)

That the shaping forces are associated with the mechanical or electrical properties of the membrane has not been universally accepted. Shrivastav and Burton (1969) presented evidence that the interior of the cell in the dimple region is birefringent and suggested that this was evidence for the existence of dynamic long-chain tactoids bridging the 1.0 μm gap and causing a force of attraction between the two opposite membranes across the interior of the cell. A difficulty with this hypothesis exists in that the evidence for form birefringence in the concave central region of the cell is not sufficient to determine whether the birefringent structure is parallel or perpendicular to the membrane.

6.2.9 Macro-models of the red cell membrane

Physical models have been constructed with various purposes in mind; however, most often a severe compromise is made in order to simulate particular cellular behavior and the model results in merely a physically empirical structure capable of mimicking a limited set of properties. Occasionally the construction of physical models to simulate the mammalian red cell is useful because the various shape transformations and constant surface area of red cells impose a set of challenging constraints which force us to acknowledge the uniqueness of membranes as a material. There is a variety of partly successful models: the tennis ball model of Pinder (1972), the rubber models of Rand (1967), and the watch-band model of Sirs (1970). However, the model of Bull and Brailsford (1973), constructed of thick deformable foam rubber blocks hinged together in a flexibly pinned network, demonstrates many properties of the red cell, including sphering, constant area and thickness, high resistance to surface stretch and a very low resistance to unidirectional strain. One major drawback is that the model tolerates maximal local extension ratios of only 15 percent, and therefore cannot account for deformation into a sausage shape, the transformation necessary for simulated entry into a micropipette. The response by Bull (1972) to the query regarding the model's similarity to a geodesic dome states clearly that the model membrane is the opposite of a geodesic structure. The model has no translational rigidity and was constructed so as to simulate a two-dimensional inextensible liquid. This particular model is not only the delight of modeling enthusiasts but is good for teaching purposes. The model, however, is not adequate in other ways, mimicking neither tear-drop formation, nor the crenated state (echinocyte II and III—Bessis, 1972). It seems unlikely that a macromodel will ever be constructed which is mechanically equivalent to the cell membrane.

6.2.10 The spiculated red cell

The red cell will adopt, often reversibly, a wide variety of configurations with projections or spicules on the surface. These have been classified according to the morphology of the spicules, whether knobby (echinocyte) or spiked (acanthocyte) and according to the basic shape of the cell (Bessis, 1972). The shapes can be produced by a wide variety of chemical agents (Ponder, 1948; Deuticke, 1968), by ATP and calcium depletion in the membrane (Weed and Chailley, 1972), and by pH alterations induced by extracellular metal electrodes (Rand, Burton and Canham, 1965). However, little has been written to explain these shapes. The blunt-ended spikes in the spheroechinocyte I would represent focal points of high surface strain and curvature, in addition to being zones of possible charge concentration. The morphology of the actual spike, i.e. the extent to which the plasma membrane enters to the tip, the curvatures at the tip, and the surface area, are not established. A single spike can be induced and is reversibly reabsorbed into the cell surface, as shown by the work of Hochmuth and Mohandas (1972). The subsequent work of Hochmuth, Mohandas and Blackshear (1973) suggested the formation of long filaments of intact membrane (up to 70 μm in length) between two hemoglobin-filled minicells. Despite the enormous tether length, the surface area and volume of the original cell were probably preserved. The approximate surface area of the tether (using the reported value of 0.1 μm for the tether diameter from scanning electron microscopy) is πdL (where d and L are tether diameter and length) or equal to 20 μm², leaving most of the surface area (116 μm²) to enclose the volume in two minicells, less than 0.6 μm³ of hemoglobin solution being contained in the tether. Assuming two equal spherical cellular elements at the end of the tether filament, each of area 58 μm², a maximal cellular volume of 2×42 μm³ can be calculated. The original volume of the average cell of area 136 μm² is approximately 95 μm³. Probably the example given by Hochmuth, Mohandas and Blackshear (1973) is from a larger cell which has an excess of area over that required to enclose its volume in two spheres. The process of filament formation was reported to be reversible and is compatible with the preservation of area and volume; however, there are enormous strains, even greater than 1000 percent, challenging the elastomer concept of Evans (1973a). No longer can the term 'reversible' be used casually, but must be used critically and quantitatively. Evans and LaCelle (1974)* have acknowledged the problem of the reversibility of shape changes associated with large strains in the membrane. The tether experiments demonstrated the feasibility of sharp spikes, but large forces are required, either as elastic shearing forces on the cross-linked surface protein matrix or by the extension of the outer membrane layer relative to the inner layer, as discussed previously. The spheroechinocyte I is, indeed, a bizarre but stable shape.

The echinocyte I and II, which is characterized by round capped bumps of red cell surface, can be induced by osmotic shrinking in Ringer's solution of 400–600 milliosmolar, and by elevated pH (greater than pH 9). The reversibility was demonstrated using metal electrodes attached to a low-voltage battery (Rand, Burton and Canham, 1965). Cells reverted from stomatocyte (cup shape) to echinocyte and back to stomatocyte by the reversal of the electrode polarity. The impression was that the crenations, or bumps, appeared repeatedly in the same location. The transformation to echinocyte is thought to be due entirely to the membrane because the effort required to aspirate an echinocyte into a 3.0 μm pipette is the same as that for a biconcave cell (Leblond, 1972). The surface area and volume of the echinocyte have not been calculated. However, the echinocytic shape appears to be a low-stress shape, whether one is using the previously discussed curvature hypothesis or the strain energy hypothesis of Evans (1973a).

6.3   MECHANICAL PROPERTIES OF OTHER MEMBRANES

Now that a cohesive body of work on the mammalian erythrocyte is emerging, some of the techniques can be applied to other cells. Obviously, conclusions and comparisons about membrane behavior will need to be made cautiously because of the nature of the interior of cells other than red blood cells, and the ability of other cells to make membrane. An interesting example of rapid membrane production is found in a fungal cell which grows in an advancing 10 μm tube. The growing tip in the hyphae of the fungus *Basidiobolus ranarum* has been observed to produce plasma membrane at a rate of 200 μm² min⁻¹ in addition to the extracellular coat of cellulose (Robinow, 1963). Admittedly the example pertains to a cell well endowed with the machinery for membrane production; however, when a cell can produce or resorb membrane its behavior in shear deformations and flow in tubes could be partly due to the potentially transient existence of plasma membrane.

* The author is most appreciative of being sent details of this work by Dr Evans prior to its publication.

An example of leukocytic rigidity, relative to the nonrigidity of the red cell, was illustrated in P.-I. Brånemark's 16 mm film shown in Paris in 1970. A capillary of approximately 10 μm diameter branching off a major vessel was seen to carry red cells at a high velocity until a leukocyte, probably a neutrophilic granulocyte, lodged in the opening to the branch. For a period of approximately 3 s the branch flow stopped and the leukocyte elongated into a long sausage-shaped cell approximately 34 μm in length. Although the leukocytes in the film gave the impression of being nearly spherical, nonsphericity was essential for traversing the capillary sidebranch. It is anticipated that an enormous amount of work will be needed to clarify these questions.

Lichtman (1973) reported that a pressure of 50 mmH_2O was sufficient for 7–9 μm leukocytes to pass through polycarbonate filters with a pore diameter of 8.0 μm. He also reported that lymphocytes will traverse pipettes with a diameter 1.5 μm smaller than the cell diameter. Possibly there is a small decrease in volume (A. W. L. Jay and P. B. Canham, unpublished work), or an increase in area, or possibly the cell was initially nonspherical. When the cellular part outside the pipette is only slightly larger than the inner diameter of the pipette, the intracellular pressure and tension in the membrane become large. Using this geometry, and Laplace's Law for a sphere ($P = 2T/R$), which relates transmembrane pressure (P) to membrane tension (T) and radius of curvature (R), the experiments by Rand (1964) on the red cell and the experiments on lymphocytes by Lichtman (1973) may be compared. For the reported pipette sucking pressure of 50 mmH_2O, a 7.5 μm pipette and a 9.0 μm lymphocyte, one arrives at an intracellular pressure of 300 mmH_2O and a tension of approximately 10 dyn cm^{-1}; the tension is the same as that tolerated for 3 min by a swollen red cell before hemolysis and entry into a pipette as shown in *Figure 6.3*. This calculation suggests that the plasma membrane of the lymphocyte is similar mechanically to the red cell membrane although more work is needed to establish this.

Lichtman (1973) interpreted the extended duration of deformation for lymphocytes upon release from the pipette as indicating that the lymphocyte possessed a higher degree of plasticity. However, differences between leukocytes and erythrocytes might be accounted for by differences in cellular interior, geometry and membrane permeability, and not necessarily by differences in the mechanical properties of the membrane.

6.4 SUMMARY AND CONCLUSIONS

The accumulation of experimental data on the red cell membrane has underscored certain mechanical properties. The ability of the red cell to adopt a wide variety of shapes reversibly is a function largely of its unique shape, which provides a membrane area in excess of that required to contain a constant volume. It is well established that the area of the membrane is constant and that the membrane possesses a very high degree of one-directional deformability combined with a high resistance to two-dimensional stretch. These properties, which are linked to the structure of the membrane, could have valid implications for other cell membranes, especially if one accepts the membrane flow hypothesis, namely that intracellular membrane systems have a common origin (Franke et al., 1971).

Investigation of these possible mechanical similarities is just beginning. For example, there are the pipette experiments on leukocytes by Lichtman (1973), as discussed previously. Also, Baker, Hodgkin and Shaw (1962) demonstrated that nerve fibers emptied of axoplasm would transmit action potentials if filled with replacement fluid under pressure, which suggests that tension might be necessary to the normal function of nerve cell membranes.

It is unfortunate that a complete discussion of the mechanical properties of cellular membranes cannot be presented at this time. While red cell investigation is growing apace, there is still a lack of related studies dealing with other cellular membranes.

Acknowledgement

The author is pleased to acknowledge the support of the Ontario Heart Foundation.

REFERENCES

ADAMS, K. H. (1973a). *Biophys. J.*, **13**:209.
ADAMS, K. H. (1973b). *Biophys. J.*, **13**:1049.
BAKER, P. F., HODGKIN, A. L. and SHAW, T. I. (1962). *J. Physiol.*, **164**:330.
BESSIS, M. (1972). *Nouv. Revue fr. Hémat.*, **12**:721.
BLAUROCK, A. E. (1973). *Biophys. J.*, **13**:281.
BRAASCH, D. (1969). *Pflügers Arch. ges. Physiol.*, 313:316.
BRÅNEMARK, P.-I. (1971). *Intravascular Anatomy of Blood Cells in Man*. Basle; S. Karger.
BRÅNEMARK, P.-I. and LINDSTRÖM, J. (1963). *Biorheology*, **1**:139.
BRECHER, G. and BESSIS, M. (1972). *Blood*. **40**:333.
BULL, B. (1972). *Nouv. Revue fr. Hémat.*, **12**:835.
BULL. B. and BRAILSFORD. J. D. (1973). *Blood*. **41**:833.
BURTON, A. C. (1970). *Permeability and Function of Biological Membranes*. Ed. L. BOLIS, A. KATCHALSKY, R. K. KEYNES, W. R. LOEWENSTEIN and B. A. PETHICA. New York; North-Holland.
CANHAM, P. B. (1969a). *Circulation Res.*, **25**:39.
CANHAM, P. B. (1969b). *J. cell. Physiol.*, **74**:203.
CANHAM, P. B. (1970). *J. theor. Biol.*. **26**:61.
CANHAM, P. B. (1972). *Nouv. Revue fr. Hémat.*, **12**:825.
CANHAM, P. B. and BURTON, A. C. (1968). *Circulation Res.*. **22**:405.
CANHAM, P. B. and PARKINSON, D. R. (1970). *Can. J. Physiol. Pharmac.*, **48**:369.
DANON, D. (1961). *J. cell. comp. Physiol.*, **57**:111.
DEUTICKE, B. (1968). *Biochim. biophys. Acta*, **163**:494.
DODGE, J. T., MITCHELL, C. D. and HANAHAN, D. J. (1963). *Archs Biochem. Biophys.*, **100**:119.
EVANS, E. A. (1973a). *Biophys. J.*, **13**:926.
EVANS, E. A. (1973b). *Biophys. J.*. **13**:941.
EVANS, E. A. and FUNG, Y. C. B. (1972). *Microvascular Res.*, **4**:335.
EVANS, E. A. and LACELLE, P. L. (1975). *Blood*, **45**:29.
EVANS, E. A. and LEBLOND, P. F. (1972). *Nouv. Revue fr. Hémat.*, **12**:851.
EVANS, E. A. and LEBLOND, P. F. (1973). *Biorheology*, **10**:1.
FRANKE, W. W., MORRÉ, D. J., DEUMLING, B., CHEETHAM, R. D., KARTENBECK, J., JARASCH, E.-D. and ZENTGRAF, H.-W. (1971). *Z. Naturf.*. **26b**:1031.
FUNG, Y. C. B. (1966). *Fedn Proc. Fedn Am. Socs exp. Biol.*, **25**:1761.
FUNG, Y. C. B. and TONG, P. (1968). *Biophys. J.*, **8**:175.
HOCHMUTH, R. M. and MOHANDAS, N. (1972). *J. Biomechanics*. **5**:501.

HOCHMUTH, R. M., MOHANDAS, N. and BLACKSHEAR, P. L., JR. (1973). *Biophys. J.*, **13**:747.
JAY, A. W. L. (1973). *Biophys. J.*, **13**:1166.
JAY, A. W. L. (1975). *Biophys. J.*, **15**:205.
JAY, A. W. L. and BURTON, A. C. (1969). *Biophys. J.*, **9**:115.
KATCHALSKY, A., KEDEM, O., KLEBANSKY, C. and DE VRIES, A. (1960). *Flow Properties of Blood and Other Biological Systems*, pp. 155–164. Ed. A. L. COPLEY and G. STAINSBY. Oxford; Pergamon Press.
KORN, E. D. (1969). *A. Rev. Biochem.*, **38**:263.
LACELLE, P. L. (1972). *Biorheology*, **9**:51.
LEBLOND, P. F. (1972). *Nouv. Revue fr. Hémat.*, **12**:815.
LESSIN, L. S. (1972). *Nouv. Rev. fr. Hémat.*, **12**:871.
LEW, H. S. (1970). *J. Biomechanics*, **3**:569.
LEW, H. S. (1972). *J. Biomechanics*, **5**:399.
LICHTMAN, M. A. (1973). *J. clin. Invest.*, **52**:350.
LOPEZ, L., DUCK, I. M. and HUNT, W. A. (1968). *Biophys. J.*, **8**:1228.
LOEWENSTEIN, W. R. (1968). *Emergence of Order in Developing Systems*, p. 151. Ed. M. LOCKE. New York; Academic Press.
MARCHESI, V. T., JACKSON, R. L., SEGREST, J. P. and KAHANE, I. (1973). *Fedn Proc. Fedn Am. Socs exp. Biol.*, **32**:1833.
NYSTROM, R. A. (1973). *Membrane Physiology*, Chapter 2. Englewood Cliffs, N. J.; Prentice-Hall.
PINDER, D. N. (1972). *J. theor. Biol.*, **37**:407.
PONDER, E. (1948). *Hemolysis and Related Phenomena*. New York; Grune & Stratton.
PONDER, E. (1961). *The Cell*, Vol. II, pp. 1–80. Ed. J. BRACHET and A. E. MIRSKY. New York; Academic Press.
RAND, R. P. (1964). *Biophys. J.*, **4**:303.
RAND, R. P. (1967). *Fedn Proc. Fedn Am. Socs exp. Biol.*, **26**:1780.
RAND, R. P. (1968). *J. gen. Physiol.*, **52**:1735.
RAND, R. P. and BURTON, A. C. (1963). *J. cell. comp. Physiol.*, **61**:245.
RAND, R. P. and BURTON, A. C. (1964). *Biophys. J.*, **4**:115.
RAND, R. P., BURTON, A. C. and CANHAM, P. B. (1965). *Nature, Lond.*, **205**:977.
RAND, R. P. and LUZZATI, V. (1968). *Biophys. J.*, **8**:125.
ROBINOW, C. F. (1963). *J. Cell Biol.*, **17**:123.
SEEMAN, P. and ILES, G. H. (1972). *Nouv. Revue fr. Hémat.*, **12**:889.
SEEMAN, P., SAUKS, T., ARGENT, W. and KWANT, W. O. (1969). *Biochim. biophys. Acta*, **183**:476.
SHRIVASTAV, B. B. and BURTON, A. C. (1969). *J. cell. Physiol.*, **74**:101.
SINGER, S. J. and NICOLSON, G. L. (1972). *Science, N.Y.*, **175**:720.
SIRS, J. A. (1970). *J. theor. Biol.*, **27**:107.
SIRS, J. A. and STOLINSKY, C. (1971). *Micron*, **2**:382.
SKALAK, K. (1973). *Biorheology*, **10**:229.
STEIN, W. D. (1967). *The Movement of Molecules Across Cell Membranes*. London; Academic Press.
STOECKENIUS, W. and ENGELMAN, D. M. (1969). *J. Cell Biol.*, **42**:613.
TRELOAR, L. R. G. (1967). *The Physics of Rubber Elasticity*, 2nd edn. Oxford; Clarendon Press.
WEED, R. I. and CHAILLEY, B. (1972). *Nouv. Revue fr. Hémat.*, **12**:775.
WEED, R. I., REED, C. F. and BERG, G. (1963). *J. clin. Invest.*, **42**:581.
WEINSTEIN, R. S. (1969). *Red Cell Membrane, Structure and Function*, pp. 36–82. Ed. G. A. JAMIESON and T. GREENWALT. Philadelphia; J. B. Lippincott.
WEINSTEIN, R. S. and KOO, V. M. (1968). *Proc. Soc. exp. Biol. Med.*, **128**:353.

7

Enzyme distribution in mammalian membranes*

R. H. Hinton and E. Reid
Wolfson Bioanalytical Centre, University of Surrey, Guildford

7.1 MEMBRANES TO BE DISCUSSED AS ENZYME SITES

Much of the following presentation is concerned with liver, insofar as this tissue is particularly amenable to subcellular investigation and is therefore better understood than other animal tissues. This chapter represents an introductory survey, complementing the more comprehensive treatment of particular topics by subsequent contributors. An acquaintance is assumed with cellular knowledge gained by the electron microscope (EM), as admirably outlined by Haggis (1966). Especial prominence is given to the plasma membrane (PM), which will serve to illustrate investigative approaches.

The classical approach is to examine fractions obtained from homogenates by 'differential pelleting', a term which is clearer than 'differential centrifugation' (Reid, 1972a; Reid and Williamson, 1974). In early studies by this approach, commonly with 0.88 M sucrose although the conclusions also hold for 0.25 M sucrose, liver homogenates centrifuged at high speed gave a fraction containing 'microsomes' (an 'operational' term), the nature of which was elucidated by EM techniques that hinged on ultrathin sectioning. As was shown by Palade and Siekevitz (1956), the pelleted material consists mainly of vesicles, which are mostly 'rough-surfaced' due to attached granules, and which seemingly have arisen during homogenization

---

* The following abbreviations are used: ER, endoplasmic reticulum (excluding the Golgi); PM, plasma membrane (the equivalent term 'cell membrane' is not favoured). The usage of the terms 'cytochemical' (here restricted to EM studies), 'microsome' and 'microsomal [or mitochondrial, lysosomal, etc.] fraction' is indicated in the text. Membranes in apposition are best termed 'paired', not 'double'. A 'vesicle' is a 'small bladder-like vessel' or 'sac' (*Oxford Dictionary*). The centrifugal procedure commonly termed 'differential centrifugation' is here termed 'differential pelleting', in contrast with 'banding' or 'zonal centrifugation'; the latter term does not necessarily imply use of a zonal rotor (Reid, 1972a; Reid and Williamson, 1974).

by fragmentation of membrane systems (*Figure 7.1*). If derived from liver, the fraction is rich in glucose-6-phosphatase. Evidently it has arisen mainly from the endoplasmic reticulum (ER), insofar as the latter (including 'smooth-surfaced' regions and the nuclear envelope) stains preferentially when intact-cell preparations are examined cytochemically for glucose-6-phosphatase using the Gomori reaction (*Figure 7.2a*).

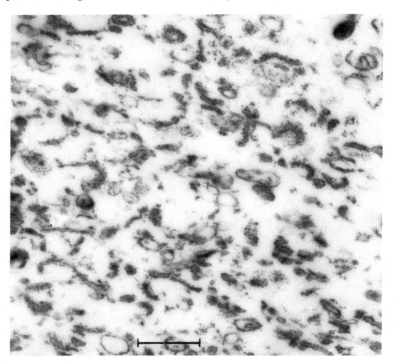

Figure 7.1 An early electron micrograph of a microsomal pellet, from the classical paper of Palade and Siekevitz (1956), courtesy of Dr P. Siekevitz, the editor of Journal of Cell Biology, and Rockefeller University Press. The pellet, from a liver homogenate prepared in 0.88 M sucrose, was fixed in OsO_4 and embedded in n-butyl methacrylate; the final appearance was less satisfactory than could be obtained by present-day techniques. The field is dominated by 'rough' (ribosome-studded) fragments which arise from the ER and which, where cut centrally, are seen to consist of vesicles. Occasional smooth vesicles are seen, e.g. a pair in the centre of the picture. The bar represents 0.5 μm

If a 5'-ribonucleotide such as UMP is used in place of glucose-6-phosphate as the cytochemical substrate, hepatocellular (as distinct from connective tissue) staining is found preferentially in the PM, although lysosomes also stain owing to their acid phosphatase content (*Figure 7.2b*; Goldfischer, Essner and Novikoff, 1964, *inter alia*). Earlier biochemical studies in several laboratories, as summarized by El-Aaser and Reid (1969a), had shown that 5'-nucleotidase activity is present both in the microsomal fraction and in the crude-nuclear fraction. Yet biochemists were slow to draw the obvious conclusion that PM fragments in centrifuged homogenates are distributed between these two fractions (Reid, 1967).

The PM fragments that sediment with nuclei consist mainly of 'sheets', and of elements that correspond to the bile canaliculi seen in electron micrographs such as *Figure 7.2b*; some membrane pairing is still evident where the homogenization has not affected an area of contact between two cells. On the other hand, the microsomal PM fragments mostly lack morphological characteristics to distinguish them from smooth-surfaced vesicles derived from the smooth ER—which imposes an experimental handicap. The microsomal PM fragments, which have been less studied than the 'nuclear' PM fragments, might arise from non-canalicular regions of the plasma membrane and conceivably have different enzymological features (Norris

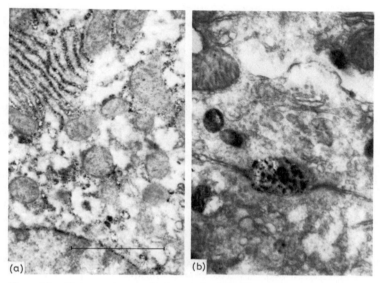

Figure 7.2 Electron micrographs showing lead phosphate deposits in hepatocytes, after incubation of the tissue with (a) glucose-6-phosphate, (b) UMP (courtesy of Dr A. A. El-Aaser). In (a) there are deposits in the nuclear membrane (bottom) and ER the PM (not shown in this field) and mitochondria are unstained. In (b) there are deposits in the PM, particularly in a bile canaliculus, and also (due to acid phosphatase) in lysosomes, but not in ER (top left and top right) or in mitochondria. The bar represents 0.5 μm

et al., 1974). In the case of intestinal mucosa and kidney tubules, the 'brush border' region of the PM yields large fragments and has received particular attention; it shows glucose-6-phosphatase cytochemically, unlike liver PM.

Recently there has been progress in isolating fractions corresponding to the Golgi apparatus—a cluster of smooth membranes that seems to be important for intracellular transport processes and enzyme storage, especially in connection with lysosome formation. Palade and Siekevitz (1956) deserve credit for having pointed out that microsomal fractions are likely to contain Golgi fragments as a minor element. They may be distinguished by drastic treatment with osmium tetroxide (Fleischer, Fleischer and Ozuwa, 1969). Other types of element that may be present in trace amount include lysosomal membranes and mitochondrial outer membranes inadvertently released *in vitro*. Caution is evidently needed in attributing an enzyme to ER merely because it is found in the microsomal fraction (Reid, 1967).

Intact cytoplasmic organelles are surveyed later in this chapter, from the viewpoint of membrane enzymology. Little has hitherto been known of the membrane that bounds peroxisomes (microbodies). The lysosomal membrane has attracted much interest, and is relevant to 'latency'. The lysosome has a single bounding membrane rather than a pair. For the paired-membrane system of mitochondria there is now much information on methodology (see Vol. 2, Chapter 4 of this series). Here it may be noted that the term 'paired' rather than 'double' is to be preferred where there are two apposed membranes.

The present introductory chapter, which is oriented towards the enzymology of the PM although this is scrutinized in more detail by P. Emmelot (Vol. 2, Chapter 1 of this series), pays some attention to subtle aspects of distribution. Thus, possible enzymological dissimilarities between different regions of the membrane are touched on, and also intramembrane location where an enzyme appears to be an intimate constituent of the membrane. Primarily, however, distribution studies entail assigning particular enzymes to particular membrane systems amongst those listed above.

7.2 ISOLATION OF MEMBRANE FRAGMENTS

The general experimental framework is as follows. For preparations other than erythrocyte 'ghosts', the initial sample is a suspension of membrane fragments together with various organelles. Other unwanted material may include unbroken cells, fragments of connective tissue, erythrocytes and other blood elements (even if the tissue has been perfused), and myofibrillar fragments in the case of muscle. Differential pelleting may be usefully applied to a suspension of cells and fragments, but only as a prelude to more subtle fractionation. Even with microsomal material it is usually unrewarding merely to manipulate differential pelleting conditions, with a uniform medium (Reid, 1967). This approach, with a 1.3 M sucrose–15 mM $CaCl_2$ medium, did nevertheless enable Eriksson and Dallner (1971) to subfractionate rough microsomes, although a stepped medium was used in other experiments where Dallner's group used cations such as Cs^+ to alter the centrifugal behaviour of particular types of vesicle (Dallner and Ernster, 1968).

Particular fractions may indeed be banded quite sharply at the interface between two steps (Cline, Dagg and Ryel, 1974); but for isopycnic separation many authors favour a continuous gradient, particularly where there is not the sharp difference in density between different types of fragment that is encountered in a mixture of smooth and rough vesicles. For the sake of resolution, as well as scale, it may be advantageous to centrifuge in a zonal rotor rather than in tubes, possibly with a gradient of subtle design. However, smooth vesicle populations typically have densities which overlap and may even be close to those for organelles such as mitochondria. Separation may then be impossible unless one type of element is selectively modified in density before centrifugation, or is not allowed to come to its isopycnic position in the centrifuge.

Sometimes it may be advantageous to introduce the sample at a high-density position rather than at the top, and then band (usually isopycnically) by flotation rather than by sedimentation. Thus, Touster *et al.* (1970) obtained quite pure PM fragments by flotation. General guidance on centrifugal approaches will be found in other publications, e.g. Reid and Williamson (1974). While a single banding procedure may give tolerably good separation of the desired element, a further stage is often needed.

Approaches for particular types of membranous element are described elsewhere (Birnie, 1972) and in subsequent chapters in the present series. Progress obviously hinges on the use of markers (*see* Section 7.3.2 below), yet in some papers there is a paucity of assay data, such as enzyme values for the original homogenate, that are vital if purity is to be assessed. Investigations on microsomal vesicles (Dallner and Ernster, 1968) have benefited from the recent practice of including data for a PM marker. All too often, authors who have prepared a fraction supposedly representing smooth ER have been seemingly oblivious to the likelihood that PM and Golgi fragments might also have been present. The allied problem of preparing microsomal PM vesicles free from smooth ER is more easily solved. In general, success in reproducibly separating particular membranous elements from rat liver is critically dependent on scrupulous adherence to particular conditions. These may be poorly documented and may also be inapplicable to other species (Lauter, Solyom and Trams, 1972) and to other tissues (Dallner and Ernster, 1968). The 'philosophy' of the design of conditions has been set down, for particular applications, by Graham (1973) and by Prospero and Hinton (1973).

7.3 METHODOLOGY FOR ASSIGNING ENZYMES TO PARTICULAR MEMBRANES

The biochemical approach, as surveyed below (Section 7.3.2), entails preparing membrane-containing fractions, the origin of which can be pin-pointed so that enzyme assignment is feasible. Even if a fraction seems to be morphologically homogeneous, there may be multiplicity of origins insofar as tissues contain diverse types of cell. The relative contribution of these types to the fraction obtained is hard to establish and they may differ, at least quantitatively, in biochemical features. An example is cited in Section 7.5.1 (hepatic adenyl cyclases). With liver, it is uncertain to what extent the traditional use of quaint types of cloth (or, nowadays, a nylon tea strainer) to filter homogenates before centrifuging results in hold-back of biliary elements. Moreover, even parenchymal cells appear to differ according to their position in the lobule (Morrison *et al.*, 1965; Reid, 1967).

Uncertainty about the morphological purity of subcellular elements present is often avoidable. Where a term such as 'mitochondria' is used, rather than 'mitochondrial fraction', the onus is on the author to show lack of contamination by elements such as vesicles and lysosomes, through morphological and/or biochemical examination.

An alternative approach to enzyme assignment, hopefully complementary to the biochemical approach, has been mentioned above. It depends on microscopic examination of intact-cell preparations incubated with an appropriate substrate, with minimal pre-fixation lest there be disastrous loss of enzyme activity. For this approach, the use of the light microscope (histochemistry) helps mainly in connection with the PM. Use of EM cytochemistry has been the key to progress.

7.3.1 The cytochemical approach

The examples already mentioned, glucose-6-phosphatase and 5'-nucleotidase, are both phosphatases, such that 'capture' of the liberated inorganic phosphate as lead phosphate furnishes the requisite electron-dense deposit for EM examination. Like 5'-nucleotidase, ATPase (usually assayed with Mg^{2+} present and, in some laboratories, K^+ and Na^+ also—see Song et al., 1969) seems to have a PM location in liver, but especially mild pre-fixation conditions also disclose mitochondrial ATPase activity (Sabatini, Miller and Barrnett, 1964).

Ingenious cytochemical procedures have been devised for a few other types of enzyme, as reviewed by Shnitka and Seligman (1971); here the electron-dense material that is finally examined is not lead phosphate but a reaction product (from an 'artificial' substrate) that is itself osmiophilic or is rendered osmiophilic by an appropriate capturing agent. Aminopeptidase and γ-glutamyl transpeptidase have been studied by this approach, but are of little interest in the present connection. For certain oxidoreductases the expected localization on the inner mitochondrial membrane (outer surface) has been confirmed and is on the outer surface (Shnitka and Seligman, 1971).

A notable limitation to cytochemical techniques is that usually they are only qualitative. The intensity of a deposit is an unreliable guide to the amount of enzyme present, even if pre-fixation damage is minimal. Although the cytochemical approach suffers from this constraint, from possible artefacts (see Section 7.4), and from limited applicability (Goldfischer, Essner and Novikoff, 1964; Shnitka and Seligman, 1971), it has been crucial in building up our knowledge of membrane enzyme distribution, as demonstrated above by glucose-6-phosphatase. Whilst the microsomal fraction is rich in this enzyme and seems, on morphological grounds, to have originated from the ER, the firm assignment of the enzyme to the ER had to await cytochemical confirmation. In the case of 5'-nucleotidase it was cytochemical rather than biochemical evidence that led to its establishment as a hepatic PM 'marker'. An alternative marker for hepatic PM (or, at least, for bile canaliculi), which some authors have adopted on cytochemical grounds, is nucleoside triphosphatase, the 'ATPase' mentioned above.

There is some justification for the use of nucleoside diphosphatase assayed with ADP or CDP as substrate, or of alkaline phosphatase with p-nitrophenyl phosphate as substrate, although Golgi elements also show activity (Goldfischer, Essner and Novikoff, 1964); but it cannot be taken for granted that any species or strain will show these activities. The particular usefulness of 5'-nucleotidase as a PM marker has not been seriously undermined by a recent claim (Widnell, 1972) that the ER also has some activity, postulated to be so susceptible to destruction in cytochemical procedures that it escapes detection when whole-cell preparations are examined. However, the reported presence of 5'-nucleotidase in the Golgi (Bergeron et al., 1973a), contrary to earlier work (e.g. Fleischer, Fleischer and Ozuwa, 1969), is a significant complication.

Widnell (1972) is one of the few investigators who have reported the use of subcellular fractions, as distinct from intact-cell preparations, in cytochemical studies. El-Aaser et al. (1973) have reviewed other literature, and were able to demonstrate abundant 5'-nucleotidase staining and sparse glucose-6-phosphatase staining in a fraction derived from a microsomal preparation and supposedly enriched in PM as distinct from ER fragments (*Figure 7.3*). The requisite cytochemical techniques (*see also* Widnell, 1972;

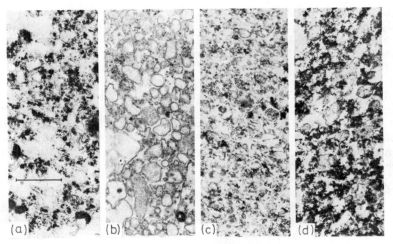

Figure 7.3 Electron micrographs showing lead phosphate deposits in subcellular fractions incubated with substrate, by the technique of El-Aaser et al. (1973) (courtesy of Dr A. A. El-Aaser). Hepatic microsomal material was incubated with (a) glucose-6-phosphate, (b) UMP. A derived fraction believed to be rich in PM was likewise incubated with (c) glucose-6-phosphate, (b) UMP; as expected, staining was sparse with the former substrate. The bar represents 0.5 μm

El-Aaser and Holt, 1973) are, however, not easy, and have yet to make substantial contributions to our knowledge of membrane enzymology. This approach has, however, recently been applied in support of the above-mentioned biochemical finding of 5'-nucleotidase in Golgi fractions (Morré et al., 1974).

The biochemist seeking to assign enzymes to subcellular elements relies on a battery of 'marker enzymes' which cytochemical techniques have, where feasible, helped to establish, together with a set of fractions each, ideally, corresponding to a particular element.

7.3.2 Characterization of subcellular fractions with the aid of 'markers'

The 'markers' used to characterize fractions are mostly enzymes (an important exception being DNA, a nuclear marker) which it is to be hoped meet the following criteria.

1. Stable and readily measurable activity towards the substrate employed, such that even feebly active fractions can be assayed.
2. Absence of endogenous inhibitors.
3. No tendency to undergo migration *in vitro* such that the cytosol fraction or a particulate fraction other than the parent one shows artefactual activity.
4. Unique in attacking the substrate employed, at least under the chosen assay conditions.
5. Unique assignment to a particular type of element in the tissue.

These and other points are considered more fully elsewhere (e.g. de Duve, 1967; Shnitka and Seligman, 1971). Obviously there are many pitfalls: one example (to which S. J. Holt has drawn our attention) is the possible presence in 'β-glycerophosphate' of sufficient α-glycerophosphate to give artefactual 'acid phosphatase' activity, really due to glucose-6-phosphatase. Particular difficulties in respect of criterion 5 (Morré *et al.*, 1974; Touster, 1974) are considered later, but some general comments now follow.

The 'one-enzyme–one-location' hypothesis implied by criterion 5 is reinforced if a unimodal distribution is demonstrated amongst the spectrum of fractions obtained from a homogenate. If a bimodal distribution is encountered, it may turn out that activity resides in a cellular element which is present in more than one fraction, as in the case of 5'-nucleotidase. The study of sharply bimodal distributions can, then, be illuminating.

The position is more difficult where an activity shows a main locus in a fraction rich in one type of cellular element and only a minor locus in another fraction which, as judged by morphological and/or marker studies, is not sufficiently contaminated with this element to account for the activity. One example is the presence of acid phosphatase activity in Golgi as well as in lysosome-containing fractions; here the problem is not serious, and there is a functional explanation, in that the Golgi is concerned in packaging enzymes for new lysosomes. Another example is β-glucuronidase: it should be shunned as a lysosomal marker (de Duve *et al.*, 1955), since a microsomal component has been shown in differential pelleting experiments. Possibly the enzyme should be assigned to a particular type of lysosome (Futai, Tsung and Mizuno, 1972), or enzyme protein destined for lysosomes may truly be present in the ER. There is, however, no simple precursor–product relationship in the case of the β-glucuronidase activities of microsomes and lysosomes; the activities are attributable to different although closely related proteins (Touster, 1974). Sometimes a bimodal distribution, or a skew distribution compared with a marker, might reflect the presence of two or more quite distinct enzyme proteins, differently located in the cell, that attack the chosen substrate.

One must choose the substrate judiciously, as in the case of 'monoamine oxidase', which was used as a marker to scan for fragments of outer mitochondrial membrane in the studies of Hinton, Norris and Reid (1971): kynuramine was a convenient substrate but is not ideal for mitochondrial studies insofar as enzymes in the inner as well as the outer mitochondrial membrane may attack it (Greenawalt and Schnaitman, 1970). Multiple forms of monoamine oxidase exist in various tissues besides liver (Youdim, 1973). Other enzymes that seemingly exist in multiple forms, one or more being located at least partly in lysosomes, include aryl sulphatases and certain glycosidases besides β-glucuronidases (Touster, 1974). Intestinal mucosa contains at least three Na^+,K^+-'ATPases', one having a preference for GTP (Mooney and McCarthy, 1973).

Table 7.1 SOME MARKER ENZYMES FOR HEPATOCELLULAR ELEMENTS
Papers cited in later sections may be consulted for assay conditions, as also given for some enzymes by de Duve et al. (1955) and Reid (1972). A simplified assay for galactosyltransferase is given by Bauer, Lukashek and Reutter (1974).

Ref. to Section	Element	Marker	Remarks
7.5.1	Plasma membrane (PM)	5'-Nucleotidase	Golgi and ER may also have some activity
		Ouabain-sensitive Na^+,K^+-ATPase	Other ATPases may handicap specific assay for this one
		Nucleotide pyrophosphatase/ phosphodiesterase I	Seemingly one enzyme with two activities
7.5.2	Endoplasmic reticulum (ER)*	Glucose-6-phosphatase Cytochrome P_{450} or b_5	
7.5.3	Golgi apparatus	Galactosyltransferase	
7.5.4	Lysosomes	Acid phosphatase	Assay with β-glycerophosphate
7.5.5	Mitochondria	Succinate dehydrogenase	In inner membrane. INT† is a convenient H acceptor
		Monoamine oxidase	Choice of substrate important. Outer membrane
7.5.6	Peroxisomes	Catalase	Measure substrate disappearance (not ideal)
		Urate oxidase	Absent from human liver

* Including nuclear outer membrane (Section 7.5.6).
† 2-(p-Iodophenyl)-3-(p-nitrophenyl)-5-phenyltetrazolium chloride.

Fortunately, there are hepatic markers (*Table 7.1*) which seem to be substantially if not uniquely located in one cellular site, even if it is really a 'family' of enzymes or isoenzymes that is being assayed. Enzymes of unknown location can then be matched up with markers in a spectrum of fractions (not forgetting the original homogenate). A minor discrepancy in matching, as in the comparison of certain hydrolases with lysosomal acid phosphatase (Futai, Tsung and Mizuno, 1972; Prospero et al., 1973, *inter alia*), may be due to intrinsic heterogeneity in the element studied, such as that known to exist in the case of mitochondria as well as lysosomes. A bimodal, polymodal or skew distribution, suggestive of a multiple location in the cell if not matchable with that for a marker, needs scrutiny from the viewpoint of possible enzyme diversity. If two enzymes or isoenzymes are concerned, differing in location and, perhaps, in substrate specificity, the 'one-enzyme–one-location' hypothesis need not be abandoned for the activity concerned.

No enzymatic markers are available to distinguish smooth and rough ER membranes. If one wishes to check biochemically whether rough ER is present in a particular fraction, one must look for evidence of ribosomes, with RNA (18 and 28 S) as a possible criterion in addition to EM appearance.

For Golgi a transferase enzyme (*Table 7.1*) is now favoured, although there has been no cytochemical study such as that which led to the earlier adoption of thiamine pyrophosphatase as marker. The latter activity may serve for some tissues, but in liver it now seems to be attributable to an enzyme protein that is located largely in the ER and has nucleoside diphosphatase activity towards GDP, IDP or UDP (Yamazaki and Hayaishi, 1968; Wattiaux-de Coninck and Wattiaux, 1969a; Shnitka and Seligman, 1971).

7.4 METHODS FOR THE STUDY OF MEMBRANE-ASSOCIATED ENZYMES

Four main questions may be asked about any membrane-associated enzyme activity. (a) With what membrane system(s) is the activity associated? (b) How is the enzyme located in relation to the structure of the membrane? (c) What is the specificity of the enzyme responsible for the observed activity? (d) What is the role of this enzyme activity in the economy of the cell? We have already dealt with (a), and will repeatedly allude to (d) later in this chapter. The methodology associated with (b) will be considered here only in outline, since this aspect of membrane enzymology has recently been reviewed comprehensively by Coleman (1973), whose excellent article is recommended as doing more justice to the subject.

As all biological membranes are 'two-faced', separating two functionally distinct spaces, so there are four distinct possibilities for the location of membrane-bound enzymes. They may be associated specifically with either face of the membrane, they may be present on both faces, or they may penetrate the membrane. No enzyme has been demonstrated to be present on both sides of a biological membrane, and only transport enzymes are known to penetrate through the lipid bilayer. Insofar as the two faces of a biological membrane are functionally distinct, one must assume, by the application of Occam's razor, a unique localization on one face or the other for any particular 'non-penetrating' enzymatic activity, until contrary evidence emerges. Moreover, there is the possibility of 'longitudinal' as well as this 'transverse' specialization, for example in the PM of liver cells between bile canalicular regions and sinusoid regions.

Three principal methods are available for the study of the intramembrane location of particular membranes: selective solubilization, cytochemical study and immunochemical study. It is also possible to label proteins exposed on the surface of membranes (e.g. Hubbard and Cohn, 1975), but there is a danger that the enzyme will be inhibited during the reaction. In selective solubilization, one cannot distinguish between the two faces of the membrane, but one can distinguish whether an enzyme is associated with the periphery or the core of the membrane (*see* Wallach, 1972a). It is generally assumed that enzymes solubilized by isotonic saline solutions, as studied by Benedetti and Emmelot (1968a) with PM preparations, are adsorbed on to the membrane, probably after cell breakage, and cannot be regarded as true membrane constituents. Enzymes solubilized in solutions of high ionic strength or by very brief digestion with proteolytic enzymes are usually taken to be associated with the external part of the membrane. Enzymes which are released only by treatment with an organic solvent or detergent would appear to be more closely associated with the core of the membrane, whilst enzymes which require lipid for their functioning, such as glucose-6-phosphatase or Na^+,K^+-ATPase, are assumed to be tightly bound to the lipid core of the membrane. An extensive list of lipid-requiring enzymes is given by Coleman (1973).

The remaining two methods for precise localization of enzymatic activity depend on the EM and, like most EM techniques, are difficult to quantitate and are subject to many artefacts. Cytochemical techniques (Section 7.3.1 above) can indeed be valuable, as in examples mentioned below: 5'-nucleotidase is on the outside of the PM, while glucose-6-phosphatase in the ER can be seen to face onto the cisterna (Goldfischer, Essner and Novikoff, 1964). However, there are two problems. First, a transport enzyme may release its product on one side of the membrane although itself located in the central part of the membrane. Secondly, the deposit seen on the EM may not be exactly at the site of activity of the enzyme. Here there are two possible sources of artefact. With dehydrogenase enzymes, it may not be the primary activity that produces the final deposit. Thus a number of dehydrogenases differing in intramitochondrial, or even intracellular, location may all be able to pass electrons to the same diaphorase and will all, therefore, be assigned the same location if cytochemical methods are used. Moreover, the deposition of the final product may not take place at the actual site of enzyme activity. Most membranes are about 12 nm thick, which is an insignificant distance for diffusion to take place. If a charged group which gave good nucleation lay close to an uncharged enzyme site, one might expect deposition to take place at the nucleation site rather than at the site of enzyme activity.

With immunochemical methods, the objections are rather different. Firstly, immunochemical localization depends crucially on the purity of the antigen used to obtain the antiserum, and there are problems when, as mentioned in Section 7.5.1.4 for 5'-nucleotidase, the purification methods of different investigators give different products. Secondly, the immune reaction will occur most readily at certain determinant sites in a protein. Thus the antibodies raised to a protein which is partly within the membrane may, in fact, be specific to groups exposed only on one side. Thus one should be careful in interpreting both cytochemical and immunochemical results and should generally use both techniques together with selective dissociation, before drawing definite conclusions about the intramembrane localization of an enzyme protein.

However, when used carefully, immunochemical methods may give very valuable results, especially in the determination of the distribution of enzymes within the PM. Two types of test are possible. First, by examining the effect of antibodies to purified membrane enzymes on the enzyme in the intact membrane one may determine whether an enzyme is exposed on the surface of the cell or buried in the membrane structure (Gurd and Evans, 1974). Secondly, one may examine the distribution of enzymes over the surface of intact cells by reacting the cells with specific antibodies coupled to a 'label' such as tobacco mosaic virus and examining the distribution using a scanning electron microscope (Nemanic et al., 1975).

Immunochemical methods can, then, be valuable in studying membrane enzymes, even without use of the EM (Blomberg and Perlman, 1971). However, the pure enzyme proteins needed as antigens may be very difficult to obtain. Those enzymes associated with the periphery of the membrane may be separated from the membrane and purified without any great change in their properties. However, as discussed by Coleman (1973), many enzymes are dependent, to a greater or lesser extent, on the lipid of the membrane for their activity, such that it may be very misleading to attempt to study their properties away from their natural lipid-rich environment. In addition, a membrane with its two-dimensional organization may well be more favourable to multi-enzyme systems than the 'soluble' phase of the cell, where diffusion in three dimensions is possible. Thus one must become chary of considering results obtained from 'purified' membrane enzyme preparations unless these are related directly to the properties of the enzyme in the intact membrane. Indeed, it will only be possible to understand membrane enzymes completely when one can study artificially assembled membranes, complete with enzymes, in an experimental situation where one can be confident about the location of the enzyme within the lipid bilayer and about its relationship to other enzymes present in the system.

7.5 ENZYMES OF PARTICULAR MEMBRANE SYSTEMS

Having outlined the characteristics of membrane-bound enzymes and the methods for their study, we now discuss the enzymes found in particular membrane systems. By way of example we first consider PM, one advantage of which is that its 'two-sidedness' helps in appraising the role of its constituent enzymes. One can be fairly confident about the purity of PM preparations and can, in some instances, correlate observations on an enzyme system *in vitro* with its function in the living cell. Later we discuss more briefly the enzymes found in the membranes of other subcellular components.

In discussing each membrane system we say little about the methods used for isolating the membrane, although clearly the confidence with which enzymes can be assigned to a given fraction will depend on the purity of the preparation studied. We deal in turn with enzymes involved in the synthesis and renewal of the membrane, with enzymes having other clearly defined functions, and with enzymes whose role is not yet clear. Finally we consider evidence on how the enzymes are bound into the structure of the membrane.

7.5.1 Plasma membrane enzymes

Looking back now, it is amazing to realize that, in 1961, an authoritative review could reasonably state that 'If the reader of this chapter has concluded that the author is not convinced about the structure, or even the necessary existence of the plasma membrane as generally described, he will not be far wrong' (Ponder, 1961). Since that time numerous methods have been developed to isolate the surface membranes of many types of mammalian cells (Hinton, 1972, *inter alia*) and these preparations have been shown to possess a distinctive chemical and enzymatic composition. The purity of these preparations has been discussed extensively elsewhere (Hinton, 1972;

Wallach, 1972b). In summary, most PM preparations are remarkably free of the other formed elements, the most important contaminant being small amounts of ER. However, considerable amounts of enzymatic and other protein can be removed by treatment with isopycnic saline, and it is likely that this protein represents soluble material adsorbed onto the PM (*see* Vol. 2, Chapter 1 of this series, by P. Emmelot).

7.5.1.1   ENZYMES CONCERNED WITH PLASMA MEMBRANE SYNTHESIS AND RENEWAL

It is unlikely that much of the PM is synthesized *in situ*. The proteins, like other membrane proteins of the cell, are presumably made on the bound ribosomes of the ER. A protein phosphokinase which acts specifically on a single endogenous protein is found in diaphragm PM (Pinkett and Perlman, 1974); it may be connected with transport across the membrane. The PM is not capable of synthesis of phosphatides *de novo* (van Golde, Fleischer and Fleischer, 1971), although it can synthesize phosphatidylglycerol from CDP–diglyceride and glycerol-3-phosphate (Victoria *et al.*, 1971). Lysolecithin acyltransferase is present in both rat liver PM (Stein, Widnell and Stein, 1968) and lymphocyte PM (Ferber *et al.*, 1972). Acyl CoA is also incorporated, although only slightly, into neutral lipid (Stahl and Trams, 1968). However, although the products of these enzymes are incorporated into the PM (Stein, Widnell and Stein, 1968), it has been suggested that the real role of the lysolecithin acyltransferase is the uptake of non-esterified fatty acids from the serum (Wright and Green, 1971). Liver PM also contains a distinctive phospholipase (van Golde, Fleischer and Fleischer, 1971; Newkirk and Waite, 1971) together with phosphatidate phosphohydrolase (Coleman, 1968) and glycerylphosphorylcholine phosphodiesterase (Lloyd-Davies, Michell and Coleman, 1972). It can actively bind chylomicrons and hydrolyse their lipids (Higgins and Green, 1967), and the cholesterol, in macrophage PM at least, is rapidly exchanged with serum lipoproteins (Werb and Cohn, 1971). It is therefore a plausible idea that the PM phospholipases are connected more with the uptake of serum lipoprotein than with turnover of the PM (Newkirk and Waite, 1971).

Galactosyltransferase found on the surface of chick neural retina cells (Roth, McGuire and Roseman, 1971) and the sialidase found in liver PM (Schengrund, Jensen and Rosenburg, 1972) are presumably concerned with the turnover of the outer carbohydrate coat of the cell. A role in promoting cell adhesion is suggested by the experiments of Webb and Roth (1974). However, little is known about the function of the cell coat, although it is an area of intense study at the present time and will be discussed in several other chapters in the present series.

7.5.1.2   ENZYMES WITH A FUNCTION RELATED TO INCOMING AND OUTGOING MOLECULES

As the PM is the peripheral barrier of the cell, it must act as a channel for all material which enters or leaves the cell. Much of the material which enters the cell, and a very large proportion of the material which leaves it, is handled in bulk by phagosomes, pinocytotic vesicles or secretion granules. Both phagocytosis (Werb and Cohn, 1971) and discharge of secretion granules (Jamieson and Palade, 1971) are active processes, requiring ATP and hence enzyme action, but nothing is known about the enzymes concerned. Much more is known about the classical 'permeases' and 'active transport' enzymes which either assist or actively pump small molecules into the cell, as exemplified respectively by the hepatic glucose permease and by the sodium pump found in almost all cells. Transport mechanisms will be reviewed by S. L. Bonting and J. de Pont in Vol. 4 of this series. By far the most important of these systems is the sodium pump, which is responsible for maintaining the level of potassium in the cytoplasm of the cell in a generally sodium-rich environment. The sodium pump can be identified with the Na^+,K^+-activated ATPase which should be demonstrable in any PM preparation. In practice, difficulty may be encountered in demonstrating a requirement for Na^+ and K^+ and also sensitivity to ouabain (Song et al., 1969). The number of pumping sites on the surface of a cell, as determined by binding of the inhibitor ouabain, varies from about 100 to 200 in the case of the red blood cell (Ellory and Keynes, 1969) to about 180 in isolated kidney cells; there are about 300 000 pump sites in HeLa cells corresponding to 3 percent of the area of the membrane (Baker and Willis, 1969).

Besides these transport enzymes, one should note those enzymes which are located on the outer surface of the PM such as invertase, lactase and isomaltase, which have been reported on the membranes of the intestinal brush border (Eichholz, 1967), as well as the lipid hydrolases and acyltransferase mentioned above. The hexokinase reported in hepatoma PM preparations (Emmelot and Bos, 1966a) should be placed in this class; but our own preparations of PM from hepatoma are contaminated by a material probably arising from necrotic areas which is not indicated by any of the usual marker enzymes (Prospero and Hinton, 1973), so that one cannot rule out such material as a possible locus.

As well as transporting actual molecules into the cell, the PM must transport 'chemical information' arriving from other cells as hormones or other carriers. It is now clear that one of the most important mechanisms for hormone action is the stimulation of the formation of cyclic 3′,5′-AMP by adenyl cyclase. Hormone-stimulated adenyl cyclases have been demonstrated in PM preparations from many tissues. The stimulation is normally very specific and confined to hormones known to affect that tissue. Thus thyroid PM possesses a thyrotropin-stimulated adenyl cyclase (Wolff and Jones, 1971), and the adrenal gland has one stimulated by ACTH (Lefkowitz et al., 1970); platelets have an adenyl cyclase stimulated by prostaglandin E (Krishna et al., 1972). Kidney contains two distinct adenyl cyclase enzymes, one activated by vasopressin C and the other by parathyroid hormone (Forte, 1972). Earlier results indicated that the vasopressin-stimulated adenyl cyclase is located in the medulla, the parathyroid-sensitive enzyme in the cortex (Chase and Aurbach, 1968).

Much work has been done on liver adenyl cyclase. The activity in isolated PM is stimulated by glucagon, and to a lesser extent by epinephrine (Ray, Tomasi and Marinetti, 1970; Pohl, Birnbaumer and Rodbell, 1971), and inhibited by insulin (Ray, Tomasi and Marinetti, 1970). There is evidence that there are two distinct enzymes (Bitensky, Russell and Blanco, 1970), the epinephrine-sensitive activity being confined to parenchymal cells while the glucagon-sensitive activity is largely located in reticuloendothelial cells (Reik et al., 1970). It is conceivable that the presence of PM in the microsomal fraction from the latter cells accounts for the observation of Wisher and Evans (1975) that the adenyl cyclases present in low-density PM fragments separated from the crude microsomal and crude nuclear fraction differ in their sensitivity to glucagon. The glucagon receptor protein is distinct from the enzyme (Tomasi et al., 1970), and hormone sensitivity can be destroyed by disrupting the membrane organization with phospholipase or digitonin (Birnbaumer, Pohl and Rodbell, 1971). A third protein molecule which is concerned with the effect of glucagon on liver is a glucagon-inactivating enzyme which is present in liver PM, but at a site different from that of the adenyl cyclase (Rodbell et al., 1971; Pohl et al., 1972). The mode of activity of this enzyme is not known. The liver PM may also play a role in the breakdown of 3',5'-AMP: the PM possesses most of the phosphodiesterase activity of the cell, as will be discussed below, and a specific phosphodiesterase acting on cyclic AMP has been shown to be present in a PM subfraction (House, Poulis and Weidemann, 1972) although other cyclic nucleotide phosphodiesterases are found in the cytosol (Russell, Terasaki and Appleman, 1973). The cyclic nucleotide phosphodiesterase in the PM would appear to be hormone-dependent (House, Poulis and Weidemann, 1972; Allan and Sneyd, 1975). A cyclic AMP phosphohydrolase appears to be present in glial cell PM (Morgan et al., 1971).

Another compound which is involved in the transfer of information between cells is acetylcholine, which may also play a role in the Na^+ pump (Wheeler, Coleman and Finean, 1972). An acetylcholinesterase is present in the outer surface of neuronal elements in the spinal cord, but is also present in the ER of the neuronal perikarya (Kokko, Mautner and Barrnett, 1969). Specific enzymes hydrolysing acetylcholine and related compounds are also present in liver PM; inhibition studies with a variety of substrates indicate there are at least two distinct enzymes (Wheeler, Coleman and Finean, 1972). In skeletal muscle, acetylcholinesterase is regarded as an ER rather than a PM enzyme (Headon, Keating and Barrett, 1974).

7.5.1.3   ENZYMES OF UNCERTAIN FUNCTION

Besides those enzymes to which a clear role can be assigned, many other enzymes have been detected in PM preparations. A number of these, such as the $NADH_2$–cytochrome c reductase, are almost certainly due to ER contamination (Hinton, 1972). Other activities can probably be explained as reactions of enzymes whose role has been discussed above, with artificial substrates. [Touster et al. (1970) and Thinès-Sempoux (1973) are amongst those who have discussed abridgement of the catalogue of true PM enzymes.] Thus, a portion of the alkaline p-nitrophenylphosphatase of the purified PM is stimulated by K^+ and inhibited by Na^+ and ouabain, and is probably due to part of the Na^+,K^+-ATPase system (Emmelot and Bos, 1966b). Similarly, the UDP–galactose-splitting galactosidase, which will act on p-nitrophenyl-β-galactoside (Fleischer and Fleischer, 1970), is probably a partial activity of the galactosyltransferase mentioned above. The nucleotide pyrophosphatase activity of the PM, demonstrable with NAD (Emmelot et al., 1964), has also been attributed to this enzyme (Skidmore and Trams, 1970) but would seem to resemble more the ATP pyrophosphatase (pyrophosphohydrolase) (Lieberman, Lansing and Lynch, 1967; Ray, 1970) or the distinctive inorganic pyrophosphatase which has been identified in rat liver PM (Emmelot and Bos, 1970). More recently, however, all these enzyme activities have been shown to be due to the phosphodiesterase which is discussed below (Evans, Hood and Gurd, 1973; Bischoff, Tran-Thi and Decker, 1975).

Besides the Na^+, K^+-activated ATPase and the ATP pyrophosphatase mentioned above, PM preparations contain a wide variety of phosphatases. All preparations appear to possess considerable nucleoside triphosphatase activity (not specific for ATP) and it is generally assumed, although with no positive evidence, that this is due to some uncoupled pump mechanism. Most PM preparations also possess considerable 5'-nucleotidase (nucleoside monophosphatase) activity, although in brain this enzyme may be confined to the membranes of glial cells (Torack and Barrnett, 1964). It should be noted that 5'-nucleotidase is not spread uniformly over the surface of the PM (see Section 7.5.1.4 below) and may not be confined to the PM (Morré, Merlin and Keenan, 1974; Johnsen, Stokke and Prydz, 1974).

Mention has already been made, in connection with early cytochemical work (Goldfischer, Essner and Novikoff, 1964), of nucleoside diphosphatase activities in PM and ER, and biochemical literature was cited at the end of Section 7.3.2. The PM enzyme was not demonstrable with ADP as substrate in Emmelot's laboratory (Emmelot et al., 1964); nor was microsomal UDPase activity demonstrable in Ernster's laboratory (Ernster and Jones, 1962). Our own results, however, accord with the conclusion (Berman, Gram and Spirtes, 1969; Wattiaux-de Coninck and Wattiaux, 1969b) that a wide-specificity PM enzyme does exist. Some of the activity in vitro may, however, be due to myokinase in the PM (Ray, Tomasi and Marinetti, 1969), and to nucleoside mono- and triphosphatases. Moreover, part at least is due to an apyrase with a very broad activity spectrum, for Blomberg and Perlman (1971) found that there was apparent identity between nucleoside di- and triphosphatase activities on immunoelectrophoresis of solubilized PM; they recognized that more specific enzymes might have been inadvertently destroyed.

Possibly connected with these nucleotide-dephosphorylating enzymes are a range of alkaline phosphatase activities detected with various artificial substrates, notably p-nitrophenyl phosphate and acetyl phosphate (Emmelot et al., 1964) and α-naphthyl phosphate (Dulaney and Touster, 1970). There would appear to be at least three distinct isoenzymes: a K^+-dependent phosphatase, an Mg^{2+}-sensitive activity and an Mg^{2+}-independent activity (Emmelot and Bos, 1966b). None of these enzymes can split β-glycerophosphate, and indeed such splitting by normal hepatocyte PM is found only in some species. Considerable β-glycerophosphatase activity is, however, found in rat-liver plasma membrane after bile duct ligation (Emmelot et al., 1964). It is possible that the enzyme responsible for this activity is the same as the alkaline phosphatase induced in cultured rat liver cells by dibutyryl cyclic AMP (Nose and Katsura, 1974). At least one of these enzymes, probably the third, Mg^{2+}-independent one, may have a role in hydrolysing phosphatidylcholine for transport of the choline into the bile (Pekarthy et al., 1972). This suggested role, analogous to a role suggested for 5'-nucleotidase (Ku and Wang, 1963), would connect the enzyme with the next group to be discussed, the phosphodiesterases.

Various authors, including Emmelot et al. (1964) and Ray (1970), have reported enzyme activity in the PM directed towards artificial substrates such as bis-*p*-nitrophenyl phosphate. In the view of Touster et al. (1970)— supported by Evans, Hood and Gurd (1973) but at variance with that of Futai and Mizuno (1967)—activity with this substrate is due to an enzyme that can also attack *p*-nitrophenyl-5'-thymidylate (Erecinska, Sierakowska and Shugar, 1969) or substrates such as NAD, i.e. the above-mentioned nucleotide pyrophosphatase. The identity of this enzyme with the ATP pyrophosphatase of Lieberman, Lansing and Lynch (1967) is doubted by Touster but appears proved by the experiments of Evans, Hood and Gurd (1973). There is, moreover, evidence to suggest identity of this enzyme ('phosphodiesterase I') with the alkaline ribonuclease (Prospero et al., 1973; but for contradictory evidence *see* Evans, Hood and Gurd, 1973) and glyceryl phosphorylcholine phosphodiesterase of PM. In the latter case the evidence from cell fractionation (Lloyd-Davies, Michell and Coleman, 1972) was thought to support the idea of an enzyme distinct from that which attacks bis-*p*-nitrophenyl phosphate. The phosphodiesterase which is active towards *p*-nitrophenyl-5'-thymidylate has recently been separated from solubilized PM and been shown to be inactive towards glycerophosphorylcholine, 3',5'-cAMP and RNA (Evans, Hood and Gurd, 1973). This phosphodiesterase would thus seem to be distinct from the cyclic-nucleotide-hydrolysing phosphodiesterase discussed earlier. Thus, the evidence so far available indicates that more than one phosphodiesterase is present in liver PM.

Blood platelets seem to have at least two phosphodiesterases, of which one (active towards bis-*p*-nitrophenyl phosphate) is located in the PM, and a second (active towards *p*-nitrophenyl-5'-thymidylate) seems to be in internal membranes (Taylor and Crawford, 1974).

Other PM enzymes need brief comment. L-Leucyl-β-naphthylamidase (LNA) activity is complicated by the presence of a family of aminopeptidases (Emmelot and Visser, 1971). While there is evidence of a very active $NADH_2$-ferricyanide reductase in neurone PM (Henn, Hansson and Hamberger, 1972), electron transport enzymes occasionally reported in PM preparations are generally attributable to contaminating ER (Hinton, 1972). The glucose-6-phosphatase present in low amounts in all liver and hepatoma PM preparations may be similarly explained, being largely confined to small vesicles (El-Aaser et al., 1973), but the activity could be partly due to a non-specific alkaline phosphatase (Emmelot and Bos, 1970).

7.5.1.4   INTRAMEMBRANE LOCATIONS

Cells in epithelial tissues, such as liver, pancreas or intestinal mucosa, have two distinct faces, one facing a blood capillary or sinusoid, the other facing the lumen of the epithelium. Similar specializations can be found in other types of cell, for example the synaptic endings of neurones, and the epithelial-cell 'pinocytotic vesicles' which dominated the lung PM preparation of Ryan and Smith (1971). One may expect differences in enzyme content in membrane preparations which would reflect both this functional specialization and also the diversity of cell types within a tissue. Cytochemical observations, which are subject to some uncertainty (as indicated in Section 7.4) point to a paucity of 5'-nucleotidase and ATPase in unconvoluted portions both of the PM of hepatocytes (*Figure 7.2b*), although this conflicts with histochemical evidence (Reid, 1967), and of endothelial cells in myocardial capillaries (Goldfischer, Essner and Novikoff, 1964). Also relevant, albeit inconclusive, is biochemical evidence from comparison of different hepatic PM preparations in the same laboratory.

Contrary to other biochemical reports pointing to heterogeneity of the PM, Touster *et al.* (1970) stated that 'nuclear' and microsomal PM preparations were of similar composition. However, their 'nuclear' product was apparently richer in 5'-nucleotidase, and their isolation procedure with either type of starting material may have led to inadvertent selection of lipid-rich PM elements (Thinès-Sempoux, 1973). In interpreting published comparisons of 'nuclear' and microsomal PM, there is the difficulty that the fragmentation pattern varies greatly from one laboratory to another as reflected in the relative yields. The homogenization conditions used by Touster *et al.* (1970) were notably mild, yet there was an unusually high recovery (60%) of 5'-nucleotidase of the homogenate in the microsomal fraction. Conversely, the microsomal fractions of Lieberman, Lansing and Lynch (1967) were almost devoid of the PM enzyme ATP pyrophosphatase.

Our own fractionation studies (Hinton, Norris and Reid, 1971; Norris *et al.*, 1974), in agreement with those of Thinès-Sempoux (1973), do show heterogeneity, as also found in respect of 'nuclear' PM enzymology by Benedetti and Emmelot (1968b), Evans (1970) and Evans and Gurd (1972). Likewise, microsomal PM insulin-binding capacity has been shown to be heterogeneous by House and Weidemann (1970). Reported biochemical differences amongst hepatic PM subfractions are quantitative rather than qualitative (Wisher and Evans, 1975), such that accepted general markers for PM need not be abandoned. Tentative correlations with morphological specializations, including bile canaliculi, junctional complexes and vesicle-forming elements, are emerging from studies by the investigators cited above. It is interesting to note that many enzymes appear to be concentrated in a fraction which apparently derived from bile canaliculi (Wisher and Evans, 1975). Fractionation studies on intestinal brush border have shown some correlations with morphology: thus ATPase and possibly invertase are features of microvillus membranes (Eichholz, 1967; Porteous, 1972; Mooney and McCarthy, 1973).

Intramembrane location has to be considered in the 'transverse' as well as the 'longitudinal' direction. Cytochemical approaches as indicated in Section 7.4 are helpful in this connection. Thus 5'-nucleotidase appears to act from the external face of the PM both *in situ* (Goldfischer, Essner and Novikoff, 1964; El-Aaser and Reid, 1969b) and in isolated vesicles (Widnell, 1972; El-Aaser *et al.*, 1973). The immunological recognition sites for both 5'-nucleotidase (Evans and Gurd, 1973) and phosphodiesterase (W. H. Evans, personal communication) are on the exterior face of the membrane. Cytochemical examination of erythrocyte membranes showed that ATP hydrolysis in the Na^+,K^+-ATPase system takes place on the cytoplasmic side of the membrane (Marchesi and Palade, 1967). The cytoplasmic side of erythrocyte plasma membranes also appears to be the site of the enzymes metabolizing diphosphoinositide and triphosphoinositide.

Biochemical investigation of membrane fractions by use of selective disruptive agents indicates that acetylcholinesterase is on the surface of erythrocytes (Wallach, 1972b) and of neurones in the spinal cord (Kokko, Mautner and Barrnett, 1969). In liver, the aminopeptidases mentioned above are contained in 5 nm 'knobs' which are located on the outer surface of the membrane and can be released by papain (Emmelot, Visser and Benedetti, 1969; Emmelot and Visser, 1971). Similar structures are the locus of arylamidase and invertase in intestine (Oda and Seki, 1968), yet in a rat hepatoma, which lacks these 'knobs', significant but reduced amounts of arylamidase are still present (Emmelot, Visser and Benedetti, 1968).

Na^+,K^+-ATPase, by reason of its true function, must penetrate through the thickness of the membrane, and would appear to be exceptionally sensitive to disturbance of the membrane (Emmelot and Bos, 1968). Phosphodiesterase and alkaline phosphatase activities are abolished if lipid is removed from the membrane, and would thus appear to be associated with the membrane core (Emmelot and Bos, 1968). Whilst 5'-nucleotidase survives destruction of the membrane by sonication and remains firmly associated with lipid (Barclay et al., 1972), the claim (Widnell and Unkeless, 1968) that the enzyme purified from detergent-solubilized membranes is a lipoprotein is disputed by Evans and Gurd (1973), who have isolated a lipid-free glycoprotein enzyme which would likewise appear to be associated with the membrane core. On the other hand, nucleoside triphosphate pyrophosphatase is a less tightly bound membrane constituent (Nakai et al., 1969).

7.5.1.5   CONCLUDING COMMENTS ON PLASMA-MEMBRANE ENZYMOLOGY

Evidently the period of 'random search' for PM-associated enzymes may be drawing to a close, and a period of more intensive study of molecular and functional aspects of these enzymes in the membrane will ensue. It would, in fact, appear that, amongst the diverse activities which have been reported in the PM, those which are not due to contaminating elements may be attributable to fewer enzymes than once seemed to be the case. It is noteworthy that all PM enzymes seem to have a pH optimum in the range 7–9, and all except leucylnaphthyl amidase are Mg^{2+}-dependent as reflected by the effect of EDTA treatment. This quite lengthy consideration of PM enzymology is justifiable as illustrating problems which arise generally with membrane enzymes. Where enzymes are integrated with membrane lipid, there is the special difficulty that the normal test of identity between two activities, namely co-purification, cannot readily be applied.

The emphasis on the PM is further justified by its relevance to clinical enzymology—a field not even touched on here. Workers in this field are becoming increasingly concerned with explaining the cellular origin of the increased levels, in certain diseases, of plasma enzymes that are normally present only in trace amounts, and the cell derangement that underlies the increase. Plasma enzymes such as alkaline phosphatase could arise by detachment from the PM or by synthesis *de novo*, but definitive evidence is so far scarce.

7.5.2 Enzymes of the endoplasmic reticulum

To give a brief but balanced review of ER enzymology is hardly feasible. On the one hand, the literature on some groups of enzymes, such as the microsomal oxidases, is vast. On the other hand it is uncertain whether some other groups of enzymes supposed to be in the ER (Reid, 1967) are indeed present therein. The reason for this uncertainty is simply that no proven method exists for the preparation of pure ER membranes, although many investigators have subfractionated the classical microsomal fraction. The latter contains fragments of membrane systems besides ER, notably the Golgi apparatus and the PM, which will contribute significantly to the enzyme spectrum. The outer nuclear membrane, which is continuous with the ER, is considered in Section 7.5.6.

The ER would appear to be the site of synthesis of most of the lipid made by the cell. The ER uniquely contains the full complement of enzymes required for the synthesis *de novo* of lecithin (Wilgram and Kennedy, 1963; van Golde, Fleischer and Fleischer, 1971) and, almost certainly, of the other phospholipids. This seems to hold for skeletal muscle as well as liver (Headon, Keating and Barrett, 1974). The ER can also synthesize lecithin by the CDP–choline pathway (van Golde, Fleischer and Fleischer, 1971) and can synthesize phosphatidyl glycerophosphate (Stuhne-Sekalec and Stanacev, 1970), although these activities are located mainly in the mitochondria (*see* Section 7.5.5). Synthesis of triacylglycerols would also seem to be localized in the ER (van Golde, Fleischer and Fleischer, 1971). These conclusions are drawn from studies of enzymes of isolated subcellular fractions but are, in general, confirmed by autoradiography at the EM level which shows that almost all glycerol incorporation occurs in the ER (Stein and Stein, 1967). Cholesterol synthesis is localized mainly in the ER (Chesterton, 1968), which also contains a distinctive phospholipase A acting preferentially on the A_1 position.

An important function of the ER in various types of cell is the synthesis of components to be exported from the cell. Hepatic lipid synthesis for export takes place in the ER (Stein and Stein, 1967). Protein synthesis takes place on membrane-bound ribosomes, and in the case of export proteins the product is passed through to the cisterna of the membrane (Redman and Sabatini, 1966), although this need not be the case with protein synthesized on membrane-bound polysomes for internal use (Andrews and Tata, 1971). In the case of glycoproteins the first sugar groups are added while the protein is still bound to the ribosome (Lawford and Schachter, 1966; Redman and Cherian, 1972) and others are added in the smooth ER (Redman and Cherian, 1972) before the chain is completed in the Golgi apparatus (*see* Section 7.5.3). Earlier reports that most of the synthesis of glycoprotein occurs in the ER (Hallinan, Murty and Grant; 1968; Hagopian, Bosmann and Eylar, 1968) must now be reconsidered in view of later studies on the role of the Golgi apparatus (*vide infra*).

It is also believed the lysosomal enzymes are synthesized on bound ribosomes and transported through the ER to the Golgi apparatus. These enzymes would not be detectable in the ER if, in view of their potency, they were synthesized as apoenzymes and activated only after segregation. However, it has been suggested that part of the β-glucuronidase activity seemingly associated with the ER is due to the enzyme being transported to the lysosomes (van Lancker and Lentz, 1970). Goldstone et al. (1975) have isolated a fraction of rough microsomes enriched in lysosomal hydrolases from rat kidney. Observations by Touster (1974) and others that bear on the topic of microsomal hydrolases are mentioned below in connection with lysosomes (Section 7.5.4).

Another major function of the ER, especially in the case of liver, is the metabolism of foreign compounds, of which only the briefest outline can now be given. Although the enzymes concerned are conventionally called 'detoxification' enzymes, many certainly play major roles in the normal metabolism of the cell. For convenience, two classes may be distinguished: ligases which conjugate water-insoluble compounds to give soluble derivatives, and oxidases.

Whilst the conjugating enzymes can act on a wide variety of compounds, most is known about the conjugation of steroids (Roy, 1970). Steroids can be conjugated with glucuronic acid, with sulphate or, to a lesser extent, with N-acetylglucosamine. Surprisingly, only the enzymes catalysing the transfer of glucuronyl groups occur in the microsomes; in spite of the lipid solubility of the substrates the other transferases are recovered in the soluble fraction. The source of the conjugating group is an activated intermediate (UDP–glucuronic acid, 3′-phosphoadenylyl sulphate, or UDP–N-acetylglucosamine), and it is clear that there is a battery of more or less specific enzymes catalysing the transfer of either glucuronyl or sulphate residues to different substrates.

Connected with these ligases is a group of corresponding hydrolases. Besides the above-mentioned β-glucuronidase, the function of which is under dispute, the ER also contains an aryl sulphatase which, interestingly, is absent from the ER of pancreas (Meldolesi, Jamieson and Palade, 1971), a tissue which, as far as is known, does not play a significant role in detoxification.

Especial confusion reigns in the field of the microsomal oxidases. What is known for certain is that the ER, especially of liver but also of other cells, can oxidize a wide variety of compounds including aliphatic, polycyclic and aromatic hydrocarbons, aniline, secondary and tertiary amines, barbiturates, DDT and related compounds, and steroids. The ER is also capable of the oxidative demethylation of N-, S- and O-methyl compounds (Gillette, Davis and Sasame, 1972). What is less clear is the enzymological explanation of these reactions. Microsomes can oxidize both $NADH_2$ and $NADPH_2$ using different flavoprotein dehydrogenases which can pass the electrons on to cytochrome b_5 or P_{450} respectively (Estabrook and Cohen, 1969). However, the reactions in the living cell are much more complex. Typically, the requirements for oxidation in vitro of any of the substrates mentioned above are microsomes, NADP and oxygen, which is directly involved in the reaction. The mechanism of the reaction is discussed by Gillette, Davis and Sasame (1972). It is not clear how many of these enzymes (termed mixed-function oxidases) are involved in the system or how the $NADH_2$–cytochrome

b_5 chain fits in, although it is clear that there is a close relationship between the two groups. Cytochrome b_5 is the receptor used by a smaller number of more specific dehydrogenases such as stearyl CoA desaturase.

In the case of skeletal muscle, as surveyed by Headon, Keating and Barrett (1974), regulation of the intracellular calcium concentration is an important functional activity of the ER reflected by Ca^{2+}-dependent ATPase activity. For other ATPase activities demonstrable with microsomes from muscle, or from liver (Ernster and Jones, 1962), there is doubt about the function, and even about their assignment to the ER (Morré et al., 1974).

Other enzymes of quite ill-defined function have been reported as present in the ER. These include alkaline phosphodiesterase I (Lloyd-Davies, Wichell and Coleman, 1972) and 5'-nucleotidase (Widnell, 1972). The latter enzyme is more closely considered in Sections 7.5.1.2 and 7.5.3, in relation to its PM locus. Moreover, microsomal fractions are the main locus of nonspecific esterase activity, despite cytochemical manifestation in lysosomes. However, if pre-fixation conditions are altered to minimize inhibition, activity is also manifest in rough ER (cisternae and membrane), smooth ER and the nuclear envelope, but not PM (Barrow and Holt, 1971; see 'Note added in proof', p. 197). Glucose-6-phosphatase, the favourite marker for liver ER membranes, is assumed to be mainly concerned with the hydrolysis of G-6-P to glucose prior to release into the blood. Since, however, the same enzyme has been claimed as responsible for the pyrophosphatase and the pyrophosphate–glucose phosphotransferase activities in ER (Nordlie, 1969) and for nucleoside triphosphate and diphosphate phosphotransferase activity, its role is still debatable.

Some studies of microsomal subfractions were based on mere differential pelleting in ion-free media (Imai, Ito and Sato, 1966; Reid, 1967); they showed some heterogeneity but are of limited value. Even with better methods (Reid, 1967; Dallner and Ernster, 1968; Tata, 1972) designed to separate 'rough' from 'smooth' ER, a problem arose that is repeatedly stressed here, viz. the multiple morphological origin of the smooth vesicles. These problems are considered in some detail by Amar-Costesec et al. (1974). Moreover, differential pelleting in the presence of Cs^+ or other cations, so as to aggregate particular types of vesicle, as practised in Dallner's laboratory, evidently gives products different from those obtained by isopycnic centrifugation.

Smooth vesicles from adult liver preferentially show a rise in the amount, if not the specific activity, of enzymes present in ER (e.g. glucose-6-phosphatase) in response to inducers such as phenobarbitone (Gram et al., 1971, inter alia). An enzymatic difference was evident in normal liver in the case of mixed-function oxidase activity; it was paralleled by a difference in $NADPH_2$–cytochrome P_{450} reductase activity, but this was attributable in the rat, as distinct from the rabbit, to a difference in the cytochrome concentration (Holtzman et al., 1968). In certain species, for example the guineapig, the distribution of drug-metabolizing enzymes between rough and smooth vesicles favours the latter (Gram et al., 1971).

From rough vesicles, subfractions can be obtained, one of which is relatively rich in glucose-6-phosphatase and other phosphatases, and another in electron-transport enzymes (Eriksson and Dallner, 1971). A similar distinction, in accordance with earlier fractionation studies in Louvain on whole microsomes (cited by Thinès-Sempoux *et al.*, 1969), has also been observed within the smooth-vesicle population (Glaumann and Dallner, 1970). Contrary to a report by Leskes, Siekevitz and Palade (1971), Lewis and Tata (1973) claim that rough and smooth microsomes can each furnish subpopulations which show differences in glucose-6-phosphatase content, although they remark that conceivably the pattern might merely reflect differences amongst different types of cell *in vivo*. The general impression remains, however, that observed disparities in ER enzymology amongst microsomal subfractions from adult liver are comparatively slight. Yet developmental studies (Leskes, Siekevitz and Palade, 1971) indicate that enzymes such as glucose-6-phosphatase appear in the rough ER in the first instance and later spread evenly. Literature on the ontogeny of membrane enzymes (El-Aaser and Reid, 1969b) cannot be surveyed here.

Subfractionation studies on microsomes from non-hepatic sources, for example brain (Tata, 1972), skeletal muscle (Headon, Keating and Barrett, 1974), and cultured mammalian cells (Graham, 1973), have hardly given conclusive evidence of enzyme heterogeneity in ER that can be correlated with morphological specializations *in vivo*.

The 'transverse' aspect of ER enzyme location within the membrane awaits full elucidation. Most of the enzymes show lipid dependence, indicating that they are firmly bound into the membrane structure. They differ in ease of solubilization and, indeed, of survival of activity when 'solubilized' by a detergent or other agent (Ernster and Jones, 1962; El-Aaser and Reid, 1969a; Thinès-Sempoux *et al.*, 1969), but this hardly defines their locations. The glycosyltransferases should, from their function, be located on the internal face of the membrane. Cytochemical studies indicate a similar location for glucose-6-phosphatase *in situ* (Goldfischer, Essner and Novikoff, 1964) and in isolated microsomes (Leskes, Siekevitz and Palade, 1971; El-Aaser *et al.*, 1973) but liberated glucose may enter the cytosol rather than the cisterna (Glaumann, Nilsson and Dallner, 1970).

7.5.3 Enzymes of Golgi membranes

Studies on the Golgi apparatus have developed very differently from studies on other cell organelles. The structure was recognized by the turn of this century (*see* Beams and Kessel, 1968); yet it proved very difficult to isolate, although an early report (Schneider and Kuff, 1954) claimed success with epididymis. Indeed, its major role in collecting of proteins destined either for export or for segregation into lysosomes was elucidated with the aid of EM techniques before any well characterized preparations of Golgi apparatus had been obtained (Neutra and Leblond, 1966, and *see* references cited by Jamieson and Palade, 1968). The former workers showed that, in intestinal mucosa, the Golgi was the site of addition of at least part of the carbohydrate side-chain of glycoproteins. The role of the Golgi has been reviewed by Cook (1973).

With the recent development of methods for the isolation of hepatic Golgi elements, knowledge of their enzymology is advancing. It is not known whether any part of the Golgi is synthesized *in situ*. The Golgi membranes from ox liver but not rat liver can synthesize lecithin from lysolecithin, and those from either species can carry out the synthesis of phosphatidylglycerol from CDP–diglyceride and glycerol-3-phosphate (van Golde, Fleischer and Fleischer, 1971). Golgi membranes have also been reported to possess some choline kinase and CDP–choline cytidyltransferase activity (Cheetham, Morré and Yunghans, 1970). The role of these enzymes and of the reported phosphodiesterases A_1 and A_2 (van Golde, Fleischer and Fleischer, 1971) is not yet clear.

In addition to organizing the movement of protein destined for segregation in either secretion vacuoles or lysosomes, the Golgi apparatus plays a role in the elaboration of the side chains of glycoproteins, as mentioned above, although the first sugar residues would appear to be added in the ER during or immediately after the synthesis of the protein. UDP–galactose: N-acetylglucosamine galactosyltransferase has been found in all Golgi preparations from mammalian tissue such as ox liver (Fleischer, Fleischer and Ozuwa, 1969), rat liver (Morré, Merlin and Keenan, 1969; Fleischer and Fleischer, 1970; Leelavathi et al., 1970), rat epididymis (Fleischer, Fleischer and Ozuwa, 1969), rat seminiferous tubule (Cunningham, Mollenhauer and Nyquist, 1971), and rat intestinal mucosa (Frot-Coutaz et al., 1973). Wagner and Cynkin (1969) have demonstrated transfer to endogenous protein acceptors, while Schachter et al. (1970) have demonstrated the presence of at least three specific glycosyltransferases capable of the sequential addition of the three terminal sugars of serum glycoproteins. Transfer of sugar prosthetic groups to serum lipoproteins has also been demonstrated (Lo and Marsh, 1970).

Prior to the assignment of galactosyltransferase activity to Golgi fragments, the most often suggested marker was thiamine pyrophosphatase. This is now discredited as a marker in the case of liver, as considerable activity is present in the ER and the PM. Nevertheless, both thiamine pyrophosphatase and nucleoside di- and triphosphatase activity are somewhat enriched in Golgi preparations, and it is unlikely that these activities are due to contaminants (Cheetham et al., 1971).

As Wagner, Pettersson and Dallner (1973) pointed out, the Golgi apparatus is morphologically heterogeneous, showing at least two separable components—peripheral small vacuoles and the large central cisternae. Very low-density lipoprotein particles destined for release to the serum were concentrated in the cisternal elements (Bergeron et al., 1973a). Novikoff et al. (1971), working with dorsal root ganglion, found thiamine pyrophosphatase to be concentrated in the cisternae whilst acid phosphatase, presumably destined for incorporation into lysosomes, was contained in a peripheral system of membranes (called the Golgi–endoplasmic reticulum–lysosome system, or GERL) continuous with the ER.

There have been few reports of other enzymes in the Golgi apparatus. An ox preparation was reported to contain more $NADH_2$–cytochrome c reductase which was insensitive to antimycin A than could be explained by ER contamination (Fleischer, Fleischer and Ozuwa, 1969), but this enzyme could not be found in preparations from rat liver that did nevertheless contain considerable amounts of cytochrome b_5 (Fleischer et al., 1971). Recently there have been reports of 5′-nucleotidase in the Golgi apparatus (Bergeron et al., 1973a). Morré et al. (1974) found the specific activity to be much lower than for ER, but ruled out the possibility that the activity is explicable by PM contamination. One explanation is that the Golgi apparatus is involved in the transport of proteins to the PM, which would also account for the presence of insulin-binding activity (Bergeron, Evans and Geschwind, 1973b). A small amount of galactosidase activity found in Golgi preparations is probably due to lysosome contamination, but Morré *et al.* (1974) consider that some glucose-6-phosphatase is present as a true Golgi constituent. A Ca^+-activated ATPase has been reported in vesicles from the Golgi apparatus of mammary gland (Baumrucker and Keenan, 1975). The vesicles actively accumulate Ca^{2+} ions, which apparently play a role in packaging the milk proteins.

To summarize, it appears at the moment that only one group of enzymes has a major locus in the Golgi apparatus of liver, the glycosyltransferases, of which the galactosyltransferase seems to be almost exclusively in the Golgi (Morré *et al.*, 1974). It is conceivable that nucleoside di- and triphosphatase activity may be largely a facet of the transferase activity. Most other activities seem to be due either to contaminants or to material being transported through the Golgi apparatus. Whilst it is unlikely that this is a full catalogue of Golgi enzymes, in Tom Lehrer's words, 'There may be many others but they haven't been discovered'.

7.5.4 Enzymes of the lysosomal membrane

Even using rats treated with Triton WR-1339 or with dextran, it is difficult to obtain reasonably pure preparations of liver lysosomes (*see* Beaufay, 1969; Reid, 1972b). Lysosomes prepared by centrifugation from untreated animals are inevitably heavily contaminated with other membranous components. This situation is most unfortunate as one cannot assume that the membranes of the large secondary lysosomes found in the livers of treated rats will be similar to those of the lysosomes found in normal animals. Triton and dextran are usually assumed to be taken up by pinocytosis, and if this were so they would carry considerable amounts of PM into the cell. In discussing the enzymes of lysosomal and especially of 'tritosome' membranes, one must be continually aware of the possibility that some activities might derive from the PM. Fortunately, methods have been described recently for the purification of lysosomes by free-flow electrophoresis (*see* Henning and Heidrich, 1974). So far no results have been reported concerning the enzymes present in the membranes of these lysosomes, but study of their lipid composition shows them to be very different from Triton WR-1339-loaded secondary lysosomes.

Even though lysosomes can be purified from Triton-treated rats, the homogeneity of the resulting 'membrane' preparation is still in doubt. The usual method of preparing 'membranes' from lysosomes entails disrupting the organelle by freeze-thawing, by sonication or by osmotic shock. The 'membranes' are then pelleted, but under these conditions undisrupted lysosomes and insoluble matrix components will also sediment. A more extensive discussion of the problems is given by Thinès-Sempoux (1973), by Henning (1974), and by R. Wattiaux (Vol. 2, Chapter 5 of this series). Added complexity in the case of certain lysosomes ('dense-body-type') comes from the apparent existence of an internal membrane system, as shown cytochemically (Smith, 1969).

Subject to these reservations, a number of enzymes have been reported in the membranes of Triton-loaded lysosomes (tritosomes). Of these the nucleoside diphosphatase (Wattiaux-de Coninck and Wattiaux, 1969b) and the phospholipase A (Rahman, Verhagen and van der Wiel, 1970) seem very similar to the corresponding PM enzymes. There has been more disagreement about the 5'-nucleotidase activity found in the lysosome membrane fraction. Kaulen, Henning and Stoffel (1970) have argued for a distinct 5'-nucleotidase, but Pletsch and Coffey (1972) claim that the tartrate-insensitive nucleotidase activity in lysosomal membranes is very similar to that in PM, although there is another distinct component. The situation is complicated by the reported presence of an acid nucleotidase in lysosomes (Touster, 1974), albeit hardly active at alkaline pH. In mouse peritoneal leukocytes, 5'-nucleotidase has been shown to be present in newly formed phagocytotic vesicles, but disappears with a half-life of only two hours (Werb and Cohn, 1972). Thus if the enzyme detected in the liver lysosomes were due to PM taken up during pinocytosis, the digestion would have to be much slower than in the leukocyte. This is quite plausible, however, with these overloaded lysosomes, and partial digestion of the enzymes might explain differences in their properties.

Similarly, it is not proven that the L-leucyl-β-naphthylamidase (Kaulen, Henning and Staufel, 1970) and the alkaline phosphodiesterase (Thinès-Sempoux, 1973) detected in lysosomal membranes could not derive from the PM. The chemical composition of the tritosome membrane is too similar to that of the PM (Thinès-Sempoux, 1973) to preclude the possibility that it has arisen largely from the PM. The absence of Na^+,K^+-activated ATPase (Kaulen, Henning and Staufel, 1970) is no proof of the absence of PM material for, as mentioned earlier, this enzyme is extremely sensitive to changes in the organization of the membrane and so could readily be lost. Immunochemical criteria indicate a considerable similarity between PM and the membranes of secondary lysosomes (see Berzins, Blomberg and Perlman, 1975). This, together with the considerable differences between the lipids in the membranes of primary and Triton WR1339-loaded lysosomes (Henning and Heidrich, 1974), suggests that the latter derive largely from the PM.

Of the undoubted lysosomal enzymes, acid phosphatase shows a notably high proportion in the insoluble 'membrane' fraction after lysosomal disruption (Thinès-Sempoux, 1973). Most work on the release enzymes has been done with lysosomes from untreated rats, and is not invalidated merely because of the common use of very impure lysosomal fractions such as are obtained by differential pelleting. Unlike acid ribonuclease, alkaline ribonuclease tends to be sedimentable after freeze-thawing (Reid and Nodes, 1959), but part of the activity in the lysosomal fraction is due to other membrane fragments as discussed in Section 7.5.1.3, while the truly lysosomal activity does not seem to be membrane-associated (M. Dobrota, personal communication). Baccino, Rita and Zuretti (1971) found that a number of enzymes, notably β-acetylglucosaminidase, were strongly associated with the 'membrane' fraction, but could be washed off by saline and were readily bound *in vitro*; evidently they are adsorbed rather than structurally bound. A portion of the acid phosphatase activity was, however, firmly bound, and acid phosphatase was not adsorbed *in vitro*.

Evidence for an acid phosphatase on the lysosomal membrane which was different from the soluble acid phosphatase of lysosomes had already been reported by Sloat and Allen (1969). More recently Dobrota and Hinton (1974) have separated disrupted lysosomes by zonal-rotor techniques and have shown the presence of a peak of acid phosphatase at a density of 1.17 g cm^{-3}, the density expected for free membranes. On functional grounds it is a reasonable supposition that the inner rather than the outer face of the lysosomal membrane is the locus of any constituent hydrolases that are membrane-bound. On the other hand one might expect that the HCO_3^--activated ATPase (Iritani and Wells, 1974), which may be part of the system responsible for maintaining the lysosome interior at a lower pH than the cytosol, would span the whole width of the membrane.

7.5.5 Enzymes of mitochondrial membranes

The anatomy of the mitochondrion is extremely complex. The mitochondrion is surrounded by two membranes which are quite different in structure and in function. In addition, the inner membrane has been shown to have morphological specializations—the cristae and their attached 'knobs'—besides being 'two-faced', but no attempt will be made to scrutinize this microheterogeneity in this section; it will be dealt with by R. A. Capaldi in Vol. 2 of this series. Enzymes may be located in the different membrane systems of the mitochondrion, in the space between the membranes, or in the internal matrix, as will be outlined below in conjunction with *Figure 7.4*. Much progress has now been made in assigning enzymes to the different loci mentioned above.

It is thought that a large part of the lipid of mitochondrial membranes is synthesized in the ER. Certainly a very rapid exchange of lipid between mitochondrial and ER membranes can be demonstrated *in vitro* (Wirtz and Zilversmit, 1968), and this may partly explain the rapid incorporation of labelled precursors *in vivo* (Beattie, 1969). Mitochondria have, however, been reported to be capable of some synthesis of lipids *de novo* (Kaiser, 1969).

Whether or not mitochondria are capable of synthesizing phospholipid *de novo*, they are certainly capable of modifying lipids obtained from other parts of the cell. An active phospholipase A_2 is present in the outer membrane (Nachbaur, Colbeau and Vignais, 1969; Waite, 1969), and the lysophospholipids so formed can be reacylated by acyltransferases of the outer mitochondrial membrane (Sarzala *et al.*, 1970). Mitochondrial enzymes can also acylate glycerophosphate, but the mitochondrial enzyme differs from the better known microsomal enzyme in giving rise to lysophosphatidic rather than phosphatidic acid, and in preferring saturated to unsaturated fatty acids (Bremer *et al.*, 1974). Nachbaur, Colbeau and Vignais (1969) have reported a transacylase in the inner mitochondrial membrane which transfers acyl residues from phospholipids to lysophospholipids. The complex phospholipid cardiolipin, which in mammalian cells is found only in mitochondrial outer membranes, is synthesized from CDP–diglyceride by a mitochondrial enzyme system (McMurray and Magee, 1972).

Some of the information for synthesis of mitochondrial proteins is contained in mitochondrial DNA. This DNA can be replicated by a special mitochondrial DNA polymerase and can be transcribed to RNA by a mitochondrial RNA polymerase. Mitochondrial ribosomes can use this RNA as a template for the synthesis of some of the proteins of mitochondria. This whole information system is contained in the inner part of the mitochondrion, but it is not yet clear how closely the enzymes are membrane-associated (Beattie, 1971).

The role of the outer mitochondrial membrane is not yet fully understood. Outer mitochondrial membrane preparations contain a heterogeneous group of enzymes including kynurenine hydroxylase, rotenone-insensitive $NADH_2$–cytochrome c reductase, part of the nucleoside diphosphate kinase activity, and monoamine oxidase (Schnaitman and Greenawalt, 1968). The use of monoamine oxidase as a marker has already been mentioned in Section 7.3.2. The true cytochrome acceptor for the reductase may be b_5.

The main role of the mitochondrial inner membrane is in energy generation. The inner membrane is, in contrast to the outer membrane, remarkably impermeable to small molecules other than those which are freely lipid-soluble. Materials such as ATP, phosphate ions and citric-acid-cycle intermediates are transferred across the membrane by specific carrier systems which show most of the properties of enzymes. For L-glutamate there are two such systems (Bradford and McGivan, 1973).

The inner surface of the inner membrane is, in fact, the main functional unit in energy generation, Classical work on ATPase-containing electron-transport particles ('subunits'), which comprise 'knobs' on the cristal inner faces and mediate oxidative phosphorylation, entailed the development of isolation methods, applicable to sources such as liver and heart muscle. Only recently were such studies complemented by studies on other membranous elements. Membrane fragments, differing ultrastructurally from phosphorylating particles, can be isolated from heart muscle or liver mitochondria after cholate or other treatment, and were claimed to be rich in citric-acid-cycle enzymes, although these were mainly in a soluble side-fraction (Allmann and Bachmann, 1967). While this soluble fraction no doubt corresponds to the mitochondrial matrix, the claim that the outer membrane is the source of the membrane fragments and mediates the cycle is probably untenable, as shown by examination of fractions obtained from mitochondria that have been allowed to swell and contract so as to loosen the outer membrane. Such fractions, corresponding to the outer and the inner membranes respectively, have been characterized by Sottocasa *et al.* (1967) and other authors including Greenawalt and Schnaitman (1970). Insofar as citric-acid-cycle enzymes are in part membrane-linked, it is the inner rather than the outer membrane that is the locus.

Although some aspects of mitochondrial membrane enzymology are controversial, the scheme shown in *Figure 7.4* represents views on which there is a wide measure of agreement. Of particular interest is the fact that there are enzyme activities common to the outer membrane and to ER membranes, compatible with the possibility that the latter give rise to the former. This would entail loss of certain enzymes, one example being glucose-6-phosphatase (Sottocasa *et al.*, 1967), though its loss may not be complete (Sarzala *et al.*, 1970). As has been indicated above, the seeming similarity in enzymes of lipid metabolism occurring in the outer membrane and in microsomal membranes may be misleading.

Racker (1970) has surveyed mitochondrial 'topography' with special reference to the inner membrane, besides considering the validity of purification studies on lipid-dependent enzymes. Conflicting lines of evidence on the orientation of cytochrome oxidase can be reconciled by assuming that it functions on *both* sides, although with different cytochrome substrates. It is particularly difficult to reconstitute isolated respiratory components so as to give a system that will carry out oxidative phosphorylation with a substrate such as succinate. Such complexities, and consideration of the oligomycin-sensitive ATPase of the phosphorylating particles, fall outside the

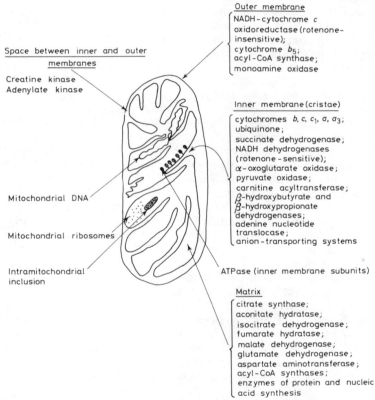

Figure 7.4 Schematic representation of mitochondrial structure and enzyme localization, from Hughes, Lloyd and Brightwell (1970), courtesy of Dr D. Lloyd, the Editor of Symposia of the Society for General Microbiology, *and Cambridge University Press. Enzymes of the citric acid cycle are possibly located not only in the matrix (as shown) but also in the inner membrane*

scope of this chapter, as does the vectorial aspect of carrier-dependent hydrogen transport into the mitochondrion. An important incentive to the study of separable elements, such as cristal and non-cristal components of the inner membrane, comes from autoimmunity to inner membranes, as encountered in the human (Berg *et al.*, 1969).

7.5.6 Enzymes of other membrane systems, including nuclear membranes

Peroxisomal membranes, hitherto little explored, are discussed by N. E. Tolbert and R. P. Donaldson in Vol. 2 of this series and will not be considered here.

Information on the membranes of storage granules is available for several tissues, but no clear picture has yet emerged. A membrane-linked ATPase is found in adrenomedullary chromaffin granules and appears to be involved in catecholamine release (Trifaró, 1972). Dopamine β-hydroxylase is largely located in the granule membrane (Hörtnagl, Winkler and Lochs, 1972). Smith (1969) has presented cytochemical evidence of acid phosphatase in the membrane of secretion granules of anterior pituitary and other glandular tissues. It is known that lysosomes can fuse with secretion granules (Farquhar, 1969), so that one cannot ascertain that the acid phosphatase is a normal constituent. However, Smith and Winkler (1969) suggest that lysosomal enzymes in the Golgi apparatus may play a role in the formation of the granules.

Meldolesi, Jamieson and Palade (1971) have isolated the membranes of pancreatic zymogen granules and found them to have an enzyme composition similar to that of the PM. This would appear to argue for 'recycling' of the secretion granule membranes after the discharge of their contents, but this has been excluded on other grounds. Meldolesi, Jamieson and Palade (1971) argue against contamination with PM fragments as an explanation, because the ratios of the enzymes differ from those for purified PM. However, another possible explanation would be contamination by a subfraction of the PM, for, as mentioned in Section 7.5.1.4, the characteristic enzymes of the PM are not distributed uniformly.

Surrounding the nucleus of eukaryotic cells there is a paired membrane system, pierced by numerous pores. The enzymology of the inner membrane awaits biochemical exploration. Although the outer membrane of the 'nuclear envelope' appears to be continuous with the ER, we deal with it here rather than in Section 7.5.2. In pioneer cytochemical studies glucose-6-phosphatase was identified in the outer nuclear membrane (Goldfischer, Essner and Novikoff, 1964). Biochemical studies on nuclear envelope preparations (Monneron, 1974) have indeed shown the presence of this and other ER phosphatases, but this does not prove that they are located on the outer membrane; a marker enzyme is lacking.

It is not yet known whether the outer nuclear membrane can synthesize lipids; but isolated nuclei do not possess acyl CoA synthetase (Lippel and Blythe, 1972), hence perhaps other lipid-synthesizing enzymes may also be absent.

On the basis of reports that DNA associated with the nuclear membrane was more rapidly labelled than the bulk of the chromatin (*see* O'Brien, Sanyal and Stanton, 1972, *inter alia*), it has been suggested that in eukaryotic cells, as well as in bacteria, the DNA polymerase is membrane-associated. This does not now seem to be so, as it has been shown that newly synthesized DNA is not preferentially associated with the nuclear membrane (Fakan *et al.*, 1972) and that purified nuclear membrane has very little DNA polymerase activity (Kay, Fraser and Johnston, 1972; Monneron, 1974). The obvious major role of the nuclear membrane is control of the passage of RNA into the cytoplasm and of the passage into the nucleus of messenger compounds which control the synthesis of DNA and RNA. Nothing is known about the enzymes involved in the first process, unless an Mg^{2+}-ATPase found by Kashnig and Kasper (1969) is involved. With regard to the second, it is interesting that about half the activity of $NADPH_2$-Δ^2-3-ketosteroid-5α oxidoreductase, which converts testosterone to 5α-dihydrotestosterone in the ventral prostate of the rat, is found in the nuclear membrane of this tissue (Moore and Wilson, 1972).

While the enzyme composition of nuclear membranes now seems to be qualitatively similar to that of the ER, there were early reports from some laboratories that glucose-6-phosphatase was almost absent from nuclear membranes (Zbarsky et al., 1969; Franke et al., 1970; Agutter, 1972). The explanation lies in the vigorous methods which were used to separate the membrane from the rest of the nucleus. Gentler methods have subsequently been developed (Price, Harris and Baldwin, 1972, 1973; Kay, Fraser and Johnston, 1972) and have confirmed that there is considerable glucose-6-phosphatase activity. Liver and thymocyte preparations also contain 5′-nucleotidase and a 3′-nucleotidase, besides Na^+,K^+-independent ATPase (Monneron, 1974).

The nuclear membrane possesses a complex and distinctive electron transport system which differs in several respects from that of the microsomes (Berezney and Crane, 1972). Like microsomes, the nuclear membrane possesses a rotenone-sensitive $NADH_2$–cytochrome c reductase (Kashnig and Kasper, 1969; Berezney and Crane, 1972) which works through cytochrome b_5. In contrast with microsomes, however, cytochrome P_{450} and $NADPH_2$–cytochrome c reductase are almost absent (Berezney, Macaulay and Crane, 1972) while a cytochrome c oxidase is present which differs in some respects from mitochondrial cytochrome oxidase (Berezney and Crane, 1972). However, other authors have failed to detect the cytochromes a and a_3 of cytochrome oxidase in purified nuclei, so that the question cannot be regarded as settled (Berezney and Crane, 1972). The absence of the microsomal $NADPH_2$ oxidase system is confirmed by experiments on the feeding of phenobarbitone, which induces $NADPH_2$–cytochrome c reductase and cytochrome P_{450} in the microsomal membrane, but has no effect on the (low) level of these components and of drug-metabolizing enzymes in the nuclear membrane (Kasper, 1971).

7.6 CONCLUDING COMMENTS

Amongst the different categories of membrane in the cells of a given tissue, there are dissimilarities in enzymatic composition which are much more striking than the morphological differences, and which in some instances are evidently related to functional differences. A functional role is particularly evident for enzymes of the mitochondrial membrane.

The ER is particularly well equipped with enzymes for synthesis and renewal of membranes, or at least of membrane lipids. In this process it seems that 'new components are inserted molecule-by-molecule into a pre-existing structural framework' (Leskes, Siekeitz and Palade, 1971), and that 'the gradual conversion of membranes from one type to another is documented by comparisons of enzymatic activities, lipid composition and progressive modification of the proteins and lipids of membranes along the . . . export routes', viz. ER→Golgi→secretory vesicle→PM, or ER→vesicle→PM (Morré et al., 1974). It is thus not surprising that, for these membranes at least, it is the exception rather than the rule to encounter a unique location for any particular enzyme.

It is, then, understandable that there is heart-searching concerning 'biochemical markers', and the phrase 'UDPase, a Golgi enzyme' that appeared in a 1968 paper would no longer pass muster. Technical difficulties in fractionating microsomal smooth vesicles, to give sub-populations that are morphologically meaningful, are a further hindrance to the elucidation of enzyme distribution amongst membranous elements in the cell.

It is likewise difficult to establish whether a particular membranous element that has morphologically different regions has corresponding enzymatic heterogeneity. Suggestive evidence for this has, however, already been obtained with the PM and the Golgi apparatus. Moreover, despite the technical difficulties it already appears that some enzymatic constituents of membranes are in the 'core' and others are associated quite loosely with the membrane or even, in some instances, artefactually. Those that are firmly bound are typically lipoproteins or glycoproteins.

Our understanding of membrane enzymes in respect of location and role is gradually improving. The catalogue of membrane enzyme activities still shows bewildering diversity, but it is to be hoped that it will become shortened through the present trend to ascribe to a single enzyme protein a range of activities, although some may have been reported in membrane preparations on the basis of assays with 'unnatural' substrates.

Acknowledgements

Our own experience in this field has largely been gained whilst pursuing studies that were supported by the Cancer Research Campaign and the Science Research Council. We acknowledge use of colleagues' unpublished data, and particularly thank Dr A. A. El-Aaser for allowing use of photographic material from his past work. The legends to the figures give further acknowledgements.

REFERENCES

AGUTTER, P. S. (1972). *Biochim. biophys. Acta*, **255**: 397.
ALLAN, E. H. and SNEYD, J. G. T. (1975). *Biochem. biophys. Res. Commun.*, **62**: 594.
ALLMANN, D. W. and BACHMANN, E. (1967). *Meth. Enzym.*, **10**: 438.
AMAR-COSTESEC, A., WIBO, M., THINÈS-SEMPOUX, D., BEAUFAY, H. and BERTHET, J. (1974). *J. Cell Biol.*, **62**: 717.
ANDREWS, T. M. and TATA, J. R. (1971). *Biochem. J.*, **121**: 683.
BACCINO, F. M., RITA, G. A. and ZURETTI, M. F. (1971). *Biochem. J.*, **122**: 363.
BAKER, P. F. and WILLIS, J. S. (1969). *Biochim. biophys. Acta*, **183**: 646.
BARCLAY, M., BARCLAY, R. K., SKIPSKI, V. P., ESSNER, E. and TEREBUS-KEKISH, O. (1972). *Biochim. biophys. Acta*, **255**: 931.
BARROW, P. C. and HOLT, S. J. (1971). *Biochem. J.*, **125**: 545.
BAUER, C. H., LUKASHEK, R. and REUTTER, W. G. (1974). *Biochem. J.*, **142**: 221.
BAUMRUCKER, C. R. and KEENAN, T. W. (1975). *Expl Cell Res.*, **90**: 253.
BEAMS, H. W. and KESSEL, R. G. (1968). *Int. Rev. Cytol.*, **23**: 209.
BEATTIE, D. S. (1969). *Biochem. biophys. Res. Commun.*, **35**: 67.

BEATTIE, D. S. (1971). *Sub-Cell. Biochem.*, **1**:1.
BEAUFAY, H. (1969). *The Lysosomes in Biology and Pathology*, Vol. 2, pp. 515–546. Ed. J. T. DINGLE and H. B. FELL. Amsterdam; North-Holland.
BENEDETTI, E. L. and EMMELOT, P. (1968a). *The Membranes*, pp. 33 *et seq.* Ed. A. J. DALTON and F. HAGENAU. New York; Academic Press.
BENEDETTI, E. L. and EMMELOT, P. (1968b). *J. Cell Biol.*, **38**:15.
BEREZNEY, R. and CRANE, F. L. (1972). *J. biol. Chem.*, **247**:5562.
BEREZNEY, R., MACAULAY, L. K. and CRANE, F. L. (1972). *J. biol. Chem.*, **247**:5549.
BERG, P. A., ROITT, I. M., DONIACH, D. and COOPER, H. M. (1969). *Immunology*, **17**:281.
BERGERON, J. J. M., EHRENREICH, J. H., SIEKEVITZ, P. and PALADE, G. (1973a). *J. Cell Biol.*, **59**:73.
BERGERON, J. J. M., EVANS, W. H. and GESCHWIND, I. I. (1973b). *J. Cell Biol.*, **59**:771.
BERMAN, H. M., GRAM, W. and SPIRTES, M. A. (1969). *Biochim. biophys. Acta*, **183**:10.
BERZINS, K., BLOMBERG, F. and PERLMAN, P. (1975). *Eur. J. Biochem.*, **51**:181.
BIRNBAUMER, L., POHL, S. L. and RODBELL, M. (1971). *J. biol. Chem.*, **246**:1857.
BIRNIE, G. D. (Ed.) (1972). *Subcellular Components: Preparation and Fractionation*, 2nd edn. London; Butterworths.
BISCHOFF, I., TRAN-THI, T.-A. and DECKER, K. F. A. (1975). *Eur. J. Biochem.*, **51**:353.
BITENSKY, M. W., RUSSELL, V. and BLANCO, M. (1970). *Endocrinology*, **86**:154.
BLOMBERG, F. and PERLMAN, P. (1971). *Biochim. biophys. Acta*, **233**:53.
BRADFORD, N. M. and MCGIVAN, J. D. (1973). *Biochem. J.*, **134**:1023.
BREMER, J., BJERVE, K. S., CRISTOPHERSEN, B. O., DAAE, L. N. W., SOLBERG, H. E. and AAS, M. (1974). *Regulation of Hepatic Metabolism; Alfred Benzon Symposium VI*, pp. 159–179. Ed. F. LUNDQUIST and N. TYGSTRUP. New York; Academic Press.
BURRISS-GARETT, R. J. and REDMAN, C. M. (1975). *Biochim. biophys. Acta*, **382**:58.
CHASE, L. R. and AURBACH, G. D. (1968). *Science, N.Y.*, **159**:545.
CHEETHAM, R. D., MORRÉ, D. J. and YUNGHANS, W. N. (1970). *J. Cell Biol.*, **44**:492.
CHEETHAM, R. D., MORRÉ, D. J., PANNEK, C. and FRIEND, D. S. (1971). *J. Cell Biol.*, **49**:899.
CHESTERTON, C. J. (1968). *J. biol. Chem.*, **243**:1147.
CLINE, G. B., DAGG, M. K. and RYEL, R. B. (1974). *Methodological Developments in Biochemistry*, Vol. 4. pp. 39–46. Ed. E. REID, London; Longmans.
COLEMAN, R. (1968). *Biochim. biophys. Acta*, **163**:111.
COLEMAN, R. (1973). *Biochim. biophys. Acta*, **300**:1.
COOK, G. M. W. (1973). *The Lysosomes in Biology and Pathology*, Vol. 3, pp. 237–277. Ed. J. T. DINGLE. Amsterdam; North-Holland.
CUNNINGHAM, W. P., MOLLENHAUER, H. H. and NYQUIST, S. E. (1971). *J. Cell Biol.*, **51**:273.
DALLNER, G. and ERNSTER, L. (1968). *J. Histochem. Cytochem.*, **16**:611.
DE DUVE, C. H. (1967). *Enzyme Cytology*, pp. 1–26. Ed. D. B. ROODYN. London; Academic Press.
DE DUVE, C. H., PRESSMAN, B. C., GIANETTO, R., WATTIAUX, R. and APPELMANS, F. (1955). *Biochem. J.*, **60**:604.
DOBROTA, M. and HINTON, R. H. (1974). *Methodological Developments in Biochemistry*, Vol. 4, pp. 177–186. Ed. E. REID. London; Longmans.
DULANEY, J. T. and TOUSTER, O. (1970). *Biochim. biophys. Acta*, **196**:29.
EICHHOLZ, A. (1967). *Biochim. biophys. Acta*, **135**:475.
EL-AASER, A. A. and HOLT, S. J. (1973). *Methodological Developments in Biochemistry*, Vol. 3, pp. 85–89. Ed. E. REID. London; Longmans.
EL-AASER, A. A. and REID, E. (1969a). *Histochem. J.*, **1**:417.
EL-AASER, A. A. and REID, E. (1969b). *Histochem. J.*, **1**:439.
EL-AASER, A. A., FITZSIMONS, J. T. R., HINTON, R. H., NORRIS, K. A. and REID, E. (1973). *Histochem. J.*, **5**:199.
ELLORY, J. C. and KEYNES, P. D. (1969). *Nature, Lond.*, **221**:776.
EMMELOT, P. and BOS, C. J. (1966a). *Biochim. biophys. Acta*, **121**:434.
EMMELOT, P. and BOS, C. J. (1966b). *Biochim. biophys. Acta*, **121**:375.
EMMELOT, P. and BOS, C. J. (1968). *Biochim. biophys. Acta*, **150**:341.
EMMELOT, P. and BOS, C. J. (1970). *Biochim. biophys. Acta*, **211**:169.
EMMELOT, P. and VISSER, A. (1971). *Biochim. biophys. Acta*, **241**:273.
EMMELOT, P., VISSER, A. and BENEDETTI, E. L. (1968). *Biochim. biophys. Acta*, **150**:364.
EMMELOT, P., BOS, C. J., BENEDETTI, E. L. and RÜMKE, PH. (1964). *Biochem. biophys. Acta*, **90**:126.
ERECINSKA, M., SIERAKOWSKA, H. and SHUGAR, D. (1969). *Eur. J. Biochem.*, **11**:465.
ERIKSSON, L. C. and DALLNER, G. (1971). *FEBS Lett.*, **19**:163.
ERNSTER, L. and JONES, L. E. (1962). *J. Cell Biol.*, **15**:563.

ESTABROOK, R. W. and COHEN, B. (1969). *Microsomes and Drug Oxidations*, pp. 95–105. Ed. J. R. GILLETTE, A. H. CONNEY, G. J. COSMIDES, R. W. ESTABROOK, J. R. FOUTS and G. J. MANNERING. New York; Academic Press.
EVANS, W. H. (1970). *Biochem. J.*, **116**:833.
EVANS, W. H. and GURD, J. W. (1972). *Biochem. J.*, **128**:691.
EVANS, W. H. and GURD, J. W. (1973). *Biochem. J.*, **133**:189.
EVANS, W. H., HOOD, D. O. and GURD, J. W. (1973). *Biochem. J.*, **135**:819.
FAKAN, S., TURNER, G. N., PAGANO, J. S. and HANCOCK, R. (1972). *Proc. natn. Acad. Sci. U.S.A.*, **69**:2300.
FARQUHAR, M. G. (1969). *The Lysosomes in Biology and Pathology*, Vol. 2, pp. 462–482. Ed. J. T. DINGLE and H. B. FELL. Amsterdam; North-Holland.
FERBER, E., RESCH, K., WALLACH, D. F. H. and IMM, W. (1972). *Biochim. biophys. Acta*, **266**:494.
FLEISCHER, B. and FLEISCHER, S. (1970). *Biochim. biophys. Acta*, **219**:301.
FLEISCHER, B., FLEISCHER, S. and OZUWA, H. (1969). *J. Cell Biol.*, **43**:59.
FLEISCHER, S., FLEISCHER, B., AZZI, A. and CHANCE, B. (1971). *Biochim. biophys. Acta*, **225**:194.
FORTE, L. R. (1972). *Biochim. biophys. Acta*, **266**:524.
FRANKE, W. W., DEUMLING, B., ERMEN, B., JARASCH, E. D. and KLEINIG, H. (1970). *J. Cell Biol.*, **46**:379.
FROT-COUTAZ, J., DUBOIS, P., BERTHELLIER, G. and GOT, R. (1973). *Expl Cell Res.*, **77**:223.
FUTAI, M. and MIZUNO, D. (1967). *J. biol. Chem.*, **242**:5301.
FUTAI, M., TSUNG, P.-K. and MIZUNO, D. (1972). *Biochim. biophys. Acta*, **261**:508.
GILLETTE, J. R., DAVIS, D. C. and SASAME, H. A. (1972). *A. Rev. Pharmac.*, **12**:57.
GLAUMANN, H. and DALLNER, G. (1970). *J. Cell Biol.*, **47**:34.
GLAUMANN, H., NILSSON, R. and DALLNER, G. (1970). *FEBS Lett.*, **10**:306.
GOLDFISCHER, S., ESSNER, E. and NOVIKOFF, A. B. (1964). *J. Histochem. Cytochem.*, **12**:72.
GOLDSTONE, A., KOENIG, H., NAYYAR, R., HUGHES, C. and LU, C. Y. (1973). *Biochem. J.*, **132**:259.
GRAHAM, J. M. (1973). *Methodological Developments in Biochemistry*, Vol. 3, pp. 205–217. Ed. E. REID. London; Longmans.
GRAM, T. E., SCHROEDER, D. H., DAVIS, D. C., REAGAN, R. L. and GUARINO, A. M. (1971). *Biochem. Pharmac.*, **20**:2885.
GREENAWALT, J. W. and SCHNAITMAN, C. (1970). *J. Cell Biol.*, **46**:173.
GURD, J. W. and EVANS, W. H. (1974). *Archs Biochem. Biophys.*, **164**:305.
HAGGIS, G. H. (1966). *The Electron Microscope in Molecular Biology*. London; Longmans.
HAGOPIAN, A., BOSMANN, H. B. and EYLAR, E. H. (1968). *Archs Biochem. Biophys.*, **128**:387.
HALLINAN, T., MURTY, C. N. and GRANT, J. H. (1968). *Archs Biochem. Biophys.*, **125**:715.
HEADON, D. R., KEATING, H. and BARRETT, E. J. (1974). *Methodological Developments in Biochemistry*, Vol. 4, pp. 279–291. Ed. E. REID. London; Longmans.
HENN, F. A., HANSSON, H. A. and HAMBERGER, A. (1972). *J. Cell Biol.*, **53**:654.
HENNING, R. (1974). *Methodological Developments in Biochemistry*. Vol. 4, pp. 187–194. Ed. E. REID. London; Longmans.
HENNING, R. and HEIDRICH, H.-G. (1974). *Biochim. biophys. Acta*, **345**:326.
HIGGINS, J. A. and GREEN, C. (1967). *Biochim. biophys. Acta*, **144**:211.
HINTON, R. H. (1972). *Subcellular Components. Preparation and Fractionation*, 2nd edn, pp. 119–156. Ed. G. D. BIRNIE. London; Butterworths.
HINTON, R. H., NORRIS, K. A. and REID, E. (1971). *Separations with Zonal Rotors*, pp. S-2.1–S-2.16. Ed. E. REID. Guildford; Wolfson Bioanalytical Centre, University of Surrey.
HOLTZMAN, J. L., GRAM, T. E., GIGON, P. L. and GILLETTE, J. R. (1968). *Biochem. J.*, **110**:407.
HÖRTNAGL, H., WINKLER, H. and LOCHS, H. (1972). *Biochem. J.*, **129**:187.
HOUSE, P. D. R., POULIS, P. and WEIDEMANN, M. J. (1972). *Eur. J. Biochem.*, **24**:429.
HOUSE, P. D. R. and WEIDEMANN, M. J. (1970). *Biochem. biophys. Res. Commun.*, **41**:541.
HUBBARD, A. L. and COHN, Z. A. (1975). *J. Cell Biol.*, **64**:438.
HUGHES, D. E., LLOYD, D. and BRIGHTWELL, R. (1970). *Symp. Soc. gen. Microbiol.*, **20**:295.
IMAI, Y., ITO, A. and SATO, R. (1966). *J. Biochem., Tokyo*, **60**:417.
IRITANI, N. and WELLS, W. W. (1974). *Archs Biochem. Biophys.*, **164**:357.
JAMIESON, J. D. and PALADE, G. E. (1968). *J. Cell Biol.*, **39**:580.
JAMIESON, J. D. and PALADE, G. E. (1971). *J. Cell Biol.*, **50**:135.
JOHNSEN, S., STOKKE, T. and PRYDZ, H. (1974). *J. Cell Biol.*, **63**:357.
KAISER, W. (1969). *Eur. J. Biochem.*, **8**:120.
KASHNIG, D. M. and KASPER, C. B. (1969). *J. biol. Chem.*, **244**:3786.
KASPER, C. B. (1971). *J. biol. Chem.*, **246**:577.
KAULEN, H. D., HENNING, R. and STOFFEL, W. (1970). *Hoppe-Seyler's Z. physiol. Chem.*, **351**:1555.

KAY, R. B., FRASER, D. and JOHNSTON, I. R. (1972). *Eur. J. Biochem.*, **30**:145.
KOKKO, A., MAUTNER, H. G. and BARRNETT, R. J. (1969). *J. Histochem. Cytochem.*, **17**:625.
KRISHNA, G., HARWOOD, J. P., BARBER, A. J. and JAMIESON, G. A. (1972). *J. biol. Chem.*, **247**:2253.
KU, K. Y. and WANG, C. T. (1963). *Shih Yen Sheng Wu Hsueh Pao (Acta Biol. exp. sin.)*, **8**(3-4): 400 (1964); *Chem. Abstr.*, **61**:2123.
LAUTER, C. J., SOLYOM, A. and TRAMS, E. G. (1972). *Biochim. biophys. Acta*, **266**:511.
LAWFORD, G. R. and SCHACHTER, H. (1966). *J. biol. Chem.*, **241**:5408.
LEELAVATHI, P. E., ESTES, L. W., FEINGOLD, D. S. and LOMBARDI, B. (1970). *Biochim. biophys. Acta*, **211**:124.
LEFKOWITZ, R. S., ROTH, J., PRICER, W. and PASTAN, I. (1970). *Proc. natn. Acad. Sci. U.S.A.*, **65**:748.
LESKES, A., SIEKEVITZ, P. and PALADE, G. E. (1971). *J. Cell Biol.*, **49**:288.
LEWIS, J. A. and TATA, J. R. (1973). *Biochem. J.*, **134**:69.
LIEBERMAN, I., LANSING, A. I. and LYNCH, W. E. (1967). *J. biol. Chem.*, **242**:736.
LIPPEL, K. and BLYTHE, D. (1972). *Biochim. biophys. Acta*, **280**:231.
LLOYD-DAVIES, K. A., MICHELL, R. H. and COLEMAN, R. (1972). *Biochem. J.*, **127**:357.
LO, L.-H. and MARSH, J. B. (1970). *J. biol. Chem.*, **245**:5001.
MARCHESI, V. T. and PALADE, G. E. (1967). *J. Cell Biol.*, **35**:385.
MCMURRAY, W. C. and MAGEE, W. C. (1972). *A. Rev. Biochem.*, **41**:129.
MELDOLESI, J., JAMIESON, J. D. and PALADE, G. E. (1971). *J. Cell Biol.*, **49**:150.
MONNERON, A. (1974). *Phil. Trans. R. Soc., Ser. B*, **268**:101.
MOONEY, P. and MCCARTHY, C. F. (1973). *Trans. biochem. Soc.*, **1**:603.
MOORE, R. J. and WILSON, J. D. (1972). *J. biol. Chem.*, **247**:958.
MORGAN, I. G., WOLFE, L. S., MANDEL, P. and GOMBOS, G. (1971). *Biochim. biophys. Acta*, **241**:737.
MORRÉ, D. J., MERLIN, L. M. and KEENAN, T. W. (1969). *Biochim. biophys. Res. Commun.*, **37**:813.
MORRÉ, D. J., YUNGHANS, W. N., VIGIL, E. L. and KEENAN, T. W. (1974). *Methodological Developments in Biochemistry*, Vol. 4, pp. 195-236. Ed. E. REID. London; Longmans.
MORRISON, G. R., KARL, I. E., SCHWARTZ, R. and SHANK, R. E. (1965). *J. Lab. clin. Med.*, **65**:248.
NACHBAUR, J., COLBEAU, A. and VIGNAIS, P. M. (1969). *FEBS Lett.*, **3**:121.
NAKAI, K., TAKEMITSU, S., KAWASAKI, T. and YAMASHINA, I. (1969). *Biochim. biophys. Acta*, **193**:468.
NEMANIC, M. K., CARTER, O. P., PITELKA, D. R. and WOFSY, L. (1975). *J. Cell Biol.*, **64**:311.
NEUTRA, M. and LEBLOND, C. P. (1966). *J. Cell Biol.*, **30**:119.
NEWKIRK, J. D. and WAITE, M. (1971). *Biochim. biophys. Acta*, **225**:224.
NORDLIE, R. C. (1969). *Ann. N.Y. Acad. Sci.*, **166**:699.
NORRIS, K. A., DOBROTA, M., ISSA, F. S., HINTON, R. H. and REID, E. (1974). *Biochem. J.*, **142**:667.
NOSE, K. and KATSURA, H. (1974). *Expl Cell Res.*, **87**:8.
NOVIKOFF, P. M., NOVIKOFF, A. B., QUINTANA, N. and HAUW, J. J. (1971). *J. Cell Biol.*, **50**:859.
O'BRIEN, R. L., SANYAL, A. B. and STANTON, R. H. (1972). *Expl Cell Res.*, **70**:106.
ODA, P. and SEKI, S. (1968). *Proceedings, 6th International Congress for Electron Microscopy, Kyoto*, Biol. Vol. p. 387.
PALADE, G. E. and SIEKEVITZ, P. (1956). *J. biophys. biochem. Cytol.*, **2**:171.
PEKARTHY, J. M., SHORT, J., LANSING, A. I. and LIEBERMAN, I. (1972). *J. biol. Chem.*, **247**:1767.
PINKETT, M. O. and PERLMAN, R. L. (1974). *Biochim. biophys. Acta*, **372**:379.
PLETSCH, Q. A. and COFFEY, J. W. (1972). *Biochim. biophys. Acta*, **276**:192.
POHL, S. L., BIRNBAUMER, L. and RODBELL, M. (1971). *J. biol. Chem.*, **246**:1849.
POHL, S. L., KRANS, M. J., BIRNBAUMER, L. and RODBELL, M. (1972). *J. biol. Chem.*, **247**:2295.
PONDER, E. (1961). *The Cell*, Vol. 2, pp. 1-84. Ed. J. BRACHET and A. E. MIRSKY. New York; Academic Press.
PORTEOUS, J. W. (1972). *Subcellular Components, Preparation and Fractionation*, 2nd edn, pp. 157-183. Ed. G. D. BIRNIE. London; Butterworths.
PRICE, M. R., HARRIS, J. R. and BALDWIN, R. W. (1972). *J. Ultrastruct. Res.*, **40**:178.
PRICE, M. R., HARRIS, J. R. and BALDWIN, R. W. (1973). *Methodological Developments in Biochemistry*, Vol. 3, pp. 159-170. Ed. E. REID. London; Longmans.
PROSPERO, T. D. and HINTON, R. H. (1973). *Methodological Developments in Biochemistry*, Vol. 3, pp. 171-186. Ed. E. REID. London; Longmans.
PROSPERO, T. D., BURGE, M. L. E., NORRIS, K. A., HINTON, R. H. and REID, E. (1973). *Biochem. J.*, **132**:449.
RACKER, E. (1970). *Essays in Biochemistry*, Vol. 6, pp. 1-22. Ed. P. N. CAMPBELL and F. DICKENS. London; Academic Press.

RAHMAN, Y. E., VERHAGEN, J. and VAN DER WIEL, D. F. M. (1970). *Biochem. biophys. Res. Commun.*, **38**:670.
RAY, T. K. (1970). *Biochim. biophys. Acta*, **196**:1.
RAY, T. K., TOMASI, V. and MARINETTI, G. V. (1969). *Fedn Proc. Fedn Am. Socs exp. Biol.*, **28**:891.
RAY, T. K., TOMASI, V. and MARINETTI, G. V. (1970). *Biochim. biophys. Acta*, **211**:20.
REDMAN, C. M. and CHERIAN, M. G. (1972). *J. Cell Biol.*, **52**:231.
REDMAN, C. M. and SABATINI, D. D. (1966). *Proc. natn. Acad. Sci. U.S.A.*, **56**:608.
REID, E. (1967). *Enzyme Cytology*, pp. 321–406. Ed. D. B. ROODYN. London; Academic Press.
REID, E. (1972a). *Sub-Cell. Biochem.*, **1**:217.
REID, E. (1972b). *Subcellular Components: Preparation and Fractionation*, 2nd edn, pp. 95–118. Ed. G. D. BIRNIE. London; Butterworths.
REID, E. and NODES, J. T. (1959). *Ann. N.Y. Acad. Sci.*, **81**:618.
REID, E. and WILLIAMSON, R. (1974). *Meth. Enzym.*, **31A**:713.
REIK, L., PETZGOLD, G. C., HIGGINS, J. A., GREENGARD, P. and BARRNETT, R. (1970). *Science, N.Y.*, **168**:382.
RODBELL, M., KRAUS, M. J., POHL, S. L. and BIRNBAUMER, L. (1971). *J. biol. Chem.*, **246**:1861.
ROTH, S., MCGUIRE, E. J. and ROSEMAN, S. (1971). *J. Cell Biol.*, **51**:536.
ROY, A. B. (1970). *Chemical and Biological Aspects of Steroid Conjugation*, pp. 74–130. Ed. S. BERNSTEIN and S. SOLOMON. Berlin; Springer-Verlag.
RUSSELL, T. R., TERASAKI, W. L. and APPLEMAN, M. M. (1973). *J. biol. Chem.*, **248**:1334.
RYAN, J. W. and SMITH, U. (1971). *Biochim. biophys. Acta*, **249**:177.
SABATINI, D. D., MILLER, F. and BARRNETT, R. J. (1964). *J. Histochem. Cytochem.*, **12**:57.
SARZALA, M. G., VAN GOLDE, L. M. G., DE KRUYFF, B. and VAN DEENEN, L. L. M. (1970). *Biochim. biophys. Acta*, **202**:106.
SCHACHTER, H. S., JABBAL, J., HUDGIN, R. L., PINTERIC, L., MCGUIRE, E. J. and ROSEMAN, S. (1970). *J. biol. Chem.*, **245**:1090.
SCHENGRUND, C. L., JENSEN, D. S. and ROSENBURG, A. (1972). *J. biol. Chem.*, **247**:2742.
SCHNAITMAN, C. and GREENAWALT, J. W. (1968). *J. Cell Biol.*, **38**:158.
SCHNEIDER, W. C. and KUFF, E. L. (1954). *Am. J. Anat.*, **94**:200.
SHNITKA, T. K. and SELIGMAN, A. M. (1971). *A. Rev. Biochem.*, **40**:375.
SKIDMORE, J. and TRAMS, E. G. (1970). *Biochim. biophys. Acta*, **219**:93.
SLOAT, B. F. and ALLEN, J. M. (1969). *Ann. N.Y. Acad. Sci.*, **166**:574.
SMITH, A. D. and WINKLER, H. (1969). *The Lysosomes in Biology and Pathology*, Vol. 1, pp. 155–166. Ed. J. T. DINGLE and H. B. FELL. Amsterdam; North-Holland.
SMITH, R. E. (1969). *Ann. N.Y. Acad. Sci.*, **166**:525.
SONG, C. S., RUBIN, W., RIFKIND, A. B. and KAPPAS, A. (1969). *J. Cell Biol.*, **41**:124.
SOTTOCASA, G. L., KUYLERSTIERNA, B., ERNSTER, L. and BERGSTRAND, A. (1967). *Meth. Enzym.*, **10**:448.
STAHL, W. L. and TRAMS, E. G. (1968). *Biochim. biophys. Acta*, **163**:459.
STEIN, O. and STEIN, Y. (1967). *J. Cell Biol.*, **33**:319.
STEIN, Y., WIDNELL, C. and STEIN, O. (1968). *J. Cell Biol.*, **39**:185.
STUHNE-SEKALEC, L. and STANACEV, N. Z. (1970). *Can. J. Biochem.*, **48**:1214.
TATA, J. (1972). *Subcellular Components, Preparation and Fractionation*, 2nd edn, pp. 185–213. Ed. G. D. BIRNIE. London; Butterworths.
TAYLOR, D. J. and CRAWFORD, N. (1974). *Methodological Developments in Biochemistry*, Vol. 4, pp. 319–326. Ed. E. REID. London; Longmans.
THINÈS-SEMPOUX, D. (1973). *The Lysosomes in Biology and Pathology*, Vol. 3. pp. 278–299. Ed. J. T. DINGLE. Amsterdam; North-Holland.
THINÈS-SEMPOUX, D., AMAR-COSTESEC, A., BEAUFAY, H. and BERTHET, J., (1969). *J. Cell Biol.*, **43**:189.
TOMASI, V., KORETZ, S., RAY, T. K., DUNNICK, J. and MARINETTI, G. V. (1970). *Biochim. biophys. Acta*, **211**:31.
TORACK, R. M. and BARRNETT, R. J. (1964). *J. Neuropath. exp. Neurol.*, **23**:46.
TOUSTER, O. (1974). *Methodological Developments in Biochemistry*, Vol. 4, pp. 247–270. Ed. E. REID. London; Longmans.
TOUSTER, O., ARONSON, N. W., DULANEY, J. T. and HENDRICKSON, H. (1970). *J. Cell Biol.*, **47**:604.
TRIFARÓ, J. M. (1972). *FEBS Lett.*, **23**:237.
VAN GOLDE, L. M. G., FLEISCHER, B. and FLEISCHER, S. (1971). *Biochim. biophys. Acta*, **249**:318.
VAN LANCKER, J. L. and LENTZ, P. (1970). *J. Histochem. Cytochem.*, **18**:529.
VICTORIA, E. J., VAN GOLDE, L. M. G., HOSTETLER, K. Y., SCHERPHOF, G. L. and VAN DEENEN, L. L. M. (1971). *Biochim. biophys. Acta*, **239**:443.

WAGNER, R. R. and CYNKIN, M. A. (1969). *Biochem. biophys. Res. Commun.*, **35**:139.
WAGNER, R. R., PETTERSSON, E. and DALLNER, G. (1973). *J. Cell Sci.*, **12**:603.
WAITE, M. (1969). *Biochemistry*, **8**:2536.
WALLACH, D. F. H. (1972a). *Biochim. biophys. Acta*, **265**:61.
WALLACH, D. F. H. (1972b). *The Plasma Membrane: Dynamic Perspectives, Genetics and Pathology*. London; English Universities Press.
WATTIAUX-DE CONINCK, S. and WATTIAUX, R. (1969a). *Biochim. biophys. Acta*, **183**:118.
WATTIAUX-DE CONINCK, S. and WATTIAUX, R. (1969b). *FEBS Lett.*, **5**:355.
WEBB, G. C. and ROTH, S. (1974). *J. Cell Biol.*, **63**:796.
WERB, E. and COHN, Z. A. (1971). *J. exp. Med.*, **134**:1545.
WERB, E. and COHN, Z. A. (1972). *J. biol. Chem.*, **247**:2439.
WHEELER, G. E., COLEMAN, R. and FINEAN, J. B. (1972). *Biochim. biophys. Acta*, **255**:917.
WIDNELL, C. C. (1972). *J. Cell Biol.*, **52**:542.
WIDNELL, C. C. and UNKELESS, J. C. (1968). *Proc. natn. Acad. Sci. U.S.A.*, **61**:1050.
WILGRAM, G. F. and KENNEDY, E. P. (1963). *J. biol. Chem.*, **238**:2615.
WIRTZ, K. W. A. and ZILVERSMIT, D. B. (1968). *J. biol. Chem.*, **243**:3596.
WISHER, M. H. and EVANS, W. H. (1975). *Biochem. J.*, **146**:375.
WOLFF, J. and JONES, A. B. (1971). *J. biol. Chem.*, **246**:3939.
WRIGHT, J. D. and GREEN, C. (1971). *Biochem. J.*, **123**:837.
YAMAZAKI, M. and HAYAISHI, O. (1968). *J. biol. Chem.*, **243**:2934.
YOUDIM, M. B. H. (1973). *Br. med. Bull.*, **29**:120.
ZBARSKY, I. B., PEREVOSHCHIKOVA, K. A., DELEKTORSKAYA, L. N. and DELEKTORSKY, V. V. (1969). *Nature, Lond.*, **221**:257.

Note added in proof (section numbers in parentheses)

In a critical survey of methods for isolating membranes (*see* Section 7.2), including Golgi fragments (7.5.3), Steck (1972) discussed the possible 'inside-out' nature of vesicles such as those formed when inner mitochondrial membranes are liberated (7.5.5). The question of artefactual orientations and of 'sided-ness' (7.4) has also been further considered, for erythrocytes, by Steck and Kant (1974). In a study of the 5'-nucleotidase of fat cells, Newby, Luzio and Hales (1975) have shown by immunological (7.4) and other approaches that the enzyme is on the external face of the PM (7.5.1.4). They cite speculations that the role of 5'-nucleotidase (7.5.1.3) may relate to the vasodilatory action of released adenosine.

Holt (1973) has clarified the localization of esterase (7.5.2), in an article which presents novel cytochemical methodology (7.3.1). Cytochemical examination of fractions (7.3.1) has been described by Farquhar, Bergeron and Palade (1974), who have given precise localizations for 5'-nucleotidase, thiamine pyrophosphatase and other enzymes in Golgi elements (7.5.3). Franke (1974) has discussed the nuclear envelope (7.5.6) with respect to constituents such as DNA and to its relationship to the ER.

ADDITIONAL REFERENCES

FARQUHAR, M. G., BERGERON, J. J. M. and PALADE, G. (1974). *J. Cell Biol.*, **60**:8.
FRANKE, W. W. (1974). *Phil. Trans. R. Soc., Ser. B*, **268**:67.
HOLT, S. J. (1973). *Electron Microscopy and Cytochemistry*, pp. 71–83. Ed. E. WISSE, W. TH. DAEMS, I. MOLENAAR and P. VAN DUIJN. Amsterdam; North-Holland.
NEWBY, A. C., LUZIO, J. P. and HALES, C. N. (1975). *Biochem. J.*, **146**:625.
STECK, T. L. (1972). *Membrane Molecular Biology*, pp. 76–114. Ed. C. F. FOX and A. D. KEITH. Stamford, Connecticut; Sinauer Associates.
STECK, T. L and KANT, J. A. (1974). *Meth. Enzym.*, **31A**:172.

8

Electrostatic control of membrane permeability via intramembranous particle aggregation

David Gingell
Department of Biology as Applied to Medicine, The Middlesex Hospital Medical School, London

8.1 INTRODUCTION

Cell membrane structure has suddenly ceased to be dull. Work in several laboratories has uncovered dynamic features which can provide a completely new way of thinking about membrane physiology. The lipid bilayer is a naturally stable configuration, not particularly responsive to the environment. Much effort devoted to studying the interaction of proteins with lipids in bilayers and cell membranes has shown that macromolecules can introduce the element of instability essential for the sensitivity to external effectors that is characteristic of living cells. The brilliant experiments of Mueller and Rudin (1967) on the evocation of action potentials from bilayers which had been treated with proteins were decisive in underlining the important role of macromolecules in lipid systems. It is natural to suppose that proteins and glycoproteins are responsible for the dynamic aspects of cell membranes. A turning point in our concept of cell membrane structures has come from recent work, summarized by Edidin (1974), which shows that biological membranes are two-dimensional hydrocarbon fluids in which 'islands' of protein and glycoprotein, which will be referred to as *intramembranous particles*, are partially embedded in the lipid and are able to diffuse in the plane of the lipid bilayer. This simple fact has important and far-reaching implications, many of which are only just beginning to be appreciated. It is the purpose of this chapter to expand a brief note (Gingell, 1973) and argue that diverse cellular actions are due to membrane permeability changes which accompany aggregation of intramembranous particles. The central point of the argument is that *the particles suffer mutual repulsion owing to their strong negative charges yet attract one another by van der Waals (electrodynamic) forces. Consequently, factors which reduce or overcome the electric repulsion cause these molecules to aggregate, setting up a region of increased permeability which acts as an information channel through the membrane.*

In a previous article (Gingell, 1972), it was suggested that many factors of the cell's environment alter the electrostatic surface potential of the membrane and cause an increase in membrane permeability leading to cellular responses. For example, pinocytosis in amoebae is stimulated by small cations and polycations which neutralize the negative surface potential of the plasma membrane. A subsequent increase in membrane permeability, as measured with microelectrodes, allows fluxes of small ions which may trigger contractile cytoplasmic events responsible for formation of pinocytosis channels. Although such transductive activity was implicated, no clear mechanism relating surface potential and permeability was evident. It now seems that the electrostatic potential of repulsion between mobile intramembranous particles may control their degree of aggregation and thereby provide the key to permeability changes.

The emphasis of this chapter will be twofold. First, following a very brief résumé of the reasons for accepting the reality of mobile intramembranous particles, an order-of-magnitude analysis of electrostatic interactions between model particles will be given; details are in the Appendix (p. 220). Secondly, reasons for believing that aggregation of the particles is a widespread phenomenon related to membrane permeability changes will be examined, taking evidence from amoeboid pinocytosis, as well as from immunoglobulin-induced capping in small lymphocytes, the cellular response to plant lectins, exocytosis and the structure of intercellular junctions.

Little attempt will be made to discuss the nature of the permeability change accompanying aggregation or the details of subsequent cytoplasmic responses, such as contraction. It has usually been assumed that an ionic channel can exist through the subunits of the close junction (McNutt and Weinstein, 1973) but it is not obvious how such channels would be opened as a result of aggregation. Another possibility is that ions pass between the aggregated particles and that the lipid properties become altered in these regions. It is even possible that, without any structural changes, ions may move easily between the membrane subunits when they are closely packed simply because it takes less energy to move an ion through lipid of low dielectric constant if part of the ion's field extends into a region of higher dielectric constant (Parsegian, 1969) caused by the proximity of protein. The possibility that adenyl cyclase activation plays a part in mediating the cytoplasmic response triggered by aggregating intramembranous particles cannot be evaluated at present. It would certainly be fashionable to assign the role of signal not to small ions but to cyclic AMP, though a Ca^{2+} ion influx is almost certainly the signal in secretory processes (Douglas, 1968); it is certainly possible that the initial fluxes of small ions are responsible for activating adenyl cyclase. Although there is little evidence either way in the case of pinocytosis, Seyberth *et al.* (1973) have shown that an earlier demonstration of increased cyclic AMP levels during phagocytosis was in fact an artefact due to the incubation medium.

8.2 GLYCOPROTEIN NATURE OF INTRAMEMBRANOUS PARTICLES

Evidence that proteins can interact with bilayers was supplemented by work (Marchesi et al., 1972) showing that glycophorin, the major glycoprotein of human red cells, has a hydrophobic end eminently suitable for insertion into membrane lipid (Morawiecki, 1964; Winzler, 1969; Marchesi et al., 1972), and by the demonstration that this molecule spans the lipid bilayer (Bretscher, 1971). Using Winzler's data (Winzler, 1969) Bretscher calculated the density of glycoprotein on the cell surface to be about one molecule per 170 nm^2, a figure of the same order as the density of $\sim$8 nm diameter particles seen in freeze-fractured preparations; the glycoprotein has a molecular weight of approximately 50 000 and carries 27 negative charges due to sialic acid. There is no evidence of other charged groups except for those of the polypeptide backbone of the molecule. The carboxyl terminal group is presented to the cytoplasm. The hydrophobic portion within the membrane is probably in helical form, while, by analogy with other proteins in solution, the portion extending into the outer aqueous phase, and bearing sugar residues in branched chains, is likely to be folded. The sugar side chains have ionizable sialic acid groups and the configuration of the molecule is influenced by their mutual repulsion.

There is strong evidence that the intramembranous particles seen in freeze-fractured red cell membranes include the charged glycoproteins. Branton's elegant demonstration that bilayers cleave down the middle (Branton and Moor, 1964; Branton, 1966) and Marchesi's more direct evidence from freeze-fractured preparations where the outside face of the membrane was previously labelled with ferritin-conjugated antibody to the glycoprotein (Marchesi et al., 1972), seem to put Branton's hypothesis that cleavage occurs between the hydrocarbon layers almost beyond doubt and make it difficult to escape the conclusion that the particles are proteins and glycoproteins. It is clear that the volume of the hydrophobic amino acid residues of glycoprotein which are believed to lie within the bilayer is considerably smaller than a freeze-fracture particle, even allowing for the thickness of the platinum-carbon replica: consequently there cannot be a one-to-one relationship between glycophorin molecules and intramembranous particles. It has been suggested that a major membrane protein and glycophorin combine to form a particle. Grant and McConnell (1974) were able to demonstrate freeze-fracture particles in phospholipid membranes treated with glycophorin, and Segrest, Gulik-Krzywicki and Sardet (1974) have succeeded in incorporating the hydrophobic segment of glycophorin (residues 12 to 34) into phospholipid bilayers. Freeze-etching showed intramembranous particles of diameter 8 nm which the authors calculate to be composed of 10 to 20 monomers. They argue that in vivo the 8 nm particle of red cell membranes is likely to have hydrophobic regions of not less than 4 to 12 glycophorin molecules. This interesting suggestion does not of course preclude the possibility that they are combinations of glycoprotein and protein, and in the electrostatic calculations which follow it will be assumed that each particle represents one glycophorin molecule.

8.3 MOBILITY OF INTRAMEMBRANOUS PARTICLES

Evidence for membrane fluidity has come largely from observations on markers attached to membrane components (*see* Oseroff, Robbins and Burger, 1973, and also Tamm and Tamm, 1974, for a novel argument). The celebrated experiments of Frye and Edidin (1970) showed that fusion of cells bearing distinctly different immunofluorescent labels resulted in dye intermingling within minutes. A comparable result attended the immunofluorescent labelling of B lymphocytes with antilymphocytic serum or anti-immunoglobulin (Raff and de Petris, 1973). Aggregation of the label into a 'cap' followed by pinocytosis of the cap is consistent with the hypothesis that membrane components are free to move in the plane of the cell surface. A high rate of lateral diffusion of spin-labelled phospholipids in sarcoplasmic reticulum vesicles has been found by Scandella, Devaux and McConnell (1972). A lateral diffusion constant of 6×10^{-8} cm^2 s^{-1} corresponds to the translation of a single phospholipid molecule a distance of 5000 nm in one second. The authors, however, stress that the motion of some membrane lipids might be considerably slower.

The relation between the distribution of membrane components, which can be labelled using extracellular markers, and the movement of intramembranous particles has recently been examined by freeze-fracture methods combined with electron-opaque molecular markers. Nicolson and Painter (1973) were thus able to show that redistribution of surface binding sites is paralleled by movement of intramembranous particles. They exposed red cell ghosts to antispectrin antibody, which binds exclusively to sites on the cytoplasmic face of the membrane, and found that the pattern of intramembranous particle clumping reflected the distribution of binding sites for colloidal ferric hydroxide at the extracellular face of the membrane. Work with *Ricinus communis* agglutinin has given the same results (Ji and Nicolson, 1974). Similar conclusions were reached by Pinto da Silva, Moss and Fudenberg (1973) using ferritin with a high isoelectric point as an extracellular surface marker for the redistribution of intramembranous particles in response to pH. The finding that the extent of particle aggregation is a function of pH and can be reversible (Pinto da Silva, 1972) suggests that electrostatic forces can play a decisive role in particle stability. Intact red cells do not, however, show this behaviour: Elgsaeter and Branton (1974) have shown that removal of spectrin from the cytoplasmic face of the membrane at low ionic strength markedly increases the response of intramembranous particles to changes in environmental pH. Further examples of constraints on the free lateral diffusion of particles will be discussed later in relation to the action of lectins.

Caution must be exercised in attributing these relatively well-documented properties of red cell ghosts to other membranes. Pinto da Silva and Martinez-Palomo (1975) found that intramembranous particles of 3T3 cells fail to show aggregation after treatment with Con A at a concentration sufficient to cluster Con A receptors at the cell surface: they concluded that intramembranous particles are distinct from Con A receptors in 3T3 cells. Further, capping induced with Con A-peroxidase in *Entamoeba histolytica* fails to aggregate intramembranous particles, whose distribution coincides far more closely with that of strongly acidic surface sites, which are presumably sialic acid (Pinto da Silva, Martinez-Palomo and Gonzalez-Robles, 1975). Pinto da Silva and co-workers conclude that the particles probably carry the acidic sites. Whether endocytosis followed capping by Con A-peroxidase was not reported: I shall argue that endocytosis is triggered by particle aggregation and it would be important to know whether the same mechanism can be triggered in its absence.

Nevertheless, red cell membranes do not provide the only system where particle movements can be correlated with labels which bind to the external face of the membrane. Jan and Revel (1974) have beautifully demonstrated that the ultrastructural localization of rhodopsin in outer segment membranes of retinal rods corresponds to that of intramembranous particles. The rhodopsin molecules apparently span the bilayer. Specific antibodies directed against rhodopsin aggregate the intramembranous particles, showing very clearly that the surface sites and particles are combined as kinetic units. In chloroplast membranes also (Ojakian and Satir, 1974), reversible aggregation of particles can be correlated with the extracellular magnesium concentration.

8.4 FORCES BETWEEN INTRAMEMBRANOUS PARTICLES

Molecules which act as bridges can hold the intramembranous particles together, as in the case of lymphocyte capping, characteristic of the binding of complete antibodies and bivalent F(ab')$_2$ fragments (Raff and de Petris, 1973). The modes of intramembranous particle aggregation are illustrated in *Figure 8.1*. However, bridging is not essential; any factors which reduce inter-subunit forces will tend to aggregate the particles. It is a simple but important fact that, *if the force of repulsion between them is overcome or diminished sufficiently, membrane subunits must tend to aggregate owing to van der Waals electrodynamic forces which are universally attractive between identical bodies* (Dzyaloshinskii, Lifshitz and Pitaevskii, 1961). This is true of colloid systems, and the membrane with intercalated particles can be thought of as a two-dimensional colloid. Having assumed the identity of the particles shown in freeze-fractured preparations with biochemically characterized molecules bearing sialic acid charges, we can say that these suffer mutual electrostatic repulsion which is responsible for keeping them in a dispersed state. A reduction in pH sufficient to reduce the acidic charge, or an increase in ionic strength which lowers the electrostatic potential of the charges, will allow the underlying electrodynamic attraction to pull the particles together. It is possible that strong aggregation can occur at molecular 'contact' (primary potential energy minimum) and weaker (reversible) aggregation at larger separations (secondary potential energy minimum) under appropriate conditions.

The energy G^{ed} of electrodynamic attraction between two bodies of identical material (type 2) separated by a different material (type 1) depends on geometric factors of their shape and separation and on the dielectric properties of the materials, denoted by ε_1 and ε_2.

$$G^{ed} \propto -\sum \left(\frac{\varepsilon_1 - \varepsilon_2}{\varepsilon_1 + \varepsilon_2}\right)^2 \tag{8.1}$$

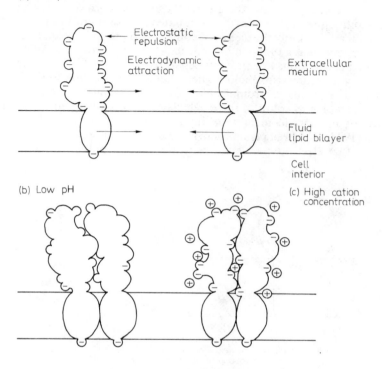
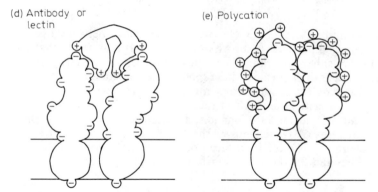

Figure 8.1 Diagrams of possible modes of interaction between mobile glycoproteins at the cell surface. (a) Except for the carboxyl-terminal group of the polypeptide chain all electrostatic charges are assumed to reside on the extracellular surface of the bilayer. Electrostatic repulsion tends to push particles apart and electrodynamic attraction tends to pull them together. (b) Low pH reduces the negative charge density on the glycoproteins. Electrodynamic forces pull them together. (c) High cation concentration shields negative charges, reduces repulsion and electrodynamic forces aggregate the particles. (d) Divalent antibody or lectin molecules can act as bridges, probably with little effect on electrostatic repulsion. (e) Polycationic molecules reduce the negative electrostatic repulsion and also act as bridges

Consequently the greater the difference in dielectric properties of the bodies and the separating medium, the greater the attraction exerted between them. The quantity $[(\varepsilon_1 - \varepsilon_2)/(\varepsilon_1 + \varepsilon_2)]^2$ is evaluated at a series of frequencies from static (where ε is simply the dielectric constant) to the ultraviolet. In principle ε can be obtained as a function of frequency from measurements of refractive index and absorption. For proteins separated by water the major contribution will come from visible and higher frequencies since the static dielectric constant of aqueous macromolecular solutions is similar to that of water. A more complete account of these calculations for planar geometries in the context of membrane–membrane interactions is given by Parsegian and Gingell (1973). The brief résumé here is intended to add some substance to the assertion that electrodynamic forces would tend to pull subunits together.

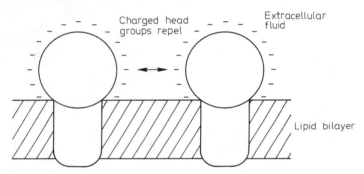

Figure 8.2 Simplified model for electrostatic interaction between mobile glycoproteins at the cell surface. The hydrophilic headgroups are assumed to be spheres of diameter 6 nm and the carboxylic charge of sialic acid groups is assumed to lie at the surface of the spheres. Anchorage by a hydrophobic polypeptide inserted into the lipid bilayer keeps the charged heads co-planar, but the distortion of spherical charge symmetry caused by this is neglected

The simplified model for electrostatic repulsion between identical negatively charged particles is shown in *Figure 8.2*. It is assumed that the glycoproteins are spheres of radius 3–6 nm (*see* Appendix) carrying a charge due to sialic acid at their surface; these spheres are embedded in the lipid by their tails but this defect in spherical symmetry is neglected. The precise shape is probably not critical as both attraction and repulsion will be affected together and only an order-of-magnitude calculation will be attempted. The repulsive energy $G^{es}(l)$ depends on the distance of centre-to-centre separation l, particle radius a, electric surface potential ψ_0, a function κ describing the ionic milieu and the dielectric constant of water (ε_w). Thus

$$G^{es}(l) = \psi_0^2 \frac{\varepsilon_w a^2}{l} e^{-\kappa(l-2a)} \qquad (8.2)$$

The energy decreases rapidly since it is negatively exponential in the separation. In the Appendix it is shown how the repulsion depends on the pH and ionic strength of the solution. Essentially, if the pH falls below the pK_a of the surface ionogenic groups, repulsion is decreased. The decrease in energy depends on ionic strength; at any given pH, increasing the ionic strength decreases repulsion.

In modelling repulsive forces between intramembranous particles no

ELECTROSTATIC CONTROL OF MEMBRANE PERMEABILITY 205

account will be taken of changes in conformation of the glycoproteins under different ionic conditions. Such second-order effects no doubt occur but it is impossible to assess their significance on even a semi-quantitative level.

8.5 RESULTS AND DISCUSSION

The electrostatic free energy of repulsion $G^{es}(l)$ is shown as a function of centre-to-centre separation and closest-approach distance $(l-2a)$ in *Figure*

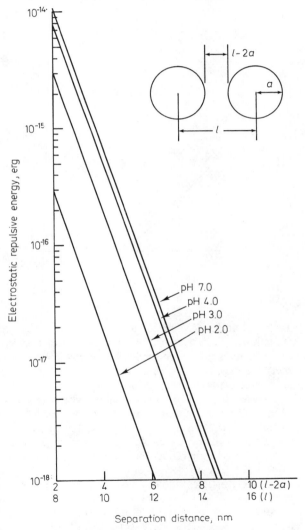

Figure 8.3 Electrostatic energy of repulsion between glycoprotein units of diameter 6 nm bathed in a solution containing a monovalent cation. The ionic strength at all pH values is 0.15. The inset diagram defines distances l and a

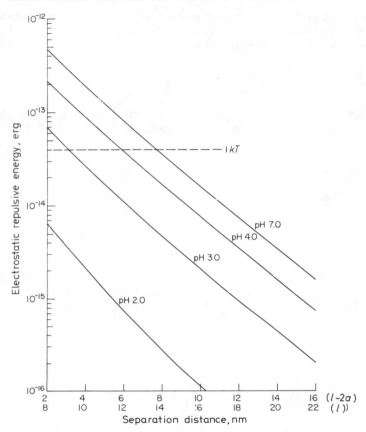

Figure 8.4 The interaction is as in Figure 8.3, but the ionic strength is 0.01

8.3 for 150 mM NaCl and in *Figure 8.4* for 10 mM NaCl. The curves for different bulk pH values lie nearly parallel on a logarithmic scale owing to the overwhelmingly exponential dependence of energy on distance.

Since the membrane particles do not undergo spontaneous and irreversible aggregation, but remain more or less randomly distributed, there is presumably a critical distance of approach at which repulsion begins to prevail over attraction. The distance must be that at which repulsion is significantly greater than the average thermal energy of collision (kT), otherwise irreversible collision would tend to aggregate the whole population. Particles separated by distances exceeding the critical distance will not be mutually adherent*. The computed curves tell us something very interesting about membrane particles when this is realized: in 150 mM NaCl the repulsive energy is equal to kT at pH 7 at some distance less than 2 nm, consequently particles in apparent 'contact' in freeze-fracture electron micrographs are not necessarily stuck together. When the pH is lowered sufficiently (just how much depends on the attractive force), particles will stick without a detectable change in separation. We may conclude that

* This is discussed further in a 'Note added in proof', p. 223.

closely apposed particles without resolvable separation seen in fracture faces fixed at physiological pH are not necessarily stuck together in the primary potential energy minimum, a fact which may account for the reversibility of particle apposition in gap junctions, discussed below.

At low ionic strength the situation is different. The electric field near the particles is higher and extends further into the bathing medium. The energy of repulsion is equal to kT at about 8 nm separation, while at pH 3 this becomes about 3 nm separation. Consequently, particles are predicted to repel each other more strongly at larger distances and should be harder to aggregate by lowering the pH at low ionic strength. The above conclusions have been stated somewhat dogmatically for clarity—details depend on the model chosen, but the general features are probably reliable guides to real behaviour.

Data with which to compare the preceding calculation are limited to the red cell ghost. The ionic conditions which initiate pinocytosis in the large free-living amoebae cannot be compared quantitatively because of the thick glycoprotein coat of unknown charge density. Direct visualization of the aggregation of membrane subunits in response to changes in the ionic milieu has been achieved by Pinto da Silva (1972) and Elgsaeter and Branton (1974). Pinto da Silva observed subunit aggregation in freeze-fractured preparations of both fixed and unfixed red cell ghosts at a series of pH values in 8 nM phosphate buffer. Particles were aggregated irreversibly below pH 4 (suggestive of primary minimum adhesion) and there was also reversible aggregation between pH 5.5 and 6 (secondary minimum adhesion), but the particles were disaggregated between pH 4 and 4.5. Elgsaeter and Branton, however, failed to find disaggregation in this region. Glutaraldehyde fixation did not influence the state of aggregation appropriate to unfixed cells at a corresponding pH. While the increase in particle aggregation with falling pH is exactly what the electrostatic model predicts, the effect sets in at a surprisingly high pH: neither chemical data from glycoproteins nor electrophoretic data from whole red cells (Heard and Seaman, 1960; Seaman and Heard, 1960; Cook, Heard and Seaman, 1961) provide evidence of surface groups which could explain particle aggregation near pH 5.5. It is clear from the calculation above that neuraminic acid could hardly be associating in this region, even at low ionic strength where the effective pK_a is probably raised nearly a whole pH unit. It is a more severe problem to account for Pinto da Silva's finding that particles disaggregate at pH 4 to 4.5 on any electrostatic basis, and it seems almost certain that the contrary result of Elgsaeter and Branton is correct, a conclusion supported by Nicolson (1973), who investigated the pH-induced aggregation of colloidal iron hydroxide binding sites on erythrocyte ghost membranes and found results closely paralleling those of Pinto da Silva with the exception that no re-dispersal occurred near pH 4.5. Trypsin, which removes a glycopeptide containing a large part of the sialic acid of glycophorin, caused irreversible clumping as judged by the behaviour of the iron label, a result clearly in accordance with electrostatic predictions. Elgsaeter and Branton (1974) also observed that intramembranous particles are strongly aggregated by trypsin and neuraminidase, both of which reduce the cell surface charge. Furthermore, these authors studied particle distribution as a function of ionic strength and found that aggregation increased smoothly with increasing ionic strength, indicating electrostatic screening and confirming the prediction that the particles experience greater repulsion at low ionic strength, for any given pH. Thus the rather limited available data on particle aggregation as a function of pH, ionic strength and enzyme action are apparently explicable in straightforward electrostatic terms, although the unexpectedly high pH for the onset of aggregation remains to be explained.

The aggregation of particles which can occur when membranes are exposed to 25% (v/v) glycerol, used as a cryoprotective agent (N.B. Gilula, personal communication), may perhaps be interpreted in terms of long-range forces. From equation 8.1 it can be seen that interparticle attraction increases as the dielectric function ε_2 of the glycoprotein headgroups becomes more dissimilar from that of the intervening medium ε_1. This would occur if the glycerol concentration were raised sufficiently, so that ε_1 (glycerol solution) $> \varepsilon_2$ (glycoprotein). In the visible part of the spectrum, for example, ε is equal to the square of the refractive index, n, so that the condition for attraction in this region would be $n_2 < n_1$ since $n > 1$ always. Spectroscopic data for constructing ε_1 and ε_2 would make it possible to calculate the critical concentration of glycerol necessary to bring about an increase in interparticle attraction.

8.6 BIOLOGICAL EVIDENCE

The foregoing sections have been essentially devoted to sketching the hypothesis and attempting to put it on a semi-quantitative footing. The following describes evidence pertinent to the idea that intramembranous particle aggregation can lead to changes in membrane permeability. This is indeed putting the horse behind the cart, but the evidence is better evaluated if the reader is already familiar with the theory.

8.6.1 Pinocytosis in amoebae

The most fully documented account of endocytosis is to be found in the case of pinocytosis by free-living amoebae. The phenomenon will be summarized fairly briefly, having already been covered elsewhere (Gingell, 1972). Amoebae exhibit pinocytosis when treated with a wide variety of proteins or other polycations which have a net positive charge near neutral pH. Negatively charged molecules or neutral molecules, such as sugars or neutral amino acids, have no detectable inducing effect. Since 'neutral' amino acids are neutral only at their pI (near pH 6), their lack of activity is presumably due to their low pK_a (near 2.1), so that even at pH 4 only 2 percent of the molecules exist in the NH_3^+ form. Low pH itself antagonizes induction, an action which may perhaps be attributable to inhibition of cytoplasmic contractility rather than a failure to alter the membrane surface potential; this requires further electrophysiological investigation.

Monovalent cations, but not anions, are potent inducers and amoebae fail to discriminate between Na^+ and K^+ in this respect. Divalent cations—calcium in particular—antagonize the inducing action of monovalent cations. It is well known, however, that external Ca^{2+} increases membrane electrical resistance and that Ca^{2+} depletion renders membranes leaky, so that Ca^{2+} may have a double action; first as a 'colloidal' counter-ion for membrane negative charges, in common with monovalent cations; secondly, in reducing permeability. Since all inducing substances which have been tested lower membrane resistance, these two functions of Ca^{2+} are mutually antagonistic.

There is, then, reasonable evidence that only positively charged ionic species induce pinocytosis. It is known that cationic macromolecules adsorb at the cell surface and can be removed by salt solutions, which is indicative of electrostatic attraction. Monovalent cations have not been shown to adsorb and their action is most likely to reduce the magnitude of the negative surface potential simply by acting as counter-ions in the double layer (*Figure 8.1*).

An exciting development in the physiology of pinocytosis was made simultaneously by Brandt and Freeman (1967) and by Gingell and Palmer (1968) (*see* Gingell, 1967, 1972), who found that inducers caused a decrease in membrane resistance, as monitored by intracellular microelectrodes. The change in membrane resistance which is caused by the adsorption of ribonuclease requires the continuous presence of the protein. Washing with salt solution, followed by removal of excess salt, returns the resistance to its initial value as shown by the removal of fluorescently labelled ribonuclease (Gingell, 1967). Similar decreases in resistance occur when pinocytosis is induced with sodium chloride. Brandt and Freeman (1967) found that calcium ions reversibly antagonize the permeability increase caused by inducers.

It is clear from these results that cationic substances which induce pinocytosis in amoebae increase membrane permeability. It is not yet known whether the ensuing ion fluxes are responsible for the morphological events of pinocytosis, but analogy with a contractile phenomenon in amphibian eggs described below makes this ripe for investigation. Amoebae are known to contain contractile actomyosin-like filaments (*see* Komnick, Stockem and Wohlfarth Botterman, 1973). If, as Bray (1973) has argued, such filaments link up with the cell membrane, perhaps with the mobile glycoproteins, a pinocytosis channel could be formed in the following manner (*Figure 8.5*): (a) the inducer reduces repulsion between the mobile subunits and they aggregate; (b) ion channels are formed, perhaps between the subunits; (c) ion fluxes trigger off contraction in actomyosin filaments connected to the units (rather like strings attached to balloons), thus pulling the membrane inwards; and (d) pinocytosis. The budding off of vesicles at the end of the pinocytosis channel may be a purely physical process like the breaking up of a tube of water at a critical length:diameter ratio. This sequence of events automatically curtails channel formation and leads to channel regression as subunits become used up by internalization.

8.6.2 **Ion permeability change in Xenopus egg membrane**

Evidence that electrostatic changes at a non-excitable cell surface can produce ion permeability changes has been obtained in eggs of the amphibian *Xenopus laevis* (Gingell and Palmer, 1968; Gingell, 1970). Adsorption of polycations such as polylysine, polyarginine and polyornithine, but not polyanions, at the cell membrane causes a decrease in resistance as measured with intracellular microelectrodes. Calcium ions move into the cell and cause contraction of actomyosin-like filaments adjacent to the cell membrane. Direct iontophoresis of calcium and, to a lesser extent, strontium and barium ions has a similar effect, producing local, spontaneously reversible, contraction; while magnesium, monovalent anions and monovalent cations have no effect. The response to calcium and magnesium is strikingly similar to that of muscle actomyosin.

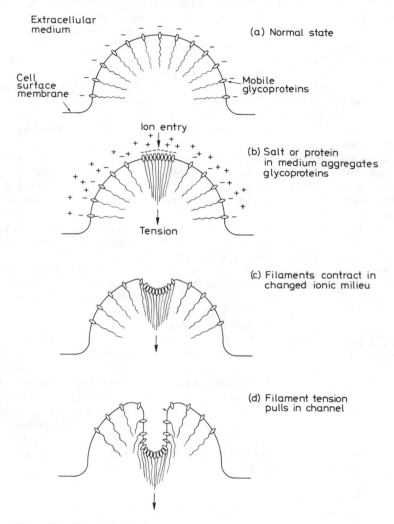

Figure 8.5 Hypothesis for formation of pinocytosis channels. (a) Mobile glycoproteins are imagined to be linked to contractile elements in the cell periphery. They are kept apart by mutual electrostatic repulsion. (b) Added salt or protein reduces repulsion between glycoproteins, which consequently aggregate. Transmembrane ionic fluxes occur at sites of aggregation. (c) The ionic fluxes initiate contraction of filaments. (d) A pinocytosis channel is formed

8.6.3 Pinocytosis in vertebrate cells

The events surrounding pinocytosis by eukaryotic cells are still less well agreed upon than in the case of pinocytosis by amoebae. This may in part be related to the use of culture media which contain serum and the fact that pinocytosis is less obvious when well-defined channels are not involved. However, as with amoebae, polycations act as inducers (Ryser, 1969), but no electrophysiological studies have yet been reported.

Within the last few years investigators at Tufts University have discovered and unravelled a fascinating sequence of biochemical events surrounding the action of a natural inducer of endocytosis in polymorphonuclear leukocytes (Nishioka *et al.*, 1973a, b). A γ-globulin called leukokinin binds as cytophilic antibody to the leukocyte surface. From this molecule a specific membrane-located enzyme then cleaves off a tetrapeptide which has been christened tuftsin and has the structure Thr–Lys(+)–Pro–Arg(+). Each molecule has two positively charged residues, making it strongly basic. Tuftsin is a strong inducer of endocytosis, both causing pinocytosis and provoking the endocytosis of particles. That tuftsin has the activity ascribed to it has been proved beyond doubt by the demonstration that the synthetic tetrapeptide perfectly mimics the action of the natural leukokinin system (Nishioka *et al.*, 1973a, b). Tuftsin has an absolute requirement for membrane sialic acid, to which it binds and thereby stimulates endocytosis (Constantopoulos and Najjar, 1973). This system very satisfyingly parallels the conditions which cause pinocytosis in amoebae and appears to act precisely as outlined in the theoretical section, though it is not yet clear whether tuftsin is simply adsorbed onto the sialic acid groups of intramembranous particles and reduces the electrostatic repulsion between neighbouring particles, or whether it is capable of physically bridging between them. Possibly both mechanisms occur. It would be of great interest to know whether intramembranous particle aggregation can be demonstrated, and whether an increase in membrane permeability also occurs.

It must be admitted that the work of Cohn (1966, 1969) has produced conclusions diametrically opposite to those of most other workers on vertebrate cell pinocytosis, in that anionic substances are considered to be potent inducers. However, his assessment of endocytosis by visual counting of a very small number of vesicles in the peripheral cytoplasm, after the lapse of more than an hour, and the use of media including calf serum which contain antibodies that are known to be inducers of pinocytosis render his conclusions questionable. More recently Seljelid, Silverstein and Cohn (1973) have described endocytosis in response to polylysine, a strongly basic polypeptide, thus confirming Ryser's observations.

8.6.4 Membrane response to antibodies

The recent discovery of the 'capping' response of small lymphocytes (Taylor *et al.*, 1971; Raff and de Petris, 1973) has provided new insight into endocytosis. Anti-lymphocytic globulin and anti-immunoglobulin labelled with a fluorescent marker rapidly show non-uniform localization ('patching') at the cell surface. The patches subsequently aggregate ('capping'), by a process that apparently requires cell movement, and the cap thus formed is eventually pinocytosed. The process bears a very close resemblance to pinocytosis in free-living cells, in which fluorescent-labelled antibodies to the cell surface, as well as less specific proteins, induce pinocytosis (Wolpert and O'Neill, 1962). However, capping is not essential to pinocytosis; Edidin and Weiss (1972) have shown that, under certain conditions, fibroblasts assume a speckled appearance in response to fluorescent antiserum before internalizing the label. It appears that aggregation of surface material into patches, but not necessarily capping, provides the trigger for pinocytosis.

Raff and de Petris (1973) reported that divalent antibody is necessary for capping although pinocytosis could be caused by monovalent Fab fragments. Similar conclusions were reached by Antoine et al. (1974): peroxidase-labelled divalent antibody directed against lymph-node cells causes patching at 4 °C and capping with pinocytosis at higher temperatures. Peroxidase-labelled Fab fragments form no discernible patches or caps at any temperature, but do induce pinocytosis. Karnovsky and Unanue (1973), however, reported that monovalent antibodies cause neither patching nor capping and that capping with divalent antibody does not necessarily lead to pinocytosis. Using the sandwich technique to prepare a compound antibody directed against H2, with the recognition sites $\sim$ 40 nm apart, they observed capping without endocytosis, from which they concluded that the sites on the membrane have to be brought into close proximity in order to stimulate pinocytosis.

Nevertheless, it would appear that sites do not need to be chemically linked for capping to occur. Stackpole et al. (1974) reported the use of hybrid $F(ab')_2$ antibody molecules which each have one specificity directed against a cell surface antigen and another directed against a foreign molecule such as ferritin, which can thus act as an electron-microscopic marker for the antibody. With one valency directed against surface IgG or H2 antigens the hybrid antibodies caused capping both with and without the addition of ferritin. Care was taken to exclude non-hybrid antibodies in these experiments as they would result in spurious cross-linking. The authors concluded that bivalent cross-bridging between intramembranous particles is not essential for aggregating them, and they speculate that combination of antibody with the surface sites reduces the negative charge on the site, causing site aggregation.

There is evidence that once antigenic sites have been aggregated by bivalent antibodies disaggregation cannot be brought about by cleaving the antibody (Loor, Forni and Pernis, 1972), indicating that electrodynamic forces, not chemical bonds, may be responsible for holding the intramembranous particles together and that the simple concept of bridging is inadequate to explain particle adhesion.

The effect of temperature on the interaction between surface antigens and antibodies resembles protein-induced pinocytosis in amoebae; at 4 °C surface binding occurs and the aggregation of mobile units is seen as patching, but neither pinocytosis nor capping occurs until the temperature is raised. Whether or not capping occurs seems to depend on the degree of cell movement and on shape changes which are temperature-dependent. That patching is not metabolically dependent is also clear from studies with metabolic inhibitors.

8.6.5 Membrane response to lectins

A remarkable experiment performed by Tosteson, Lau and Tosteson (1973) provides the most direct evidence in favour of the hypothesis presented in this chapter. These authors incorporated human red cell glycophorin into bilayers made from sheep red cell lipids and, in the course of testing their preparation for physiological viability, discovered that the agglutinin concanavalin A (Con A) caused an eightfold increase in the conductance of the voltage-clamped black film. Without glycophorin, no change in conductance occurred. It is now possible to offer an explanation for these results: the electrostatic repulsion between glycophorin molecules is overcome by the bridging action of tetravalent Con A, and the ensuing aggregation of glycoprotein particles in the bilayer is the direct cause of the increase in membrane conductance. Furthermore, it seems possible that the spontaneous fluctuations in membrane current which were observed at constant transmembrane potential difference are a manifestation of thermal diffusion of glycophorin molecules in the plane of the bilayer, forming transient clusters with associated transient excursions of membrane permeability. If this explanation for the action of Con A is correct, polycations would be predicted to have the same effect and it would be of interest to observe the response of such a preparation to polylysine.

This experiment stands in acute contrast to the confusion and disagreement characteristic of work on the interaction of lectins with living cells. The prime areas of uncertainty concern the qualitatively different effects of lectins at low and high concentrations, the relations between capping, particle aggregation and pinocytosis, as well as between patching, capping, and the action of anti-microtubule and anti-microfilament agents.

Yahara and Edelman (1973) found that Con A caps lymphocyte surface receptors at low concentration (~ 10 µg ml^{-1}) but not at high concentrations, where it inhibits capping induced by anti-immunoglobulins, presumably by acting as a competitive inhibitor. Lack of capping alone at high concentrations would at first sight appear to indicate that bridging cannot occur owing to site saturation. If this is the case, binding of Con A without bridging (analogous to hybrid antibody binding) does not reduce the charge on the intramembranous particle sufficiently to allow aggregation. However, a more complex picture emerges, since cells labelled at 4 °C with 100 µg ml^{-1} Con A, then washed and warmed to 37 °C, show capping. At 37 °C capping by Con A requires colchicine, while at 4 °C it is induced by Con A without colchicine. Since both low temperature and colchicine disrupt microtubules, the authors argue from these results that a colchicine binding system is largely responsible for holding Con A binding sites more or less rigidly at the cell surface. Their conclusion that microtubules are involved in the regulation of lectin binding-site motility is supported by the observation (Yahara, Cunningham and Edelman, 1974) that in the absence of colchicine, receptors of ferritin-labelled Con A are randomly dispersed, whereas in the presence of colchicine they are aggregated by ferritin-labelled Con A, the total number of receptors remaining constant.

The action of Con A in relation to anti-microtubule and anti-microfilament reagents has recently been carefully investigated by de Petris (1975). He finds, using conditions similar to those of Yahara, Cunningham and

Edelman (1974), the very different result that Con A strongly cross-links receptors on the lymphocyte surface and that vinblastine does not potentiate the effect, from which he concludes that microtubules are not mechanically related to Con A binding sites. Unlike vinblastine, cytochalasin B almost completely inhibits Con A capping, although some patching is visible and some pinocytosis occurs. Once caps have formed, they can be dispersed by adding cytochalasin B. These results are consistent with a process where initial clustering (patching) requires only limited surface movements which can take place at 4 °C, and is independent of cellular metabolic energy. In terms of the electrostatic model proposed earlier, this phase may correspond to the initial neutralization of the charge and results in aggregation of neighbouring Con A receptor sites. The next stage, which cannot proceed at low temperature, involves the contraction of microfilaments sensitive to cytochalasin, which by direct connection with Con A receptors, or indirectly, pull the clustered patches into larger aggregates (capping). Microtubules are known to have a cytoskeletal function, so that if they do play a role it may simply be passive resistance to cell shape changes due to forces exerted by microfilaments as they draw patches together to form caps.

A difficulty in interpreting Con A binding sites as intramembranous particles is evident from the work of Podolsky et al. (1974), who claim that Con A binds exclusively to a surface galactosyltransferase, which is probably not a component of the particles. Guérin et al. (1974) found that Con A binds almost equally to two lines of plasmocytoma cells, yet only induced clustering of intramembranous particles in one type. Karnovsky and Unanue (1973), on the other hand, were unable to find a correlation between Con A binding and intramembranous particle distribution during capping of mouse lymphocytes, and a similar conclusion was reached by Pinto da Silva and Martinez-Palomo (1975) regarding lack of particle aggregation by concentrations of Con A sufficient to cause capping of 3T3 cells.

The experiments which have been described were performed with normal tetrameric Con A. Dimers can be prepared by succinylation, yielding a derivative which binds with the same avidity as native Con A, but without causing capping (Gunther et al., 1973). However, if the degree of succinylation is reduced, producing a less strongly negatively charged dimer, it causes a response similar to the tetramer. This interesting observation would seem to warrant a simple electrostatic explanation.

On the basis of the available evidence, therefore, it seems very likely that surface receptors for Con A normally experience electrostatic repulsion which can be overcome by the molecular bridging action of Con A. The receptors may be mechanically linked to cytoplasmic contractile elements. Although capping can apparently occur without particle aggregation, suggesting that the Con A sites are not invariably on the particles, it is not clear whether pinocytosis can occur in the absence of particle aggregation, nor is it known whether permeability is related to particle aggregation, although the experiment of Tosteson, Lau and Tosteson (1973) strongly suggests that this is so. More complete evaluation of the proposed model in relation to the action of lectins must await the answers to these questions.

In relation to permeability changes and particle aggregation, there is no evidence either in amoeboid pinocytosis or in the response of lymphocytes or fibroblasts to immunoglobulins or lectins that external divalent cations enter the cell. Taylor *et al.* (1971) showed that 4 mM EDTA or EGTA had no effect on cap formation, or pinocytosis. There is indirect evidence of a permeability increase following lectin binding: Romeo, Zabrucchi and Rossi (1973) have shown that the binding of Con A to polymorphonuclear leukocytes and macrophages stimulates respiration and that blocking the binding sites for lectins prevents stimulation. They also found that Con A causes release of lysosomal enzymes into the extracellular fluid. These facts are interpreted by the authors in terms of subunit aggregation caused by lectin cross-bridging, giving rise to a change in membrane permeability.

8.6.6 Exocytosis and close junctions

The structural features of membranes immediately prior to exocytosis, and the structure of close (gap) junctions between adjacent membranes, share attributes of great interest: both show aggregations of particles similar to those seen in a more dispersed state in the rest of the membrane, and both membrane specializations can be linked with increased permeability.

Satir, Schooley and Satir (1972, 1973) have shown by freeze-fracture methods that mucus-containing vesicles are found in close proximity to the inner surface of the cell membrane, apparently just prior to exocytosis of their contents, while aggregations of particles ('rosettes') can be found both in the vesicles and in the cell membrane at the region of closest membrane–membrane approach. In the next stage, the vesicle contents swell and are expelled as the vesicle becomes confluent with the cell membrane. The authors consider that the structural rearrangement observed is responsible for a permeability change which allows an inrush of water that rapidly hydrates and swells the mucus. Remarkably similar morphological events have been found in trichocyst discharge in *Paramecium* (Plattner, 1974). Exocytosis was triggered by an influx of Ca^{2+}, using the calcium ionophore A23187. While the precise sequence of events leading to spontaneous exocytosis is not yet known, it is clear that a profound change in membrane properties goes hand in hand with subunit aggregation.

In the case of the gap junction (Gilula, Reeves and Steinbach, 1972; Bennett, 1973; Raviola and Gilula, 1975), there is good reason to believe that increased intercellular ionic permeability occurs at these sites. Freeze-fracture studies (McNutt and Weinstein, 1973) show arrays of particles of diameter 6–8 nm, similar to those disposed in the non-junctional areas, but having centre-to-centre spacings of 9–10 nm. Evidence that the particles may normally be subject to a degree of mechanical constraint comes from the studies of Raviola and Gilula (1973). They found particle aggregates surrounded by regions devoid of particles at gap junctions between rod and cone cells in the vertebrate retina. This corresponds exactly with the observation that particles stream centripetally towards cluster centres during gap junction genesis, leaving particle-free halos (Decker, 1973). Benedetti, Dunia and Bloemendal (1974) also conclude that the junctions are formed by lateral movement of pre-existing pools of membrane particles. That the clustered distribution is not necessarily fixed irreversibly is shown by the redistribution of previously aggregated particles after denervation at neuromuscular junctions (Rash, Ellisman and Staehelin, 1973). These observations are similar to those made on the distribution of acetylcholine receptor clusters in cultured muscle cells, using labelled bungarotoxin as a specific marker (Sytkowski, Vogel and Nirenberg, 1973). Denervation causes redistribution of the receptors. Freeze-fracture shows that the receptors are indeed 6–7 nm intramembranous particles (Nickel and Potter, 1973) which are denser by a factor of ten in the clusters than outside them.

In addition to the facilitation of electrical conduction at neuromuscular junctions, it is conceivable that transient particle aggregation could be responsible for the permeability changes accompanying the transmission of nerve impulses. That conformational changes in clustered protein subunits might be the basis for excitability has been suggested by Neumann, Nachmansohn and Katchalsky (1973). Peracchia (1974) has described globular arrays in freeze-etched axon surface membranes: it would indeed be fascinating to know whether the distribution of subunits is functionally related to the passage of an action potential.

8.6.7 Spontaneous aggregation

It was suggested earlier that cations from the environment can cause particle aggregation. But why particles should aggregate spontaneously during the formation of gap junctions and during exocytosis is completely unknown. There seem to be three possible mechanisms: first, their proximity to each other affects both membranes in some way which results in particle aggregation on corresponding surfaces. A mechanism of this type is apparently consistent with the results of Johnson et al. (1974), who found that particles in developing gap junctions in hepatoma cells progressively aggregate in both membranes as the distance between the membranes decreases from around 10 nm; the whole process takes about two hours and is accompanied by a corresponding increase in electrical communication. Secondly, aggregation occurs 'spontaneously' on one membrane and this induces aggregation in the opposing membrane; thirdly, aggregation proceeds independently on both membranes and the aggregated regions of different membranes make adhesive contact when chance apposition occurs. It is conceivable that 'spontaneous' aggregation (second mechanism) could be caused if several actomyosin strands leading to a group of surface subunits fan out from a common intracellular attachment point. Contraction of the strands (or movement of the attachment point) would cause centripetal surface motion of the subunits in the plane of the bilayer, thus bringing them into a cluster. Once together, having done work to overcome their mutual repulsion, the subunits might remain adherent in primary or secondary potential energy minima. So far there is no direct evidence for the attachment of intramembranous particles to contractile or skeletal elements in the cortical cytoplasm, but there is much circumstantial evidence. Bray (1973) has argued that cell movement depends on such a relationship. That contractile proteins are found in the vicinity of the cell membrane is well authenticated. Peptide sequence analysis has shown that actin is associated with the plasmalemma of 3T3 and HeLa cells (Gruenstein, Rich and

Weihing, 1973), but the authors were unable to prove chemical attachment. That some actomyosin-like material projects outside the lipid barrier is suggested by the binding of specific antibodies directed against smooth muscle actomyosin (Groschel-Stewart, Jones and Kemp, 1970).

Although there seems a definite case for the translocation of particles by molecular ropes, it is perhaps worth considering that the first mechanism proposed, whereby proximity alone causes aggregation, might be brought about by a phenomenon allied to concentration-dependent colloid flocculation; as surfaces come into contact the effective concentration of membrane subunits is raised, thus increasing the probability of collisions sufficiently energetic to result in inter-subunit adhesion*. The geometry of opposing surfaces and the diffusion constant of subunits in the bilayer would be important factors.

8.7 SYNOPSIS AND CONCLUSION

The plasma membranes of eukaryote cells include intramembranous particles of about 8 nm diameter which can diffuse in the plane of the lipid bilayer. Diffusion may in certain circumstances be limited by microfilaments attached to the particles, by means of which their distribution may be controlled endogenously by the cell. The intramembranous particles carry a strong negative charge due to sialic acid, and this is responsible for their mutual electrostatic repulsion, the size of which has been estimated. The particles however are always subjected to mutual attraction by van der Waals forces: consequently, in situations where they are sufficiently free of mechanical constraint, the particles aggregate when the electrostatic repulsion is lowered. This is thought to happen in the induction of pinocytosis in amoebae and in vertebrate cells, as well as in lymphocyte capping. The first identifiable response of large amoebae to pinocytosis inducers is an increase in ionic permeability of the membrane and there is evidence that in other situations, including exocytosis and at close junctions, aggregation of intramembranous particles is the physical basis for a similar permeability increase. In pinocytosis the cytoplasmic response to the charged ionic fluxes is cytoplasmic contraction, perhaps caused by the very mechanochemical linkages between cytoplasm and particles that can be responsible for limiting their lateral diffusion.

After considering the electrostatic model in relation to the available facts, a number of questions will occur to the reader, positive answers to any of which would be capable of casting doubts on the correctness of various aspects of the model.

1. Can intramembranous particles be made to aggregate without causing a permeability change?
2. Can a permeability change be caused by the aggregation of surface receptors *without* the concomitant aggregation of intramembranous particles?

---

* This was pointed out to the author by Dr Israel Miller of the Weizmann Institute of Science, Israel.

3. Can pinocytosis be initiated without the aggregation of intramembranous particles?
4. Can pinocytosis take place in conditions where the contraction of all microfilament classes is absolutely inhibited?

It is not suggested that the hypothesis put forward provides a rigorously watertight case. It may be argued with some justification that selected examples have been extracted from diverse experimental situations and in no single case has the complete sequence of events proposed been rigorously demonstrated on a single cell type. Nevertheless, it is felt that there exists a unifying idea linking apparently dissimilar phenomena and that out of this very diversity valuable generalizations can be extracted.

REFERENCES

ANTOINE, J.-C., AVRAMEAS, S., GONATAS, N. K., STEIBER, A. and GONATAS, J. O. (1974). *J. Cell Biol.*, **63**:12.
BENEDETTI, E. L., DUNIA, I. and BLOEMENDAL, H. (1974). *Proc. natn. Acad. Sci. U.S.A.*, **71**:5073.
BENNETT. M. V. L. (1973). *Fedn Proc. Fedn Am. Socs exp. Biol.*. **32**:65.
BRANDT, P. W. and FREEMAN, A. R. (1967). *Science, N.Y.*, **155**:582.
BRANTON, D. (1966). *Proc. natn. Acad. Sci. U.S.A.*. **55**:1048.
BRANTON, D. and MOOR, H. (1964). *J. Ultrastruct. Res.*, **11**:401.
BRAY, D. (1973). *Nature, Lond.*, **244**:93.
BRETSCHER, M. (1971). *Nature, New Biol.*, **231**:229.
COHN, Z. A. (1966). *J. exp. Med.*, **124**:557.
COHN, Z. A. (1969). *Cellular Recognition*, p. 39 *et seq*. Ed. R. T. SMITH and R. A. GOOD. Amsterdam; Appleton–Century–Crofts.
CONSTANTOPOULOS, A. and NAJJAR, V. A. (1973). *J. biol. Chem.*, **248**:3819.
COOK, G. M. W., HEARD, D. H. and SEAMAN, G. V. F. (1961). *Nature, Lond.*, **191**:44.
DANIELLI, J. F. (1937). *Proc. R. Soc. Ser. B*, **122**:155.
DECKER, R. S. (1973). *J. Cell Biol.*, **59**:74A.
DE PETRIS, S. (1975). *J. Cell Biol.*, **65**:123.
DOUGLAS, W. W. (1968). *Br. J. Pharmac. Chemother.*, **34**:451.
DZYALOSHINSKII, I. E., LIFSHITZ, E. M. and PITAEVSKII, L. P. (1961). *Adv. Phys.*, **10**:165.
EDIDIN, M. (1974). *A. Rev. Biophys.*. **4**:179.
EDIDIN, M. and WEISS, A. (1972). *Proc. natn. Acad. Sci. U.S.A.*, **69**:2456.
ELGSAETER, A. and BRANTON, D. (1974). *J. Cell Biol.*, **63**:10018.
FRYE, L. D. and EDIDIN, M. (1970). *J. Cell Sci.*, **7**:319.
GILULA, N. B., REEVES, O. R. and STEINBACH, A. (1972). *Nature, Lond.*, **235**:262.
GINGELL, D. (1967). *Ph.D. Thesis*, London University.
GINGELL, D. (1970). *J. Embryol. exp. Morph.*, **23**:583.
GINGELL, D. (1972). *Membrane Metabolism and Ion Transport*, Vol. 3, p. 317 *et seq*. Ed. E. E. BITTAR. London; John Wiley.
GINGELL, D. (1973). *J. theor. Biol.*, **38**:677.
GINGELL, D. and PALMER, J. F. (1968). *Nature, Lond.*, **217**:98.
GRANT, C. W. M. and MCCONNELL, H. M. (1974). *Proc. natn. Acad. Sci. U.S.A.*, **71**:4653.
GROSCHEL-STEWART, U., JONES, B. M. and KEMP, R. B. (1970). *Nature, Lond.*. **227**:280.
GRUENSTEIN, E., RICH, A. and WEIHING, R. (1973). *J. Cell Biol.*, **59**:127A.
GUÉRIN. C., ZACHOWSKI. A., PRIGENT. B., PARAF. A., DUNIA. I., DIAWARA, M.-A. and BENEDETTI, E. L. (1974). *Proc. natn. Acad. Sci. U.S.A..*, **71**:114.
GUNTHER, G. R., WANG, J. L., YAHARA, I., CUNNINGHAM, B. A. and EDELMAN, G. M. (1973). *Proc. natn. Acad. Sci. U.S.A.*, **70**:1012.
HEARD, D. H. and SEAMAN, G. V. F. (1960). *J. gen. Physiol.*, **43**:635.
JAN, L. Y. and REVEL, J. P. (1974). *J. Cell Biol.*, **62**:257.
JI, T. H. and NICOLSON, G. L. (1974). *Proc. natn. Acad. Sci. U.S.A.*, **71**:2212.
JOHNSON, R., HAMMER, M., SHERIDAN, J. and REVEL, J. P. (1974). *Proc. natn. Acad. Sci. U.S.A.*, **71**:4536.

KARNOVSKY, M. and UNANUE, E. R. (1973). *Fedn Proc. Fedn Am. Socs exp. Biol.*, **32**:55.
KOMNICK, H., STOCKEM, W. I. and WOHLFARTH BOTTERMAN, K. E. (1973). *Int. Rev. Cytol.*, **34**:169.
LOOR, F., FORNI, L. and PERNIS, B. (1972). *Eur. J. Immun.*, **87**:447.
MARCHESI, V. T., TILLACK, T. W., JACKSON, R. L., SEGREST, J. P. and SCOTT, R. E. (1972). *Proc. natn. Acad. Sci. U.S.A.*, **69**:1445.
MCNUTT, N. S. and WEINSTEIN, R. S. (1973). *Prog. Biophys. molec. Biol.*, **26**:45.
MORAWIECKI, A. (1964). *Biochim. biophys. Acta*, **83**:339.
MUELLER, P. and RUDIN, D. O. (1967). *Nature, Lond.*, **213**:603.
NEUMANN, E., NACHMANSOHN, D. and KATCHALSKY, A. (1973). *Proc. natn. Acad. Sci. U.S.A.*, **70**:727.
NICKEL, E. and POTTER, L. T. (1973). *J. Cell Biol.*, **57**:246A.
NICOLSON, G. L. (1973). *J. Cell Biol.*, **57**:373.
NICOLSON, G. L. and PAINTER, R. G. (1973). *J. Cell Biol.*, **59**:395.
NISHIOKA, K., CONSTANTOPOULOS, A., SATOH, P. S., MITCHELL, W. M. and NAJJAR, V. A. (1973a). *Biochim. biophys. Acta*, **310**:217.
NISHIOKA, K., SATOH, P. S., CONSTANTOPOULOS, A. and NAJJAR, V. A. (1973b). *Biochim. biophys. Acta*, **310**:230.
OJAKIAN, G. K. and SATIR, P. (1974). *Proc. natn. Acad. Sci. U.S.A.*, **71**:2052.
OSEROFF, A. R., ROBBINS, P. W. and BURGER, M. M. (1973). *A. Rev. Biochem.*, **42**:647.
PARSEGIAN, V. A. (1969). *Nature, Lond.*, **221**:844.
PARSEGIAN, V. A. and GINGELL, D. (1973). *J. Adhesion*, **4**:283.
PERACCHIA, C. (1974). *J. Cell Biol.*, **61**:107.
PINTO DA SILVA, P. (1972). *J. Cell Biol.*, **53**:777.
PINTO DA SILVA, P. and MARTINEZ-PALOMO, A. (1975). *Proc. natn. Acad. Sci. U.S.A.*, **72**:572.
PINTO DA SILVA, P., MARTINEZ-PALOMO, A. and GONZALEZ-ROBLES, A. (1975). *J. Cell Biol.*, **64**:538.
PINTO DA SILVA, P., MOSS, P. S. and FUDENBERG, H. H. (1973). *Expl Cell Res.*, **81**:127.
PLATTNER, H. (1974). *Nature, Lond.*, **252**:722.
PODOLSKY, D. K., WEISER, M. M., LAMONT, J. T. and ISSELBACHER, K. J. (1974). *Proc. natn. Acad. Sci. U.S.A.*, **71**:904.
RAFF, M. C. and DE PETRIS, S. (1973). *Fedn Proc. Fedn Am. Socs exp. Biol.*, **32**:48.
RASH, J. E., ELLISMAN, M. H. and STAEHELIN, L. A. (1973). *J. Cell Biol.*, **59**:280A.
RAVIOLA, E. and GILULA, N. B. (1973). *Proc. natn. Acad. Sci. U.S.A.*, **70**:1677.
RAVIOLA, E. and GILULA, N. B. (1975). *J. Cell Biol.*, **65**:192.
ROMEO, D., ZABRUCCHI, G., and ROSSI, F. (1973). *Nature, New Biol.*, **243**:111.
RYSER, H. J. P. (1969). *Bull. schweiz. Akad. med. Wiss.*, **24**:363.
SATIR, B., SCHOOLEY, C. and SATIR, P. (1972). *Nature, Lond.*, **235**:53.
SATIR, B., SCHOOLEY, C. and SATIR, P. (1973). *J. Cell Biol.*, **56**:153.
SCANDELLA, C. J., DEVAUX, P. and MCCONNELL, H. M. (1972). *Proc. natn. Acad. Sci. U.S.A.*, **69**:2056.
SEAMAN, G. V. F. and HEARD, D. H. (1960). *J. gen. Physiol.*, **44**:251.
SEGREST, J. P., GULIK-KRZYWICKI, T. and SARDET, C. (1974). *Proc. natn. Acad. Sci. U.S.A.*, **71**:3294.
SELJELID, R., SILVERSTEIN, S. C. and COHN, Z. A. (1973). *J. Cell Biol.*, **57**:484.
SEYBERTH, H., SCHMIDT-GAYK, H., JAKOBS, K. H. and HACKENTHAL, E. (1973). *J. Cell Biol.*, **57**:567.
STACKPOLE, C. W., DEMILIO, L. T., HAMMERLING, U., JACOBSON, J. B. and LARDIS, M. P. (1974). *Proc. natn. Acad. Sci. U.S.A.*, **71**:932.
SYTKOWSKI, A. J., VOGEL, Z. and NIRENBERG, M. W. (1973). *Proc. natn. Acad. Sci. U.S.A.*, **70**:270.
TAMM, S. L. and TAMM, S. (1974). *Proc. natn. Acad. Sci. U.S.A.*, **71**:4589.
TAYLOR, R. B., DUFFUS, W. P. H., RAFF, M. C. and DE PETRIS, S. (1971). *Nature, New Biol.*, **233**:225.
TOSTESON, M. T., LAU, F. and TOSTESON, D. C. (1973). *Nature, New Biol.*, **243**:112.
VERWEY, E. J. W. and OVERBEEK, J. TH. G. (1948). *Theory of the Stability of Lyophobic Colloids.* Amsterdam; Elsevier.
WINZLER, R. J. (1969). *The Red Cell Membrane, Structure and Function*, pp. 157–171. Ed. G. A. JAMIESON and T. J. GREENWALT. Philadelphia; J. B. Lippincott.
WOLPERT, L. and O'NEILL, C. H. (1962). *Nature, Lond.*, **196**:1261.
YAHARA, I., CUNNINGHAM, B. A. and EDELMAN, G. M. (1974). *Fedn Proc. Fedn Am. Socs exp. Biol.*, **33**:765.
YAHARA, I. and EDELMAN, G. M. (1973). *Expl Cell Res.*, **81**:143.

APPENDIX

Electrostatic interaction between particles

It is assumed that all the charge is on the outside of the membrane and the glycoprotein is in the form of spheres of diameter 6 nm (100 amino acid residues would occupy an extended length of ∼50 nm) and that there are about 27 sialic acid (carboxyl) groups on each unit; in the model these reside on the surface of an impenetrable sphere with a smeared-out distribution but the true conformation may be a less compact sphere, loosened by mutual repulsion between carboxyl groups, or a more cylindrical form. Some counter-ion penetration will also doubtless occur, reducing the external field. Consequently the interparticle force of repulsion to be calculated will be an upper estimate.

For spheres whose radius, a, is comparable with the double layer thickness of Debye length $(1/\kappa)$ Verwey and Overbeek (1948) derived the approximate interaction potential for small values of κa ($\kappa a < 3$) at constant charge as

$$G^{es}(l) = Q[\psi_0(l) - \psi_0(\infty)] \simeq \psi_0^2(l) \frac{\varepsilon a^2}{l} e^{-\kappa(l-2a)} \qquad (8.A.1)$$

where Q is the surface charge, $\psi_0(l)$ is the surface potential at interparticle centre-to-centre separation l, $\psi_0(\infty)$ is the surface potential for infinite separation and ε is the dielectric constant of water. The Debye length $(1/\kappa)$ is defined by

$$\kappa^2 = \frac{8\pi I e^2}{\varepsilon k T} \quad \text{and} \quad I = \frac{1}{2}\sum_i c_i z_i^2.$$

I is the ionic strength and c_i the concentration of ions of type i of valency z_i. Equation 8.A.1 holds when $\kappa(l-2a) \gg 1$. In order to calculate $G^{es}(l)$ we need to know the relation between surface charge and surface potential. For small values of κa,

$$\psi_0(l) = \frac{zne}{a\varepsilon}\left[1 + (1-e^{-2\kappa a})\frac{e^{-\kappa(l-2a)}}{2\kappa l}(1+\alpha)\right] \Big/ (1+\kappa a)\{1-\delta(1-\alpha)\} \qquad (8.A.2)$$

where z is the valency and n the number of ionized groups on the glycoprotein.

where

$$\alpha = \lambda_1\left(1 + \frac{1}{\kappa l}\right) + \lambda_2\left(1 + \frac{3}{\kappa l} + \frac{3}{(\kappa l)^2}\right)$$

$$\lambda_1 = -\frac{\kappa^2 a^3}{l} e^{-\kappa(l-2a)}$$

and

$$\lambda_2 = \frac{\lambda_1}{9}$$

$$\delta = \frac{e^{-\kappa(l-2a)}}{2\kappa l}\left(\frac{\kappa a - 1}{\kappa a + 1} + e^{-2\kappa a}\right)$$

It is found that the expressions for λ_1 and λ_2 hold when the distance of closest approach $(l-2a)$ exceeds the Debye length $(1/\kappa)$. When $l/a > 4$ (i.e.

$l > 12$ nm) equation 8.A.2 can be usefully simplified to a function independent of l, viz.

$$\psi_0 = zne/a\varepsilon(1+\kappa a) \qquad (8.A.3)*$$

This is because the potential becomes a strong function of separation only when the closest approach equals a Debye length or so. In 10 mM NaCl and 100 mM NaCl the Debye lengths are approximately 3 and 1 nm respectively.

The surface potential of the glycoprotein headgroup at infinite separation from any other is found from equation 8.A.3 to be -45 mV in 10 mM NaCl, -22.5 mV in 100 mM NaCl and -18.9 mV in 150 mM NaCl. This result supports the rather arbitrary model of the glycoprotein as a sphere of 3 nm radius, insofar as it suggests that the surface charge density is not likely to be lower (i.e. the surface area is unlikely to be larger), since the zeta potential of human red blood cells in 0.145 M NaCl is about -12.7 mV. While the calculation of zeta potentials takes no cognizance of unevenly distributed charge, the surface potential of the glycoprotein must be greater in magnitude than the zeta potential. Differentiation of equation 8.A.1 gives the force

$$F^{es}(l) = \varepsilon a^2 \psi_0^2(l) \left(\frac{1}{l^2} + \frac{\kappa}{l}\right) e^{-\kappa(l-2a)} \qquad (8.A.4)$$

pK_a as a function of ionic strength and pH

Before the above equations can be used to calculate interaction forces and energies it is necessary to know how the surface charge density varies with pH. The pH near a charged surface differs from that of the bulk liquid (Danielli, 1937). The difference is a strong function of ionic strength while the pK_a of dissociable groups responsible for the surface charge is a function of surface pH and ionic strength. Near a biological membrane, dissociated carboxyl groups of sialic acid give a negative electrostatic potential ψ_0 which attracts hydrogen ions according to the Boltzmann relation

$$[H^+]_{\text{surface}} = [H^+]_{\text{bulk}} e^{-ze\psi_0/kT} \qquad (8.A.5)$$

The potential ψ_0 set up by ionized groups is related by equation 8.A.3 to the density of such groups, σ, on a spherical surface of radius a, to a first approximation. Consequently the magnitude $|-\psi_0|$ is increased as that of $|-\sigma|$ increases and as the ionic strength decreases. From equation 8.A.5 it follows that $[H^+]_{\text{surface}}$ increases as $|-\sigma|$ increases and as ionic strength decreases. The dissociation of surface groups is described by a mass-action equation of the type

$$[RH] \underset{}{\overset{K_a}{\rightleftharpoons}} [R^-] + [H^+]_s \qquad (8.A.6)$$

where K_a is the thermodynamic dissociation constant. From this the Henderson–Hasselbach equation follows immediately

* When $a \to \infty$ the expression describes the linearized potential of a flat plate, since $n = 4\pi a^2 \sigma/e$, where σ is the charge density.

$$\mathrm{pH_s} = \mathrm{p}K_a - \log_{10}\frac{[\mathrm{RH}]}{[\mathrm{R}^-]} \qquad (8.\mathrm{A}.7)$$

where [RH] and [R$^-$] are surface concentrations. If α is the degree of dissociation, then from equation 8.A.7

$$\alpha = \frac{1}{1+[\mathrm{H}^+]_s/K_a} \qquad (8.\mathrm{A}.8)$$

Writing the concentration of dissociable acid groups (assuming only one kind for simplicity) per unit area as σ_0 we have $\sigma = \sigma_0 \alpha$. Substituting equation 8.A.3 into 8.A.5, we obtain

$$[\mathrm{H}^+]_s = [\mathrm{H}^+]_{\mathrm{bulk}} \exp\left[-4\pi z e \sigma_0 \alpha/kT\kappa\varepsilon\left(1+\frac{1}{\kappa a}\right)\right] \qquad (8.\mathrm{A}.9)$$

Equation 8.A.8 expresses the fact that the degree of dissociation is a function of surface pH while equation 8.A.9 expresses the fact that surface pH is a function of the degree of dissociation, ionic strength and bulk pH. [H$^+$]

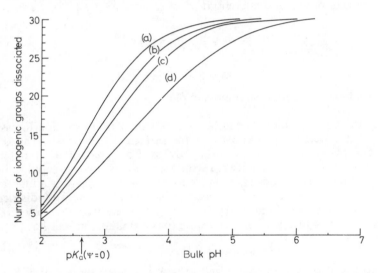

Figure 8.6 Number of ionogenic groups dissociated (COO$^-$) per glycoprotein unit as a function of pH in the bulk phase. The maximum is 30. (a) Ionic strength = 0.1, sphere diameter = 12 nm; (b) ionic strength = 0.01, sphere diameter = 12 nm; (c) ionic strength = 0.1, sphere diameter = 6 nm; (d) ionic strength = 0.01, sphere diameter = 6 nm

is eliminated between these two equations and the resulting expression is solved for α. The surface potential ψ_0 is then calculated from equation 8.A.1 or 8.A.3 using the relationship $n = 4\pi a^2 \sigma_0 \alpha/e$. Finally, interaction force and energy are calculated from equations 8.A.4 and 8.A.1.

Figure 8.6 shows the calculated number of dissociated surface charges per glycoprotein molecule, as a function of pH and ionic strength, assuming that each molecule bears 30 dissociable carboxyls of sialic acid. It will be seen that the optimum pH for the half-charged state can be displaced more than a whole unit due to the effect of surface potential shifting the pK_a. Surface potential as a function of pH and ionic strength is illustrated in *Figure 8.7*. These curves demonstrate that association of charges can be appreciable below pH 6 and that increasing the ionic strength markedly reduces the electric potential. The size of the particle also determines the charge density since the number of dissociable groups per particle is considered fixed.

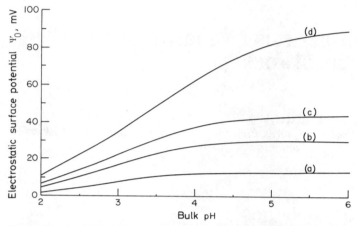

Figure 8.7 Surface electrostatic potential of glycoprotein unit as a function of pH in the bulk phase. (a)–(d) as in Figure 8.6

Note added in proof (*see* p. 206)

A 'secondary minimum' adhesion (adhesion at finite separation) can occur only at a distance where the attraction alone exceeds thermal energy, kT. Bodies would move closer than this, but repulsion at this distance, which must also exceed kT, prevents it. So a *stable* secondary minimum must be closer than the separation where repulsion alone equals kT. Computations which include electrodynamic attraction in salt solutions of concentration 150 mM or less at pH 7 show that in fact there will not be stable secondary minimum adhesion between glycoproteins. The electrostatic repulsion dominates to such an extent that it alone is a good approximation to the net interaction energy.

9
Biogenesis of mammalian membranes

S. K. Malhotra
Biological Sciences Electron Microscopy Laboratory, University of Alberta, Edmonton

9.1 INTRODUCTION

The rapid progress in our understanding of the biogenesis of cellular membranes during the last ten years has been stimulated by developments in cytoplasmic genetics. The presence of DNA in mitochondria was suggested in the late 1950s on the basis of autoradiography and Feulgen staining (*see* Roodyn and Wilkie, 1968). The biochemical confirmation of this in the early 1960s (Luck and Reich, 1964) has resulted in the application of biochemical genetics to the understanding of the problem of the biogenesis of mitochondria (Getz, 1972).

The mitochondrial genetic system is relatively simple, and could be utilized for the study of basic problems in cell biology, for example replication of DNA, synthesis of RNA, structure and function of ribosomes, and biosynthesis of cellular membranes (Borst, 1972).

The assembly of functional membranes may take place at a site removed from the site(s) of initial synthesis and an understanding of the mechanism of translocation may require the investigation of the passage of membrane components within and/or across membranes.

Transport within membranes depends upon the interrelationship between chemically and/or morphologically characterizable membranes of a cell. The process of assembly would provide information on the interactions amongst molecular components that regulate the various vital phenomena associated with membranes.

Some of the questions pertinent to the biogenesis of membranes are thus:

1. Are membranes formed *de novo* in nature?
2. How are functional membranes assembled—is it a single-step or a multiple-step process?

3. Is the assembly of membranes an ordered process or does it occur by random aggregation of components?
4. How does a membrane grow?
5. Can an 'old' membrane be distinguished from a 'new' membrane?
6. Where are the constituents synthesized?
7. Is there a functional mosaicism within a single structurally characterizable membrane?
8. How does a cell control the biogenesis of its membranes?
9. What is the mechanism of cellular transformation produced by viral or chemical agents (Singer and Rothfield, 1973)?

Some of these questions have been answered; information about others is rapidly becoming available.

The assembly of membranes could be the most complex of the supramolecular cellular processes (Fox, 1972). Most cellular membranes are composed of many molecular species which interact in a specific manner to determine the functions of the membrane. However, the chemical composition of a membrane can be easily altered by changing the composition of the culture medium (Luck, 1965) or the diet (Dallner, Siekevitz and Palade, 1966b; Siekevitz, 1972). There is apparently no concomitant change in the function of these membranes or in their structure in thin sections examined by electron microscopy (EM). However, freeze-fracturing techniques have detected some alterations in the core of the membrane in *Mycoplasma* (Tourtellotte, Branton and Keith, 1970) as a consequence of changes in chemical composition.

9.2 EXAMPLES OF MODEL SYSTEMS FOR MEMBRANE BIOGENESIS

Microorganisms have been extensively used for investigations related to the biogenesis of membranes; such organisms include bacteria, yeast, *Neurospora* and *Chlamydomonas*. Their obvious advantages are: (a) the ease of growing large homogeneous populations of cells to obtain pure membranes; (b) the ability to apply the techniques of genetics to obtain mutants with specified alterations in the composition and/or function(s) of a particular membrane; (c) the ability to control growth by altering the conditions of the culture; (d) the ability to study synthesis *in vivo* by the application of labeled precursors. Studies on these organisms have been valuable in investigating the formation of mitochondria in yeast from 'promitochondria' (Criddle and Schatz, 1969), multiplication of mitochondria by growth and division in *Neurospora* (Luck, 1965), and the dual genetic control of biosynthesis of mitochondria in yeast (*see* Roodyn and Wilkie, 1968; Beattie, 1971; Schatz et al., 1972).

Amongst mammalian cells, liver cells have been extensively used to investigate biogenesis; they provide a normal system for studying the development and differentiation of membranes since they proliferate just before, and within a few days after, birth.

The smooth endoplasmic reticulum (SER), which is either lacking or scanty before birth (*Figures 9.1a* and *9.2*), develops within two to three days of

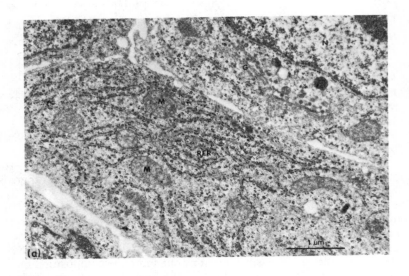

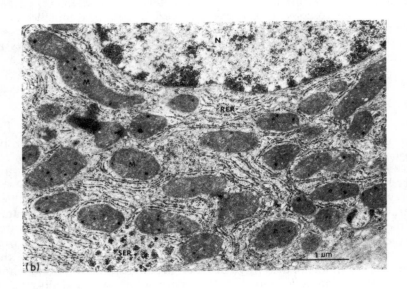

Figure 9.1 (a) Electron micrograph from a thin section of rat liver 3 days before birth. Note the irregularly dispersed cisternae of rough endoplasmic reticulum (RER) in the hepatocytes. (b) A similar electron micrograph 3 days after birth. The following changes are apparent in the structure of the hepatocytes when compared with the structure before birth: the appearance of the smooth endoplasmic reticulum (SER), the increased number of mitochondria, and formation of dense granules in them, which are thought to be lipid in nature (Afzelius, 1973). N, nucleus; G, Golgi body; M, mitochondrion

birth almost to its adult level (*Figure 9.1b*; Dallner, Siekevitz and Palade, 1966a, b; Siekevitz, 1972). Also, there is marked mitochondrial activity a few days before (*Figure 9.1a*) and after birth (*Figure 9.1b*); for example, the level of cytochrome oxidase rises rapidly during the last four days of gestation (Herzfeld, Federman and Greengard, 1973), the mitochondrial content doubles in the immediate postnatal period (Jakovcic *et al.*, 1971), and the number of mitochondria reaches the adult level by the end of the second postnatal day (Herzfeld, Federman and Greengard, 1973).

Mitochondria in liver at this period of their activity may be a useful system to investigate biogenesis, and this may be extended to the differentiation of the hepatocyte. More specifically, an answer could be sought to the question of whether there is a direct continuity between the endoplasmic reticulum (ER) and the outer mitochondrial membrane (MM) (*see* Roodyn and Wilkie, 1968; Tewari and Malhotra, 1973).

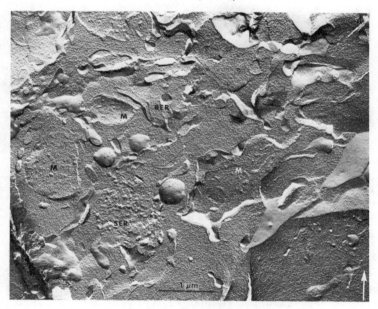

*Figure 9.2   Freeze-fracture replica of rat 1 day before birth showing the formation of smooth endoplasmic reticulum (*SER*). The arrow in the bottom right-hand corner indicates the direction of shadowing in all freeze-fracture preparations. M, mitochondrion*

Besides differentiating hepatocytes, other mammalian cells have also been utilized to obtain information on membrane biogenesis. For example, in rat heart a twofold increase in cytochrome oxidase occurs during the first ten days of development (Warshaw, 1972) and the respiratory enzyme content of adult rat brain is three and a half times as great as that of the neonate (Gregson and Williams, 1969). However, the interpretation of data from such heterogeneous organs could be complicated by the fact that the mitochondrial activity may differ greatly in different cells (Hamberger, Blomstrand and Lehninger, 1970) and also in different parts of the same cell type, as between the perikaryon and nerve ending (Hajos and Kerpel-Fronius, 1969).

Another approach that has been extensively employed for studies of membrane biogenesis is to induce alteration in membranes by drugs such as phenobarbital (Higgins and Barrnett, 1972; Kuriyama *et al.*, 1969; Siekevitz, 1972) and ethionine (Goldblatt, 1972), or hormones (Gustafsson *et al.*, 1965; Tata, 1967; Gross, 1971), or to induce regeneration by partial hepatectomy (Tata, 1970). Again, the mammalian hepatocyte has been widely used as a model system since, following induction by drugs, there is a marked proliferation of the SER and an increased accumulation of liver lipid and of microsomal enzymes, particularly those involved in detoxification of drugs (*see* Goldblatt, 1972). In hepatic regeneration and hormone-induced growth, there is a coordinated formation of rough endoplasmic reticulum (RER) (associated with ribosomes) and an increase in protein synthesis, which implies acceleration of phospholipid and ribosome synthesis (Tata, 1970).

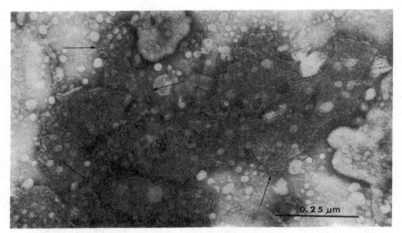

Figure 9.3 Preparation of isolated beef heart mitochondria negatively stained with sodium phosphotungstate to show the stalked particles (arrows) on the inner mitochondrial membrane

Acute exercise has also been employed to alter the normal activity of membranes. For example, Hamberger, Gregson and Lehninger (1969) observed that mitochondria isolated from heart and brain of rabbits subjected to swimming showed an increased incorporation of labeled amino acid into the mitochondrial protein.

Another model system that has been extensively explored by EM is the myelin sheath, formed from extension of the plasma membrane (PM) of a glial cell. However, myelin sheath is highly specialized, as it contains exceptionally high levels of lipid (O'Brien and Sampson, 1965; *see* Malhotra, 1970), and very low levels of protein (O'Brien, 1967). Thus, this could be a useful model for particular studies, such as lipid synthesis (Benes, Higgins and Barrnett, 1973), but not of general problems in membrane biogenesis.

The reconstitution of membranes *in vitro* has been successfully employed for partial reconstitution of oxidative phosphorylation from components of the inner mitochondrial membrane (Fessenden-Raden and Racker, 1971;

Kagawa, 1972; Kagawa, Kandrach and Racker, 1973). For example, several protein factors which seem to be required to restore oxidative phosphorylation have been isolated. Only one of these coupling factors (F_1) has been shown to have catalytic activity; it is an ATPase (Fessenden-Raden and Racker, 1971). It should be remarked that this F_1 ATPase corresponds to the well-known 9 nm stalked particles, first described by Fernández-Morán (1962) (see Malhotra and Eakin, 1967), and which are characteristically seen in negatively stained preparations on one side of the inner membranes of mitochondria in isolated fractions (Figure 9.3). Another coupling factor confers oligomycin sensitivity on the F_1 ATPase.

It has also been reported that both phosphatidylcholine and phosphatidylethanolamine are required to reconstitute membranous vesicles which are active in oxidative phosphorylation, and that these phospholipids should carry unsaturated fatty acid groups (Kagawa, Kandrach and Racker, 1973). It is apparent that this type of reconstitution of mitochondria and of other membranes, for example vesicles of sarcoplasmic reticulum (Meissner and Fleischer, 1973), has been valuable in resolving some of the problems of membrane organization, as have the studies on artificial black lipid membranes pioneered by Mueller and Rudin (see Malhotra, 1970; Finkelstein, 1972; Wallach, 1972). However, further discussion in this chapter will focus on the synthesis, assembly and interrelationships between the membranes of a cell.

9.3 GENERAL REMARKS ON THE BIOGENESIS OF MEMBRANES

Although black membranes, which resemble natural membranes in some of their properties (Finkelstein, 1972), have been prepared from lipid with or without nonlipid additives, it is generally believed that all cellular membranes are assembled by inserting newly synthesized components into preexisting membranes (Singer and Rothfield, 1973). Although it has not been ruled out, there is no adequate evidence of de novo synthesis of membranes in vivo. Even in the case of membrane-bearing virus particles, such as influenza virus, the membrane of the virus buds off from the PM of the host cell (Choppin, 1973).

The biogenesis of a membrane begins with the synthesis of its components, and information is rapidly emerging about the sites of synthesis of the components of various cellular membranes. For example, the bulk of the mitochondrial proteins are synthesized in relation to the extramitochondrial ribosomes under the genetic control of the nucleus (Roodyn and Wilkie, 1968; Beattie, 1971), while most of the protein of the membranes of the ER may be synthesized in situ on membrane-bound ribosomes (Leskes, Siekevitz and Palade, 1971a, b; Siekevitz, 1972; Eriksson, 1973).

Lipids may be synthesized at a site some distance from the site of membrane formation. For example, the major nitrogen-containing phospholipids of mammalian membranes such as phosphatidylcholine and phosphatidylethanolamine appear to be synthesized in relation to the ER (Stein and Stein, 1969; Higgins and Barrnett, 1972; Dennis and Kennedy, 1972). However, it has also been reported that isolated rat liver mitochondria rapidly incorporate labeled precursors into phospholipids (Kaiser, 1969). PM fractions of rabbit reticulocytes have been thought to be active in the synthesis of phosphatidylcholine and phosphatidylinositol. These findings are of interest because reticulocytes lose their intracellular organelles during their transformation to erythrocytes, in which *de novo* synthesis of these two lipids ceases (Percy et al., 1973). Also, acyltransferase activity involved in the biosynthesis of phospholipids has been detected histochemically and biochemically in the PM of the Schwann cell during formation of the myelin sheath (Benes, Higgins and Barrnett, 1973). The PM also possesses the capability to synthesize phosphatidic acid and phosphatidylglycerol (Victoria et al., 1971). Apart from the difficulties arising from contaminants in the isolated membrane fractions, analysis of lipid synthesis is complicated by the fact that interchange of lipids may take place between the various membranes in a cell (Murray and Dawson, 1969; Wirtz and Zilversmit, 1969; Williams and Bygrave, 1970). Irrespective of these reports, it is likely that the mitochondria play a role in the synthesis of lipids in some cells; for example, they provide the only system for fatty acid synthesis in the heart (Warshaw and Kimura, 1973) and have been reported to synthesize their own cardiolipin in the liver of guinea-pig (Davidson and Stanacev, 1971) and rat (Dennis and Kennedy, 1972). They are capable of synthesizing glycolipids, and separation of the outer and inner MMs indicates that the incorporation of [^{14}C]mannose and -glucose into the glycolipids occurs at the outer MM (Bosmann and Case, 1969). It is concluded from the above that the lipids of the membranes are synthesized at various sites in the cell, although the major phospholipids are provided by the ER in most mammalian cells.

Apart from the small, but important, contribution to protein synthesis by the mitochondria it is assumed that all membrane proteins are synthesized by the cytoplasmic ribosomes, whether membrane-bound or free (Lodish, 1973; Lowe and Hallinan, 1973). The proteins may become modified by removal of amino acids upon incorporation into the membrane as proposed by Lodish (1973). However, one report which should be further investigated, indicates that the PM fraction isolated from mouse L-cells contains 'ribosome-like particles' and that the incorporation of labeled amino acids into large polypeptides in this fraction is inhibited by those inhibitors which block protein synthesis in the microsomal fraction (Glick and Warren, 1969). These results are of interest if, indeed, the PM fraction is pure and devoid of microsomes derived from the ER. Another possibility is that the incorporation of [^{3}H]leucine into the PM of rat liver cells *in vivo* continues at undiminished rates for at least three hours following the inhibition of protein synthesis by cycloheximide (Ray, Lieberman and Lansing, 1968), suggesting that a precursor protein is formed long before its incorporation into PM. Furthermore, there is the possibility that this protein synthesis may be independent of the bulk of cellular protein synthesis and not inhibited by cycloheximide, as is true for some of the proteins synthesized by mitochondria (Korn, 1969).

The formation of membranes and the increase in demand for protein synthesis seem coordinated through an unknown mechanism. The enhanced protein synthetic activity induced by thyroid hormone or by partial hepatectomy is most noticeable after the appearance of additional cytoplasmic ribosomes attached to the ER (Tata, 1967, 1970). At the same time, there is an increase in the amount of RER as estimated by uptake of labeled choline and

^{32}P into phospholipids. A similar proliferation of cellular membranes is a feature in normal development of mammalian liver (Dallner, Siekevitz and Palade, 1966a, b; Siekevitz, 1972) and other situations that demand increased synthesis of cellular proteins not meant for export, such as viral infection and renal hypertrophy following unilateral nephrectomy (*see* Tata, 1971). Treatment by drugs, such as phenobarbital, also results in an increased incidence of hepatocyte membranes, particularly SER (Goldblatt, 1972; Siekevitz, 1972), and in some enzymes such as NADPH–cytochrome *c* reductase, with a parallel decrease in NADH–cytochrome *c* reductase (Siekevitz, 1972; Eriksson, 1973). The increase in enzymatic activity may result from actual increase in the rate of synthesis or a decrease in degradation (Kuriyama *et al.*, 1969).

Information is scarce concerning the events that take place following the initial synthesis of membrane components, and that result in the production of functional membranes. One of these events is likely to be translocation of the components to the site(s) of assembly of the membrane, when different from the site of synthesis, and the other is the insertion of the components into the existing membrane.

The transport of lipid and protein from the site of synthesis may be facilitated by rapid lateral diffusion of the newly incorporated molecules within the membrane (Palade, 1973); lipid molecules are known to migrate laterally in membranes on account of their highly fluid state (Kornberg and McConnell, 1971), as do fluorescent labeled antigens in PMs of heterokaryons produced from mouse and human cells (Frye and Edidin, 1970) and in cultured muscle fibers (Edidin and Fambrough, 1973). In this respect, continuity between the ER and the outer MMs seen in electron micrographs of thin sections (*Figure 9.4a*; Roodyn and Wilkie, 1968; Tewari and Malhotra, 1973) and replicas of freeze-fractured preparations (*Figure 9.4b*; Tewari and Malhotra, 1974) is of interest. Such a continuity may serve to transfer, to the mitochondria, phospholipids and proteins synthesized in relation to the RER and destined for incorporation into MMs. The components of the inner MM will have to be somehow transported from the outer MMs to their site of functional assembly. If a continuity between two such biochemically distinct membranes as ER and outer MM is indeed a natural phenomenon, it is of interest to know the mechanism that keeps the components of membranes from intermixing. In those cases where membrane fusion–fission has been used to explain transport of secretory proteins between distinct compartments, only the secretory proteins may be transported, and not integral membrane components (Palade, 1973). This suggestion is consistent with the report that labeled phosphatidylinositol is confined to the ER throughout the entire secretory cycle and beyond, during stimulation of enzyme secretion in guinea-pig pancreas (Gerber, Davies and Hokin, 1973).

From the data presently available on the process of assembly, it can only be stated that a cellular membrane is dynamic, with each of its various components capable of being replaced, there being no harmony in the rates of turnover of the components. The proteins and lipids turn over at rates characteristic of each type of protein and lipid (Siekevitz, 1972; Eriksson, 1973; Schimke, 1973). These rates of turnover can be modified, as is illustrated by the increase in NADPH–cytochrome *c* reductase and the decrease in the NADH–cytochrome *c* reductase in the liver following

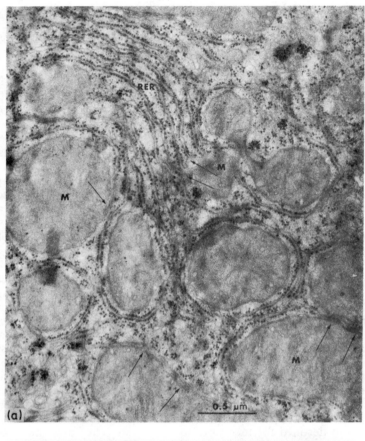
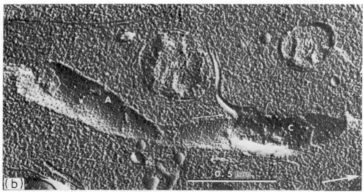

Figure 9.4 (a) Thin section of an adult rat hepatocyte showing possible association (arrows) between rough endoplasmic reticulum (RER) and the outer mitochondrial membrane. (b) Freeze-fracture replica of a rat hepatocyte 1 day before birth showing a direct continuity (arrow) between the endoplasmic reticulum and the outer mitochondrial membrane. A and C, fractures of the outer and inner mitochondrial membranes respectively; M, mitochondrion

phenobarbital treatment of the animal (Kuriyama *et al.*, 1969; Eriksson, 1973). The proliferation of membranes following drug treatment may also represent a decrease in breakdown of phospholipid compared with the normal hepatocyte (Goldblatt, 1972). The phospholipid and protein components of the MMs have been reported to have similar patterns of incorporation, suggesting that the building units are assembled as lipoprotein complexes in these membranes (Beattie, 1969). The newly synthesized components appear

Figure 9.5 Freeze-fracture replicas of rat hepatocytes (a) 3 days before birth, and (b) adult. Note that the density of particles is higher on fracture of the plasma membrane (PM) from adult rat (b) than on the corresponding fracture from a rat 3 days before birth (a). GJ, gap junction

to be inserted at random into the preexisting membranes so that the newly formed membranes are dispersed amongst the old membranes (Leskes, Siekevitz and Palade, 1971a, b; Siekevitz, 1972; Storrie and Attardi, 1973a). If discrete growth centers exist in the membranes, they have not yet been characterized. The apparently random insertion of newly synthesized components may be understandable if they rapidly diffuse laterally upon incorporation into membranes, owing to the latter's fluidity (Palade, 1973).

In relation to biogenesis, it should be mentioned that changes in the structure of cellular membranes during developmental differentiation of rat hepatocyte have been demonstrated in electron micrographs of freeze-fractured preparations (*Figure 9.5*; Tewari and Malhotra, 1974), although no such differences may be apparent in micrographs of thin sections (*see* Siekevitz, 1972, 1973). Major components of the PM (protein, phospholipid, carbohydrate) may double in quantity during interphase in cultured mouse and hamster cells to revert to the original level at cytokinesis (C.A. Pasternak, personal communication).

9.4 EVIDENCE FOR MOSAICISM IN MEMBRANES

Does a membrane represent an assembly of discrete functional areas, or a random assembly of diverse functional units? In organized tissues there is a discrete functional and structural differentiation of the PM in different locations. For example, in the proximal convoluted tubules of the kidney, where 80 percent of the glomerular filtrate is reabsorbed, the apical PM's functions are different from those of the PM at the base of the cell, where Na^+ and glucose are actively transported (Latta, Maunsback and Osvaldo, 1967; Hamburger, Richet and Grunfeld, 1971). By using freeze-fracturing techniques it has been shown that there is a higher density of macromolecules on the fractured faces of the basal PM than on the apical PM (*Figure 9.6a, b, d*; Tewari and Malhotra, 1974). This increased content may be related to a greater enzymatic activity at the basal region. The structure of the PMs of these two regions is also different from the nonjunctional PM on the lateral side (*Figure 9.6c*), as shown by the distribution of particulate material in a freeze-fractured preparation. Structural differences between luminal and lateral PM in pancreatic acinar cells have been reported by application of freeze-fracturing techniques (Camilli, Peluchetti and Meldolesi, 1974). Similarly, this type of differentiation is well known in the case of the RER and SER. Although the nuclear membrane is morphologically continuous with the ER there are distinct chemical and functional differences between these two membrane systems (Monneron, Blobel and Palade, 1972), the former having lower activities of glucose-6-phosphatase (G-6-Pase), Na^+,K^+-ATPase, cytochrome b_5 and cytochrome P_{450} (Franke *et al.*, 1970) and the latter having an esterified cholesterol content four times as high as that of the microsomal fraction in pig liver (Kleinig, 1970).

Is there any further functional differentiation within the two categories of ER? There does not seem to be a simple answer to this question. There may be differentiation in respect of a particular component or a particular functional unit but the same membrane system may lack any apparent differentiation with respect to another function. This is illustrated by several examples. The distribution of G-6-Pase *in situ* and in the microsomal fraction of developing hepatocytes of rats has been studied by using a cytochemical test involving precipitation of phosphate with lead ions (Leskes, Siekevitz and Palade, 1971a, b). The reaction product (lead phosphate) was distributed evenly over the entire surface of the RER following the first appearance of G-6-Pase and throughout the entire SER, which is formed just before or during the first 24 hours after birth. After incubating the

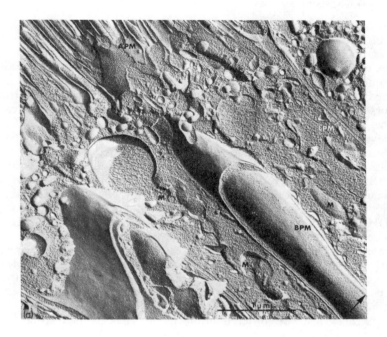

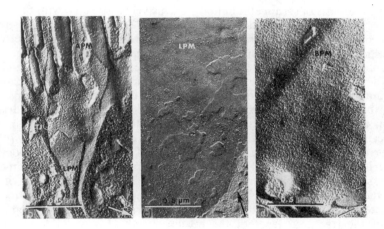

Figure 9.6 Freeze-fracture replicas of a proximal convoluted tubule cell of mouse kidney. Micrographs (b)–(d) show magnified views of the fractures of (a) apical (APM), (b) and (c), lateral (LPM) and (d) basal (BPM) plasma membrane. Note the heterogeneity in organization of the plasma membrane in these three locations. Fracture of BPM shows the highest density of particles, followed by the corresponding fractures of APM and LPM. Also note that large areas of the fracture of LPM are devoid of particles. TJ, tight junction

microsomes in the medium for the cytochemical test, attempts were made to subfractionate them into two categories, one containing lead—indicating localization of G-6-Pase—which would have higher density than the microsomes without lead. The results indicated that almost all the microsomes contained lead deposits, while at the same time most of the hepatocytes showed G-6-Pase distributed throughout the entire ER *in situ*. These results can best be interpreted to mean that the ER lacks differential areas in respect of the distribution of G-6-Pase (Leskes, Siekevitz and Palade, 1971b). However, as discussed by these authors, this apparent lack of differentiation in the microsomes may be due to the rather large size of the microsomal vesicles (average 120 nm). By repeated sonication of the liver microsomes, Dallman *et al.* (1969) produced microsomal vesicles one hundredth of their original volume, and attempted a separation of NADH-linked enzymes from the NADPH-linked group. Their initial findings suggest that the NADH-linked enzymes (NADH–cytochrome c reductase, NADH–ferricyanide reductase and cytochrome b_5) are more concentrated in the layer of rapidly sedimenting vesicles, whereas the NADPH-linked enzymes (NADPH–cytochrome c reductase and cytochrome P_{450}) are concentrated in the small, slowly sedimenting vesicles. Thus, the membranes of the ER may be differentiated into distinct functional areas in respect of the electron transport chains.

There may also be differentiation in the RER with respect to its ribosomes and their role in protein synthesis. Of the two types of membrane-bound ribosomes reported in liver cells, one could be dissociated easily from the membranes of the ER by treatment with RNase while the other could not, and the synthesis of serum albumin was exclusively associated with the latter type of ribosome (Tanaka and Ogata, 1972). Moreover, ribosomes attached to the ER in close proximity to the mitochondria may be selectively concerned with the synthesis of proteins destined for incorporation into mitochondria (Tata, 1971). Consistent with this speculation is the finding that the organization of portions of the ER which are in close association with mitochondria differs from that of cisternae dispersed in the cytoplasm. Also, the organization of cisternae in association with the mitochondria is similar to that of the outer MM (Tewari and Malhotra, 1973).

Recent evidence seems relevant to the functional differentiation within the two nuclear membranes and in the outer MM of developing rat hepatocytes (Tewari and Malhotra, 1974). Freeze-fractured preparations of these membranes show extensive areas which are markedly different from the rest of the same membrane because they lack any apparent particulate organization so characteristic of the rest of the membrane (*Figure 9.7a, b*), and are reminiscent of the fractured faces of the myelin sheath (*Figure 9.7c*), which is known to contain relatively little protein and contains as much as 80 percent of lipid (O'Brien, 1967). The nature and function of these smooth areas in the mitochondrial and nuclear membranes are not known. However, it would be of interest to know whether they represent essentially a lipid, or lipid–protein, matrix onto which enzymes and other membrane components are added to produce functional membrane.

Cytochemical tests on MM for electron transport enzymes frequently show a patchy appearance, indicating an uneven distribution of the enzyme on the inner MM (Ogawa and Barrnett, 1964; Storrie and Attardi, 1973b),

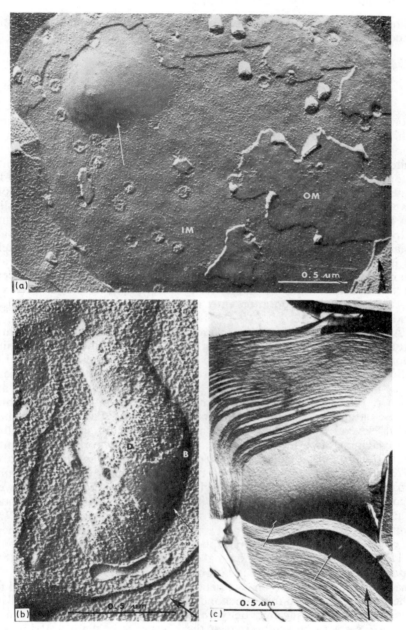

Figure 9.7 (a) *Freeze-fracture replica of a rat hepatocyte 3 days before birth showing a nucleus with a large area devoid of particles (arrow) on the fractured face of the inner membrane,* IM *(*OM, *outer nuclear membrane). (b) Freeze-fracture of a rat hepatocyte 1 day before birth. Note the presence of a large area devoid of particles (arrow) on the convex fracture of the outer mitochondrial membrane, B (D, convex fracture of the inner mitochondrial membrane). (c) Freeze-fracture of the sciatic nerve of frog showing fractures of the membranes of myelin sheath which are devoid of particles (arrows). Compare with the smooth areas in fractures of the nuclear (a) and outer mitochondrial (b) membranes*

although the uneven distribution could arise from diffusion of the reaction product(s) in these histochemical tests.

9.5   ROLE OF MITOCHONDRIA IN PROTEIN SYNTHESIS

That mitochondria synthesize protein *in vitro* and *in vivo* now seems certain. For the purposes of this chapter, it should suffice that a protein component which is associated with the inner MM is actually synthesized on mitochondrial ribosomes which are within the mitochondria, and is coded by the mitochondrial genome; this protein should, perhaps, integrate the mitochondrial proteins of the respiratory chain and/or lipids that make up the inner membrane. A recent suggestion that the apparent *in vitro* incorporation of amino acids into mitochondrial proteins reflects specific binding of labeled amino acids to the mitochondrial protein–lipid structures (Hochberg *et al.*, 1972) has not been confirmed (Ibrahim, Burke and Beattie, 1973); not only was radioactive leucine incorporated into proteins of intact isolated mitochondria of rats at a rate nearly twentyfold greater than that reported by Hochberg *et al.*, but a greater part (75 percent) of this incorporation was inhibited by chloramphenicol and there was essentially zero incorporation into mitochondria denatured by heat or acid.

An excellent comprehensive review on the synthesis of mitochondrial proteins (Beattie, 1971) discusses earlier doubts as to the ability of mitochondria to synthesize proteins. The nature of the protein synthesized by mitochondria is still an intriguing problem. The radioactive label appears to be incorporated into proteins of the inner MM (Neupert, Brdiczka and Bucher, 1967), and is recovered in insoluble membranous proteins whose precise nature and function is still uncertain (Beattie, Patton and Stuchell, 1970). Several hydrophobic amino acids, viz. leucine, isoleucine, valine, phenylalanine and methionine, were actively incorporated into mitochondrial proteins of HeLa cells *in vivo* and *in vitro* (Costantino and Attardi, 1973) in preference to the charged polar amino acids (alanine, arginine, aspartic acid, cystine, glutamic acid, glutamine, glycine and lysine) that were incorporated in substantial amounts by the polysomes associated with RER. It may be recalled that one of the subfractions (CFo) of Racker (1970) contains hydrophobic proteins and is part of the coupling device in successful reconstitution of oxidative phosphorylation.

This insoluble protein fraction makes up less than 10 percent of the total mitochondrial protein (Beattie, Patton and Stuchell, 1970). Apart from this small, but important, protein synthetic activity within mitochondria, all the other proteins required for mitochondrial functions, such as proteins of the respiratory chain (succinic dehydrogenase, cytochrome *c*, cytochrome oxidase), seem to be synthesized at the ribosomes in the cytoplasm, and subsequently transferred to the mitochondria (Roodyn and Wilkie, 1968; Beattie, 1971; Borst, 1972). Again, the direct evidence for the nuclear control of one of the mitochondrial proteins, viz. cytochrome *c*, comes from genetic analysis of mutants of yeast (Sherman *et al.*, 1970), and it can only be assumed that a similar genetic control is operative for the synthesis of these proteins in higher organisms, including mammalian cells.

9.6 EVIDENCE FOR DIVISION OF MITOCHONDRIA IN MAMMALIAN CELLS

Although a direct proof of the type provided by Luck (1965) for multiplication of mitochondria in *Neurospora* is yet to come, it is reasonable to conclude from the currently available data that the mitochondria grow and multiply by division in mammalian cells. However, a recent report that there is a single enormously long and highly branched mitochondrion in each cell of yeast and possibly also in mammalian cells (Hoffmann and Avers, 1973), if confirmed, should revolutionize our current thinking on biogenesis of mitochondria.

Mitochondria can be induced to increase markedly in size by interfering with the normal physiological conditions of animals, for example by feeding riboflavin-free diets (Tandler et al., 1969), or with cuprizone, a copper-chelating agent (Suzuki, 1969; Tandler and Hoppel, 1973), or by treating the animals with cortisone (Kimberg and Loeb, 1972).

Enlarged mitochondria can be reduced to their usual size by returning them to normal conditions. Mitochondria of normal size appear within three days of injecting riboflavin into riboflavin-deficient animals. This normalization appears to take place by division, a conclusion which is based upon analysis of micrographs of enlarged mitochondria showing transverse partitions running across them. The plasticity of mitochondria is illustrated by the fact that these partitions are often eccentric in enlarged mitochondria, and their site of formation is also variable. It is not certain how these partitions are formed, although they may be formed either by elongation or by fusion of preexisting cristae with the outer membrane pushing in between the component membranes of the septum so that two mitochondria eventually separate (Tandler et al., 1969).

When cuprizone is removed from the diet, the mitochondria attain normal size and there is an increase in their number. It is thought that the normalization in this case takes place by fission (Tandler and Hoppel, 1973), since elongated, constricted mitochondria have been seen in micrographs.

The best available evidence consistent with the division of mitochondria in mammalian cells (Storrie and Attardi, 1973b) is based on the fact that the mitochondrial protein-synthesizing system can be selectively inhibited by chloramphenicol, while this system is required for the formation of an active cytochrome oxidase. The number of mitochondria per cell remained constant for up to four generations and, in the absence of mitochondrial protein synthesis, there was no significant decrease in the number of cytochrome-oxidase-positive mitochondrial profiles visualized cytochemically. Moreover, there was a uniform decrease in the intensity of the enzymatic reaction along the inner MM. While other interpretations of these findings are possible, they strongly support the notion that mitochondria grow by random incorporation of components into the membranes of preexisting mitochondria, which then divide.

This is also indicated by the reaction of mitochondria to oxygen deficiency and their recovery upon oxygenation in cultured human (Hakami and Pious, 1967; Pious, 1970) and mouse (King and King, 1971) cells. Upon deprivation of oxygen, the cellular contents of cytochrome oxidase are greatly reduced and the mitochondrial cristae become disorganized. Upon supply of oxygen, the cytochrome oxidase activity returns to its normal level, and the cristae acquire their normal form. These findings are analogous to the formation of mitochondria from promitochondria upon adaption from anaerobic to aerobic growth in yeast (Criddle and Schatz, 1969; Schatz et al., 1972).

It should be stressed that mitochondria seem to be highly plastic membranous organelles, as shown by the varying modes of division, e.g. by transverse septa or fission, which may take place in the same cell, but perhaps under different conditions. This is illustrated in the hepatocytes of mice fed on a riboflavin-free diet, or a diet with the addition of cuprizone. The plasticity of mitochondria is also suggested in the living cell when they are seen to undergo changes in shape prior to fusing with one another and subsequently dividing by fission.

Under abnormal conditions, the division of mitochondria may produce two nonidentical progeny, one of which is selectively degraded. Two categories of mitochondria were found after prolonged treatment with ethidium bromide, where the proportion of smaller organelles increased as the drug treatment was prolonged. This could indicate resistance to ethidium bromide on the part of the small mitochondria, which EM results suggest may arise from the large mitochondria (Soslau and Nass, 1971). Somewhat on similar lines are the findings that there appear to be two populations of mitochondria in livers of rats treated with thyroid hormone (Gross, 1971), based on the observation that DNA which is synthesized following administration of the hormone turns over at a slower rate. Gross speculated that the hormone treatment might induce preexisting mitochondria to undergo division to produce two different populations, the integrity of one being maintained whilst the other undergoes degradation.

9.7 CONCLUSIONS

Our understanding of the biology of cellular membranes has markedly advanced since the 1930s when Danielli and Davson (1935) first proposed their model for the PM. This advance has been essentially in the chemical analysis of the membrane components and much less in the area of molecular interaction or in the assembly of these components into functional membranes. The discovery of mitochondrial DNA and of the dependence of mitochondrial components on the messages from the nuclear DNA has stimulated investigation of the synthesis and assembly of functional mitochondria from constituents which arise as a consequence of interactions between the mitochondrial and the nuclear genomes. Important advances have been made in the structure of DNA, ribosomes and in the mechanism of protein synthesis in mitochondria of eukaryotic cells (Roodyn and Wilkie, 1968; Borst, 1972; Storrie and Attardi, 1972, 1973a). It has also been suggested that membranes may provide a mechanism, not yet understood, for transport of membrane components from their site of synthesis to the site where they are integrated into preexisting membranes. The transport of molecules may be achieved through temporary continuities between membranes of morphologically and chemically characterizable categories, such as between ER and outer MM, and ER and PM. A direct continuity is well recognized in some cases, for example between ER and nuclear membrane, yet there may be important chemical differences between these two systems. Continuity between some other membranes is not so generally accepted, such as between ER and outer MM. Also, connections between the peroxisomes and ER have been demonstrated in mammalian cells (Black and Bogart, 1973). In this particular case, the peroxisomes are formed from the dilated regions of the ER. A direct continuity between the PM and the nuclear membrane has been reported in electron micrographs of *Blasia pusilla* (Carothers, 1972). The regulatory control needed for functional compartmentalization within the membranous system of a cell is a subject likely to be pursued actively by cell biologists. The membranes may, at times, be continuous across morphologically and biochemically recognizable entities, as was recognized in the beginning of the era of biological EM (Robertson, 1959). This is particularly true in view of the fairly recent finding that a membrane may present a 'fluid' matrix in which molecules move about, shown by the intermixing of antigens in the PM of heterokaryons produced from mouse and human tissue-culture cell lines (Frye and Edidin, 1970). Also, it has been reported that large particles ($\sim$30 nm) may be able to move in the interior of membranes, shown by studies of the PM of germinating spores of the fungus *Phycomyces* (Malhotra and Tewari, 1973). These spores have a controlled dormancy, and show uniformly dispersed particles of size 30 nm in freeze-fractured preparations. Upon initiation of germination, the particles aggregate, probably owing to movement of dispersed particles. While these observations are consistent with the fluid mosaic model proposed for functional biological membranes (Singer and Nicolson, 1972; *see* Wallach, 1972, and Chapter 10 in this volume for a detailed discussion of membrane models), further investigations are necessary to elucidate the mechanism of transport of molecules within and across membranes. Such studies are likely to advance our understanding of the molecular interactions amongst membrane constituents, which should in turn facilitate understanding of the molecular organization and biogenesis of cellular membranes.

Acknowledgements

This article has been prepared with the invaluable joint effort of my colleague, Dr J. P. Tewari. All the micrographs included in this article are previously unpublished and were prepared in the author's laboratory. The valuable technical assistance was provided by Mr M. Kobalcik. I am grateful to Dr R. S. Smith for providing the specimen of sciatic nerve and to Dr C. A. Pasternak for providing unpublished data.

The author's research project is supported by grants from the National Research Council of Canada.

REFERENCES

AFZELIUS, B. A. (1973). *The Generation of Subcellular Structures*, pp. 307–340. Ed. R. MARKHAM, J. B. BANCROFT, D. A. HOPWOOD and R. W. HORNS. Amsterdam; North-Holland.
BEATTIE, D. S. (1969). *Biochem. biophys. Res. Commun.*, 35:67.
BEATTIE, D. S. (1971). *Sub-Cell. Biochem.*, 1:1.

BEATTIE, D. S., PATTON, G. M. and STUCHELL, R. N. (1970). *J. biol. Chem.*, **245**:2177.
BENES, F., HIGGINS, J. A. and BARRNETT, R. J. (1973). *J. Cell Biol.*, **57**:613.
BLACK, V. H. and BOGART, B. I. (1973). *J. Cell Biol.*, **57**:345.
BORST, P. (1972). *A. Rev. Biochem.*, **41**:333.
BOSMANN, H. B. and CASE, K. R. (1969). *Biochem. biophys. Res. Commun.*, **36**:830.
CAMILLI, P. D., PELUCHETTI, D. and MELDOLESI, J. (1974). *Nature, Lond.*, **248**:245.
CAROTHERS, Z. B. (1972). *J. Cell Biol.*, **52**:273.
CHOPPIN, P. W. (1973). *Neurosci. Res. Prog. Bull.*, **11**:49.
COSTANTINO, P. and ATTARDI, G. (1973). *Proc. natn. Acad. Sci. U.S.A.*, **70**:1490.
CRIDDLE, R. S. and SCHATZ, G. (1969). *Biochemistry*, **8**:322.
DALLMAN, P. R., DALLNER, G., BERGSTRAND, A. and ERNSTER, L. (1969). *J. Cell Biol.*, **41**:357.
DALLNER, G., SIEKEVITZ, P. and PALADE, G. E. (1966a). *J. Cell Biol.*, **30**:73.
DALLNER, G., SIEKEVITZ, P. and PALADE, G. E. (1966b). *J. Cell Biol.*, **30**:97.
DANIELLI, J. F. and DAVSON, H. (1935). *J. cell. comp. Physiol.*, **5**:495.
DAVIDSON, J. B. and STANACEV. N. Z. (1971). *Biochem. biophys. Res. Commun.*, **42**:1191.
DENNIS, E. A. and KENNEDY, E. P. (1972). *J. Lipid Res.*, **13**:263.
EDIDIN, M. and FAMBROUGH, D. (1973). *J. Cell Biol.*, **57**:27.
ERIKSSON, L. C. (1973). *Acta path. microbiol. scand.*, **239**:1.
FERNÁNDEZ-MORÁN, H. (1962). *Circulation*, **26**:1039.
FESSENDEN-RADEN, J. M. and RACKER, E. (1971). *Structure and Function of Biological Membranes*, pp. 401–438. Ed. L. I. ROTHFIELD. New York; Academic Press.
FINKELSTEIN, A. (1972). *Archs intern. Med.*, **129**:229.
FOX, C. F. (1972). *Membrane Molecular Biology*, pp. 345–385. Ed. C. F. FOX and A. D. KEITH. Stamford, Connecticut; Sinauer Associates.
FRANKE, W. W., DEUMLING, B., ERMEN, B., JARASCH, E.-D. and KLEINIG, H. (1970). *J. Cell Biol.*, **46**:379.
FRYE, L. D. and EDIDIN, M. (1970). *J. Cell Sci.*, **7**:319.
GERBER, D., DAVIES, M. and HOKIN, L. E. (1973). *J. Cell Biol.*, **56**:736.
GETZ, G. S. (1972). *Membrane Molecular Biology*, pp. 386–438. Ed. C. F. FOX and A. D. KEITH. Stamford, Connecticut; Sinauer Associates.
GLICK, M. C. and WARREN, L. (1969). *Proc. natn. Acad. Sci. U.S.A.*, **63**:563.
GOLDBLATT, P. J. (1972). *Sub-Cell. Biochem.*, **1**:147.
GREGSON, N. A. and WILLIAMS, P. L. (1969). *J. Neurochem.*, **16**:617.
GROSS, N. J. (1971). *J. Cell Biol.*, **48**:29.
GUSTAFSSON, R., TATA, J. R., LINDBERG, O. and ERNSTER, L. (1965). *J. Cell Biol.*, **26**:555.
HAJOS, F. and KERPEL-FRONIUS, S. (1969). *Expl Brain Res.*, **8**:66.
HAKAMI, N. and PIOUS, D. A. (1967). *Nature, Lond.*, **216**:1087.
HAMBERGER, A., BLOMSTRAND, C. and LEHNINGER, A. L. (1970). *J. Cell Biol.*, **45**:221.
HAMBERGER, A., GREGSON, N. and LEHNINGER, A. L. (1969). *Biochim. biophys. Acta*, **186**:373.
HAMBURGER, J., RICHET, G. and GRUNFELD, J. P. (1971). *Structure and Function of the Kidney*. Philadelphia; W. B. Saunders.
HERZFELD, A., FEDERMAN, M. and GREENGARD, O. (1973). *J. Cell Biol.*, **57**:475.
HIGGINS, J. A. and BARRNETT, R. J. (1972). *J. Cell Biol.*, **55**:282.
HOCHBERG, A. A., STRATMAN, F. W., ZAHLTEN, R. N. and LARDY, H. A. (1972). *FEBS Lett.*, **25**:1.
HOFFMAN, H.-P. and AVERS, C. J. (1973). *Science, N.Y.*, **181**:749.
IBRAHIM, N. G., BURKE, J. P. and BEATTIE, D. S. (1973). *FEBS Lett.*, **29**:73.
JAKOVCIC, S., HADDOCK, J., GETZ, G. S., RABINOWITZ, M. and SWIFT, H. (1971). *Biochem. J.*, **121**:341.
KAGAWA, Y. (1972). *Biochim. biophys. Acta*. **265**:297.
KAGAWA, Y., KANDRACH, A. and RACKER, E. (1973). *J. biol. Chem.*, **248**:676.
KAISER, W. (1969). *Eur. J. Biochem.*, **8**:120.
KIMBERG, D. V. and LOEB, J. N. (1972). *J. Cell Biol.*, **55**:635.
KING, M. E. and KING, D. W. (1971). *Lab. Invest.*, **25**:374.
KLEINIG, H. (1970). *J. Cell Biol.*, **46**:396.
KORN, E. D. (1969). *A. Rev. Biochem.*, **38**:263.
KORNBERG, R. D. and MCCONNELL, H. M. (1971). *Proc. natn. Acad. Sci. U.S.A.*, **68**:2564.
KURIYAMA, Y., OMURA, T., SIEKEVITZ, P. and PALADE, G. E. (1969). *J. biol. Chem.*, **244**:2017.
LATTA, H., MAUNSBACH, A. B. and OSVALDO, L. (1967). *Ultrastructure of the Kidney*, pp. 1–56. Ed. A. J. DALTON and F. HAGUENAU. New York; Academic Press.
LESKES, A., SIEKEVITZ, P. and PALADE, G. E. (1971a). *J. Cell Biol.*, **49**:264.
LESKES, A., SIEKEVITZ, P. and PALADE, G. E. (1971b). *J. Cell Biol.*, **49**:288.

LODISH, H. F. (1973). *Proc. natn. Acad. Sci. U.S.A.*, **70**:1526.
LOWE, D. and HALLINAN, T. (1973). *Biochem. J.*, **136**:825.
LUCK, D. J. L. (1965). *Am. Nat.*, **99**:241.
LUCK, D. J. L. and REICH, E. (1964). *Proc. natn. Acad. Sci. U.S.A.*, **52**:931.
MALHOTRA, S. K. (1970). *Prog. Biophys. molec. Biol.*, **20**:67.
MALHOTRA, S. K. and EAKIN, R. T. (1967). *J. Cell Sci.*, **2**:205.
MALHOTRA, S. K. and TEWARI, J. P. (1973). *Proc. R. Soc., Ser. B*, **184**:207.
MEISSNER, G. and FLEISCHER, S. (1973). *Biochem. biophys. Res. Commun.*, **52**:913.
MONNERON, A., BLOBEL, G. and PALADE, G. E. (1972). *J. Cell Biol.*, **55**:104.
MURRAY, W. C. and DAWSON, R. M. C. (1969). *Biochem. J.*, **112**:91.
NEUPERT, W., BRDICZKA, D. and BUCHER, T. R. (1967). *Biochem. biophys. Res. Commun.*, **27**:488.
O'BRIEN, J. S. (1967). *J. theor. Biol.*, **15**:307.
O'BRIEN, J. S. and SAMPSON, E. L. (1965). *Science, N.Y.*, **150**:1613.
OGAWA, K. and BARRNETT, R. J. (1964). *Nature, Lond.*, **203**:724.
PALADE, G. E. (1973). *Neur. Res. Prog. Bull.*, **11**:39.
PERCY, A. K., SCHMELL, E., EARLES, B. J. and LENNARZ, W. J. (1973). *Biochemistry*, **12**:2456.
PIOUS, D. A. (1970). *Proc. natn. Acad. Sci. U.S.A.*, **65**:1001.
RACKER, E. (1970). *Membranes of Mitochondria and Chloroplasts*, pp. 127–171. Ed. E. RACKER. New York; Van Nostrand–Reinhold.
RAY, T. K., LIEBERMAN, I. and LANSING, A. I. (1968). *Biochem. biophys. Res. Commun.*, **31**:54.
ROBERTSON, J. D. (1959). *Biochem. Soc. Symp.*, **16**:33.
ROODYN, D. B. and WILKIE, D. (1968). *The Biogenesis of Mitochondria*. London; Methuen.
SCHATZ, G., GROOT, G. S. P., MASON, T., ROUSLIN, W., WHARTON, D. C. and SALTZGABER, J. (1972). *Fedn Proc. Fedn Am. Socs exp. Biol.*, **31**:21.
SCHIMKE, R. T. (1973). *Neurosci. Res. Prog. Bull.*, **11**:54.
SHERMAN, F., STEWART, J. W., PARKER, J. H., PUTTERMAN, G. J., AGRAWAL, B. B. L. and MARGOLIASH, E. (1970). *Symp. Soc. exp. Biol.*, **24**:85.
SIEKEVITZ, P. (1972). *A. Rev. Physiol.*, **34**:117.
SIEKEVITZ, P. (1973). *Gosp. Delo (Hosp. Practice)*, **8**:91.
SINGER, S. J. and NICOLSON, G. L. (1972). *Science, N.Y.*, **175**:720.
SINGER, S. J. and ROTHFIELD, L. I. (1973). *Neurosci. Res. Prog. Bull.*, **11**:1.
SOSLAU, G. and NASS, M. M. K. (1971). *J. Cell Biol.*, **51**:514.
STEIN, O. and STEIN, Y. (1969). *J. Cell Biol.*, **40**:461.
STORRIE, B. and ATTARDI, G. (1972). *J. molec. Biol.*, **71**:177.
STORRIE, B. and ATTARDI, G. (1973a). *J. Cell Biol.*, **56**:819.
STORRIE, B. and ATTARDI, G. (1973b). *J. Cell Biol.*, **56**:833.
SUZUKI, K. (1969). *Science, N.Y.*, **163**:81.
TANAKA, T. and OGATA, K. (1972). *Biochem. biophys. Res. Commun.*, **49**:1069.
TANDLER, B. and HOPPEL, C. L. (1973). *J. Cell Biol.*, **56**:266.
TANDLER, B., ERLANDSON, R. A., SMITH, A. L. and WYNDER, E. L. (1969). *J. Cell Biol.*, **41**:477.
TATA, J. R. (1967). *Nature, Lond.*, **213**:566.
TATA, J. R. (1970). *Biochem. J.*, **116**:617.
TATA, J. R. (1971). *Sub-Cell. Biochem.*, **1**:83.
TEWARI, J. P. and MALHOTRA, S. K. (1973). *Sub-Cell. Biochem.*, **2**:287.
TEWARI, J. P. and MALHOTRA, S. K. (1974). *Cytobios*, **9**:55.
TOURTELLOTTE, M. E., BRANTON, D. and KEITH, A. (1970). *Proc. natn. Acad. Sci. U.S.A.*, **66**:909.
VICTORIA, E. J., VAN GOLDE, L. M. G., HOSTETLER, K. Y., SCHESPHOF, G. L. and VAN DEENEN, L. L. M. (1971). *Biochim. biophys. Acta*, **239**:443.
WALLACH, D. F. H. (1972). *The Plasma Membrane*. New York; Springer-Verlag.
WARSHAW, J. B. (1972). *Devl. Biol.*, **28**:537.
WARSHAW, J. B. and KIMURA, R. E. (1973). *Devl Biol.*, **33**:224.
WILLIAMS, M. L. and BYGRAVE, F. L. (1970). *Eur. J. Biochem.*, **17**:32.
WIRTZ, K. W. A. and ZILVERSMIT, D. B. (1969). *Biochim. biophys. Acta*, **187**:468.

10

A perspective on models of membrane structure*

John Lenard
Department of Physiology, CMDNJ–Rutgers Medical School, Piscataway, New Jersey

Frank R. Landsberger
The Rockefeller University, New York

10.1 INTRODUCTION

The depiction of models of membrane structure has long been among the objectives of individuals engaged in membrane research. Danielli has humorously described his well-known model (Danielli and Davson, 1935) as a 'cartoon', and subsequent models have been depicted in a similar schematic fashion. These pictures, of course, serve to encapsulate the particular arrangement of lipids and proteins to which the originator's arguments have led him, and thus represent generalizations of observations usually made on one or a few types of membranes, studied from a particular experimental perspective.

It is now apparent that such cartoons can no longer represent our knowledge adequately. There are two reasons for this, both reflecting the enormous increase in knowledge of membrane structure which has accumulated in the last few years. First, it is now known that at least some lipids and proteins in membranes are in constant motion. In fact so many different (though doubtless interrelated) very rapid motions have been identified in lipid bilayer structures alone that they could be represented adequately only by an *animated* cartoon, which, to be properly appreciated, would have to be projected in slow motion. A static cartoon depicting such a structure on a printed page would inevitably be misleading. The second reason that a general model cannot be depicted in this chapter is that information about the disposition of specific proteins in specific membranes,

* Literature survey concluded October 1973, with minor additions April 1975.

though still very limited, clearly shows that a great variety of different arrangements exist. Thus, there are specific instances of proteins which traverse the membrane from side to side, of proteins which penetrate the membrane but do not traverse it, and of proteins which are associated with a single membrane surface, but do not appear to penetrate at all. The number of well-documented instances of each of these types of membrane proteins is still quite small, and it is not yet known whether each of these cases represents a large class of membrane proteins. Further, there is very little known of the functional correlates of this variety, so that one cannot generalize or predict from the few facts currently at hand.

One is, then, at an intermediate stage, when the traditional method of depicting a membrane model on a printed page can no longer provide new insights, while the information needed to propose a dynamic model for the structure of any specific membrane is not yet available. However, a great deal of important information has recently been obtained about many different membranes, by utilization of an almost bewildering variety of approaches and techniques. Many of these separate approaches give results which have converged to yield satisfyingly consistent conclusions. Others still present us with apparent discrepancies which must be resolved. We have tried to discuss these results in an integrated fashion, and to emphasize conclusions rather than specific methodology or results. We discuss experimental details and controversies where it seems justified by the importance of the underlying principles. This chapter should therefore be read not as a detailed description of particular experimental results, but as a highly subjective analysis of some recent studies which may serve as signposts pointing toward generalizations which will have to be incorporated into future models of membrane structure. Accordingly, the references cited were chosen to provide the reader with an entry into the subject under discussion, and are not exhaustive. A more detailed discussion of many topics will be found in other chapters of this treatise.

10.2 SOME BASIC CONSIDERATIONS

In this section we will recapitulate some generally accepted facts about the composition and disposition of membrane components which form the basis for topics which are considered in later sections. We also present a brief qualitative analysis of the forces which hold the membrane together.

10.2.1 Lipids

The arrangement of membrane lipids in a bilayer was first proposed over fifty years ago (Gorter and Grendel, 1925) and was an unquestioned feature of membrane models for the ensuing forty years. The realization that the earlier evidence for a bilayer structure was at best circumstantial led to several membrane models which did not include a bilayer, or indeed any significant amount of lipid–lipid interaction as a stabilizing force in the membrane (Benson, 1966; Green and Perdue, 1966). Other models retained lipid–lipid interactions, but were based upon various alternatives to the bilayer configuration (Lucy, 1964; Kavanau, 1966).

Since the time when these models were proposed, persuasive evidence has been collected that the bilayer structure represents the major organizational principle of the lipids in virtually all membranes. Early evidence came from the calorimetric measurements of Steim and his co-workers in membranes of *Mycoplasma laidlawii* (Steim *et al.*, 1969). Confirmation came from X-ray diffraction measurements, which showed that the profile of electron density across the thickness of several different membranes was very similar to that of a synthetic bilayer (Levine and Wilkins, 1971; Wilkins, Blaurock and Engelman, 1971). Further proof was supplied by the demonstration that fatty acid and steroid spin labels assumed an orientation and motional freedom in membranes closely similar to those seen when the same spin labels were incorporated into synthetic bilayers (Libertini *et al.*, 1969; Hubbell and McConnell, 1969).

The lipid composition of many different membranes has been analyzed in detail. Each membrane seems to be characterized by a heterogeneous population of lipids. Cholesterol is generally present in mammalian plasma membranes, while intracellular mammalian membranes and bacterial membranes possess little or none. The relative proportion of glycolipids and phospholipids varies considerably from one type of membrane to another. Considerable variation is found in the distribution of the different phospholipid classes in each membrane. In addition, each phospholipid class has a wide range of chain lengths of both saturated and unsaturated hydrocarbon chains, linked to the glycerol backbone by ester, vinyl ether and ether bonds. The lipid compositions of some mammalian membranes have been listed by Ansell, Hawthorne and Dawson (1973).

The lipid composition of many membranes can be altered by environmental influences such as diet or temperature, so that individual membrane functions appear to be relatively insensitive to the precise lipid composition. Some physical correlates of a few defined changes in bulk-phase lipid composition, for example cholesterol:phospholipid ratio, saturated versus unsaturated fatty acid content and fatty acid chain length have been described on the basis of studies on model systems (Section 10.3).

10.2.2 Proteins

Proteins generally comprise about one-half to two-thirds of the total material in a membrane. Membrane proteins are characteristically insoluble, forming intractable aggregates in aqueous environments. As a consequence of this inconvenient property, membrane proteins remained uncharacterized during a period when knowledge of the structure of soluble proteins was increasing rapidly. The development of suitable analytical and preparative techniques for these proteins—most notably polyacrylamide gel electrophoresis in the presence of the anionic detergent sodium dodecyl sulfate—has resulted in rapid progress in the last few years. These techniques quickly established the fact that most membranes contain an assemblage of proteins possibly even more diverse than their lipids. With the exception of a few specialized membrane structures such as rod outer segment membranes and myelin, the number of recognizable proteins in each membrane is limited chiefly by the resolving power of the technique used to study them. Furthermore, different membranes, even from the same organism, contain practically no recognizable common components (Kiehn and Holland, 1968). A thorough and readable review on membrane proteins has appeared (Steck and Fox, 1972).

10.2.3 Carbohydrates

Carbohydrates are present in membranes mainly as glycoproteins and glycolipids. Compounds consisting exclusively of carbohydrate are not thought to play an important structural role, although heparan sulfate has been identified as a surface component in several lines of cultured cells (Kraemer, 1971a, b). Evidence has been presented that carbohydrates are localized exclusively on the outer surface of erythrocyte and other membranes (Section 10.5).

10.2.4 Forces holding the membrane together

Singer (1971) has presented a lucid qualitative discussion of the forces which stabilize membrane structure. These forces are similar to those which stabilize other macromolecular systems in an aqueous environment. For purposes of illustration we can consider a pure phospholipid dispersed in water. The structure of maximum stability for such an amphiphilic molecule would be one which maximizes the interactions of the polar headgroups with water, and which minimizes the interactions of the nonpolar regions with water by sequestering them away from the aqueous environment. Any association of the nonpolar or hydrophobic groups with water will cause ordering of water molecules, which will decrease entropy and increase free energy. The most stable conformation of phospholipids *in water* is therefore a bilayer, since this conformation maximizes both hydrophilic and hydrophobic interactions. Considerations of the free energy of association of pure lipids (Smith and Tanford, 1972; Haberland and Reynolds, 1973), as well as of models for the hydrophobic and hydrophilic portions of the molecules (Singer, 1971; Tanford, 1972), suggests that the free energy of stabilization is quite large.

The maximization of hydrophilic and hydrophobic interactions is also required for stabilization of membrane proteins. This maximization is achieved in soluble proteins by sequestering hydrophobic residues in the protein interior, while the hydrophilic residues are located on the surface. In the case of membrane proteins which penetrate into or through the nonpolar portion of the lipid bilayer, a quite different arrangement of hydrophilic and hydrophobic surfaces must be required if these interactions are to be maximized. These considerations therefore can explain, in a general way, why many membrane proteins require lipid-like agents (e.g. detergents) for solubilization.

It may be useful to consider the stabilization of the lipid bilayer in a related but still qualitative manner so as to provide a framework in which the reader might more easily visualize the many different molecular motions which are discussed in Section 10.3. All forces stabilizing the bilayer may be separated into two 'components', one parallel and one perpendicular to the bilayer surface. Each of these can be loosely viewed as the sum of opposing hydrophobic and hydrophilic forces. The hydrophilic interaction perpendicular to the surface tends to pull the polar headgroup out into the aqueous environment, while the opposing hydrophobic interaction tends to pull the hydrocarbon chain inward, away from the aqueous environment and deeper into the surrounding hydrocarbon phase. This would suggest that a minimum hydrocarbon chain length and a minimum degree of polarity of the hydrophilic group are both required for bilayer stability. At equilibrium the two opposing forces must be equal, but a transient imbalance between the forces could cause a lipid molecule to be pulled slightly out of or into the bilayer. The molecule may then undergo an oscillatory motion about its equilibrium position, and by interaction with neighboring molecules might induce waves propagating parallel to the membrane surface. It may be speculated that such a motion of molecules in bilayers could provide a mechanism for transmitting information within the plane of the membrane.

In a similar manner, one can consider a pair of opposing forces acting between the molecules in a direction parallel to the bilayer surface. These opposing forces are (a) the hydrophobic and van der Waals interactions which act to pull the hydrocarbon chains together and thus exclude water from the hydrocarbon region, and (b) the solvation of the polar headgroups by water, and repulsion between charges in the lipid headgroups, which tend to push the molecules apart. When these opposing forces on an individual molecule do not precisely cancel, at any given instant there is a net force in one direction, and lateral displacement of the molecule results.

10.3 DYNAMIC STRUCTURE OF THE LIPID BILAYER

The conclusive demonstration that the bilayer structure is a predominant one in biological membranes (Section 10.2.1) sparked a renewed interest in synthetic bilayers as relevant and important models of membrane structure. The bilayer has proved to be quite different from the static, rigid entity that was previously supposed, and to possess a quite unexpected dynamism and fluidity. Several different intra- and intermolecular motions characteristic of lipids in the bilayer configuration have been described, and such motions clearly underlie the motions of membrane proteins discussed in Section 10.6. These motions in turn provide the basis for several recent models of membrane structure (Singer and Nicolson, 1972; Wallach, 1972).

10.3.1 Motions within phospholipid molecules

It was demonstrated by spin labeling studies (Hubbell and McConnell, 1969, 1971; Jost et al., 1971) and confirmed by NMR (Metcalfe et al., 1971; Chan, Seiter and Feigenson, 1972; Horwitz, Horsley and Klein, 1972; Lee et al., 1972; Levine et al., 1972; Godici and Landsberger, 1974) that bilayers are characterized by a relatively fluid interior, with a progressive increase in rigidity toward the glycerol moiety. Spin labels with nitroxide groups attached to carbons high on the fatty acid chain, close to the glycerol group, appear to be in a highly constrained, rigid environment, while those further down, close to the middle of the bilayer, are in a much more fluid environment. The characteristics of this progressive change in fluidity, or flexibility gradient, depend upon the degree of unsaturation of the fatty acids (Rottem et al., 1970; Tourtellotte, Branton and Keith, 1970; Levine et al., 1972; Godici and Landsberger, 1974), the presence of cholesterol (see below), the presence of proteins (Section 10.4), and whether the temperature is above or below the gel to liquid-crystalline transition for the system (Chapman, 1973).

The effect of cholesterol on the physical properties of bilayers has been characterized by recent studies. Cholesterol abolishes the gel to liquid-crystalline thermal transition, although disagreement exists over whether this occurs at cholesterol:phospholipid ratios above 1:2 (Hinz and Sturtevant, 1972) or 1:1 (Chapman, 1973). At high cholesterol concentrations, the state of fluidity of the fatty acyl chains is intermediate between that of the more rigid gel phase and the less rigid liquid-crystalline phase. Thus, above the thermal transition temperature of the phospholipid, cholesterol increases the rigidity of the bilayer. The largest increase in rigidity is in the middle of the fatty acyl chains, with a relatively small effect in the deep hydrocarbon region (Darke et al., 1972; McConnell and McFarland, 1972; Keough, Oldfield and Chapman, 1973; Godici and Landsberger, 1975). Below the transition temperature, cholesterol renders the lecithin bilayers more fluid (Oldfield and Chapman, 1972). It has not yet been determined whether or not cholesterol forms complexes or phases with phospholipids in a bilayer. It has been suggested that complexes which have phospholipid:cholesterol ratios of 1:1 (Chapman, 1973) or 2:1 (Darke et al., 1972; Phillips and Finer, 1974; Engelman and Rothman, 1972; Hinz and Sturtevant, 1972) are formed in mixed bilayers, and that these complexes coexist with the free phospholipid. On the other hand, other workers find no evidence for such complexes above the transition temperature of the pure phospholipid (Shimshick and McConnell. 1973a; Metcalfe et al., 1974; Godici and Landsberger, 1975).

The choline groups in lecithin bilayers have been found to undergo appreciable motion (Davis and Inesi, 1971; Metcalfe et al., 1971; Horwitz, Horsley and Klein, 1972; Lee et al., 1972; Levine et al., 1972; Sheetz and Chan, 1972). The existence of a second flexibility gradient has been shown to exist using high-resolution ^{13}C-NMR spectroscopy, which demonstrated that the magnitude of the motion of the carbon atoms increases with increasing distance from the glycerol moiety (Godici and Landsberger, 1974). Cholesterol has no detectable effect upon this flexibility gradient (Godici and Landsberger, 1975).

The existence of two different flexibility gradients, one above and one below the glycerol moiety, suggests that glycerol acts as a relatively immobile backbone which forms the anchoring structure for the remainder of the lipid molecule. This raises the interesting question of whether this relative immobility arises from stabilizing interactions between the glycerol groups, or from the net effect of the hydrophobic and hydrophilic stabilizing forces acting above and below the glycerol group (Section 10.2.4). To our knowledge no experimental data are available to suggest an answer to this question.

10.3.2 Translational motion of phospholipid molecules

Lateral diffusion of phospholipid molecules parallel to the plane of the bilayer was first demonstrated in model systems by Kornberg and McConnell (1971b). These workers measured the extent of broadening of the PMR peaks of the phospholipid upon introduction of a phospholipid spin label into the bilayer. The rate of lateral diffusion of the spin label proved to be so rapid that only a lower estimate of the root mean square displacement of 0.05 μm s^{-1} could be made using this technique. In subsequent experiments, spin label vesicles were fused to sarcoplasmic reticulum membranes, and diffusion on the order of 5 μm s^{-1} was measured (Scandella, Devaux and McConnell, 1972; Devaux, Scandella and McConnell, 1973). Similar rapid lateral diffusion of spin labels was demonstrated in membranes of *Mycoplasma laidlawii* (Grant and McConnell, 1973).

Shimshick and McConnell (1973a, b) have developed a technique, based on the temperature-dependent partitioning of a small spin label between the hydrocarbon and aqueous phases of a lipid dispersion, from which the phase diagram of a binary mixture of lipids can be deduced. The phase diagrams thus obtained agree with those determined by calorimetric methods (Phillips, Ladbrooke and Chapman, 1970). By the use of this technique it has been demonstrated that in both synthetic bilayers and natural membranes there exists a fairly broad temperature range over which both solid-like and fluid-like regions coexist (Shimshick and McConnell, 1973a, b; Linden *et al.*, 1973). In many microorganisms, physiological growth temperatures fall within this temperature range (Steim, 1972; Linden *et al.*, 1973). Linden *et al.* (1973) suggest that the coexistence of the two phases increases the lateral compressibility of the bilayer, thus facilitating insertion of newly synthesized membrane proteins and lipids. The coexistence of the two phases requires lateral diffusion of the phospholipid molecules, and thus provides evidence that the natural phospholipids, like the spin label derivatives, undergo rapid translational motions in bilayers and natural membranes (Shimshick and McConnell, 1973a, b; Shimshick *et al.*, 1973). Using the spin label technique, Shimshick and McConnell (1973a) have also obtained evidence for coexistence of two phases of different composition in systems containing cholesterol. Edidin (1974) has presented a comprehensive review of lateral diffusion of lipids in biological membranes.

10.3.3 Bilayers and membranes

The intra- and intermolecular motions of phospholipid molecules discussed above have been amply demonstrated in both bilayer and membrane systems. The molecular basis of membrane 'fluidity' is a combination of these motions. It has been suggested that a decrease in the intramolecular motions correlates with a decrease in the intermolecular motions (Devaux and McConnell, 1972; Lee, Birdsall and Metcalfe, 1973). However, the detailed relationship between the two has not yet been demonstrated. The coexistence of solid-like and liquid-like phases in bilayers of binary lipid mixtures implies that there exist local differences in molecular motions of both types.

These local differences may be more diverse in biological membranes with their heterogeneous lipid populations, containing proteins with differing affinities for specific lipids (Sections 10.4 and 10.5).

10.4 INTERACTIONS BETWEEN LIPIDS AND PROTEINS

The nature of the protein–lipid interactions is presently one of the least understood and most actively studied aspects of membrane structure. It is well known that a variety of membrane-associated enzymes require lipids for activity. However, most of the enzymes which have been characterized are strikingly insensitive to the type of lipid required for reactivation. In many cases, membrane lipids can be dispensed with altogether, and replaced by detergents of various types. In others, there may be a requirement for acidic phospholipids or fatty acids, suggesting some kind of charge interaction with the membrane protein. An absolute requirement for a single class of phospholipid is unusual, although a few instances have been reported. Often, when phospholipids of any kind are required for activity, the addition of detergent is found to activate the enzyme further. The overall impression from the sum of these observations is that one major action of lipid or detergent is to solvate those parts of the protein which normally form an interface with the nonpolar regions of the bilayer.

These considerations suggest that the physical state of the lipid or detergent may be more important than its chemical structure in affecting enzyme activity. A considerable amount of recent research has been directed toward testing this hypothesis. Much work has been carried out with bacterial mutants requiring externally added unsaturated fatty acids for growth. This permits experimental variation of the lipid composition of genetically defined microorganisms. Esfehani, Barnes and Wakil (1969) found that incorporation of *cis*-unsaturated fatty acids into the bacterial membranes increased as the growth temperature was lowered, suggesting the existence of a regulatory mechanism which functions to control the physical state of the membrane within narrow limits. Thus, changes in growth temperature are compensated by changes in the fatty acid composition of the membrane.

When the logarithms of the rates of a number of different membrane enzyme and transport processes are plotted as a function of reciprocal temperature (Arrhenius plots), biphasic, or in some cases triphasic, curves are obtained. The linear portions of these curves intersect at a point (or points) termed the transition temperature(s). For a given transport or enzyme function, this transition temperature was found to change as the fatty acid composition of the membrane was altered (*see* Machtiger and Fox, 1973). Thus the thermodynamic state of the membrane lipids plays a major role in determining the functional state of the enzyme or carrier proteins.

It has been of great interest to correlate these functional transition temperatures with the thermally induced phase changes in the membrane and in isolated membrane lipids. In general, the transition temperatures of transport functions have been shown to correlate well with the phase transitions in membranes and isolated lipids, decreasing with increasing unsaturation of the externally added fatty acid (Wilson, Rose and Fox, 1970; Overath, Schairer and Stoffel, 1970; Esfehani and Wakil, 1972). On the other hand, the transition temperatures of various enzyme functions did not correlate with the bulk phase properties of the membrane (Esfehani et al., 1971; Esfehani and Wakil, 1972; Mavis and Vagelos, 1972), suggesting that these enzymes may be interacting with specific lipids, or may occur in localized regions containing a lipid composition different from that of the bulk membrane.

The two transition temperatures seen in the Arrhenius plots for glucoside and galactoside accumulation in *Escherichia coli* have been correlated by Linden et al. (1973) with the onset and completion of the gel to liquid-crystalline transition, as measured by spin label methods. Another study of membrane vesicles of *E. coli* using spin labels has also demonstrated two thermally induced transitions (Baldassare, McAfee and Ho, 1973), corresponding quite closely to the transition temperatures observed by Linden et al. (1973). However, Baldassare, McAfee and Ho (1973) showed that the transition occurring at higher temperature involved lipid–protein interactions, since it was elevated 8 °C by denaturation of the membrane protein. A similar effect of protein on phase behavior was observed in *Mycoplasma laidlawii* membranes using a calorimetric technique (Melchior et al., 1970). These observations suggest that at least one of the transition temperatures of sugar transport might involve specific lipid–protein interactions.

Spin labeling studies of hepatic microsomes (Eletr, Zakim and Vessey, 1973) and sarcoplasmic reticulum (Eletr and Inesi, 1972) show two thermal transitions. In both cases, the higher-temperature phase change appeared to depend upon the presence of native protein in the membrane, while the lower one was characteristic also of the isolated lipid. Arrhenius plots of the microsomal enzyme UDP–glucuronyltransferase showed two transition points, at the same temperatures as those found in the spin label studies (Eletr, Zakim and Vessey, 1973). This suggested that both bulk-phase lipid properties and specific lipid–protein interactions played a role in the activity of this enzyme. On the other hand, the plot for the microsomal enzyme glucose-6-phosphatase showed only one transition, at the lower temperature (Eletr, Zakim and Vessey, 1973). Thus, this enzyme apparently depends upon the bulk-phase properties of the membrane lipid, but is independent of the interactions which give rise to the higher-temperature transition.

Membrane proteins also alter the properties of membrane lipids. Vesicles prepared from extracted membrane lipids are generally more fluid and have lower thermal transitions than the parent membranes. This has been demonstrated for mitochondria (Steim, 1972), microsomes (Eletr, Zakim and Vessey, 1973), sarcoplasmic reticulum (Seelig and Hasselbach, 1971), several different microorganisms (Rottem et al., 1970; Tourtellotte, Branton and Keith, 1970; Esfehani et al., 1971; Melchior et al., 1970; Esser and Lanyi, 1973) and enveloped animal viruses (Sefton and Gaffney, 1974; Altstiel, Landsberger and Compans, 1975).

Spin-label studies of reconstituted lipid–protein systems suggest several different ways in which proteins can influence lipid structure. Hong and Hubbell (1972) have incorporated lipid-free rhodopsin into synthetic bilayers and have shown that the protein recovers its biological property of photoreactivation. The order parameter of phospholipid spin labels increases in direct proportion to the amount of rhodopsin incorporated in the bilayer. This effect is reminiscent of the effect of cholesterol on bilayer rigidity, and suggests that rhodopsin may simply act as a rigid structure to decrease the motion of the fatty acid chains. On the other hand, addition of glycophorin, the major glycoprotein of the red cell membrane, lowers and broadens the thermal transition of phosphatidylcholine (PC) bilayers (Grant and McConnell, 1974).

Jost et al. (1973a, b) have presented evidence that a layer of 'boundary lipid' is associated with cytochrome oxidase particles incorporated into bilayers. The spectrum obtained from fatty acid spin labels in this system is a composite of two different lipid phases. At a high protein: lipid ratio most of the spin label is in a highly immobilized phase, the 'boundary lipid'. Above a level of about 0.2 g lipid/1 g protein, however, the amount of 'boundary lipid' remains constant, and additional lipid shows the fluidity characteristic of the protein-free bilayer. These observations suggest that in this case the protein acts directly to bind a fixed amount of lipid, while the remainder of the lipid behaves essentially as it does in a protein-free bilayer.

It is clear from these studies that lipids and proteins do interact in membranes to modify each others' structural and functional properties in diverse ways. The variety of thermal effects presumably arises from the fact that neither the proteins nor the lipids of a membrane are homogeneously distributed throughout the membrane. The asymmetric distribution of proteins and lipids (Section 10.5) implies that profoundly different types of lipid–protein interactions must occur on each side of the membrane. Lateral segregation of both lipids and proteins occurs also, in a similar manner to that already described for binary lipid systems (Section 10.3). The complexity of lipid–protein interactions is further reflected in the wide range of different conditions required to solubilize different membrane proteins. Further work on a number of chemically defined systems of simpler composition will have to be undertaken before generalizations can emerge which might be of use in constructing models of membrane structure.

10.5 MEMBRANE ASYMMETRY

The structural asymmetry of biological membranes has long been assumed as a necessary correlate of the functional sidedness evident from transport and enzyme studies. It is now clear that this structural asymmetry is extensive, and involves all the major chemical components of the membrane.

Carbohydrates have been found exclusively on the outer surface of erythrocytes and other membranes (Nicolson and Singer, 1971). Carbohydrates are generally accessible to enzymes and antibodies directed against the outer surface of cells and of enveloped viruses, suggesting that localization on the outer surface is a general property of the carbohydrate groups in membranes. This generalization necessarily implies that those lipids and proteins which contain carbohydrates must likewise be asymmetrically arranged in the membrane.

The asymmetrical distribution of lipids and proteins which do not contain carbohydrates has also been firmly established for the red cell membrane, and numerous observations in the literature imply that a similar type of asymmetry characterizes other membranes as well. Evidence for the red cell membrane has come from two completely different experimental approaches—labeling experiments and modification with purified phospholipases. The conclusions from these two approaches are wholly consistent with one another; together they yield a detailed picture of the asymmetric distribution of proteins and lipids which exists in the red cell membrane.

Treatment of intact erythrocytes with relatively nonpenetrating, radioactive, group-specific reagents such as diazosulfanilic acid (Berg, 1969) or formylmethionylsulfone methyl phosphate (FMMP) (Bretscher, 1971a) labels essentially only those proteins which can be attacked by proteolytic enzymes. These are very limited in number, consisting of the major glycoprotein (called glycophorin by Marchesi et al., 1972) and another major protein of molecular weight ca. 105 000 daltons (Bretscher's component a). On the other hand, virtually all the membrane proteins of isolated ghosts are labeled under the same conditions (Bender, Garan and Berg, 1971; Carraway, Kobylka and Triplett, 1971; Bretscher, 1971a, b). When the nonpenetrating enzyme lactoperoxidase is used to label the intact cell with radioactive iodine, the label is found only in glycophorin and one smaller glycoprotein, but not in component a. Again, all the proteins of the isolated ghost are readily labeled (Phillips and Morrison, 1970, 1971a, b). Surprisingly, similar results were also obtained using a penetrating label, dansyl chloride (Schmidt-Ullrich, Knüfermann and Wallach, 1973). This has been interpreted to mean that the arrangement of protein in the isolated membrane may be different from that in the intact cell membrane, and labeling results simply reflect these differences (Zwaal, Roelofson and Colley, 1973). However, such a gross alteration of membrane structure during isolation seems unlikely since the isolation conditions are quite gentle, and enzyme and transport systems are generally preserved even with more rigorous isolation procedures commonly used to prepare other membranes. Further, no differences in lipid structure between isolated ghosts and intact cells could be demonstrated by spin labeling (Landsberger, Paxton and Lenard, 1971). A more likely explanation of the labeling results is that the interior of the intact cell differs from the exterior in pH, in concentrations of specific small molecules, and in hemoglobin, and that quite different reaction conditions are, therefore, present on each side of the membrane. Ghosts are freely permeable to both small and large molecules, so reaction conditions on both surfaces will be similar. Therefore, the asymmetric localization of proteins may be correctly revealed by the labeling techniques, even though the results do not depend upon the impermeability of the label. A further implication is that resealed ghosts need not have the same reactivity difference on their two sides as is found in intact cells, unless the label is *completely* nonpenetrating, a condition which is apparently not fulfilled by any of the small labels which have been employed (Schmidt-Ullrich, Knüfermann and Wallach, 1973).

Assuming, then, that the arrangement of membrane proteins is not grossly altered by the isolation of ghosts, the labeling experiments demonstrate that only two of the major polypeptides of the membrane—glycophorin and component a—are exposed on the outer surface, while virtually all the other major membrane polypeptides are exposed only on the inner surface. There is thus an asymmetry not only of the kind of polypeptide exposed on each surface, but also of the amount, since the glycoprotein and component a together comprise no more than about 35 percent of the total membrane protein.

Bretscher (1971a) has extended these labeling experiments to demonstrate that both glycophorin and component a are exposed on both surfaces of the membrane, and must therefore span the membrane from side to side. By radioautography of the fingerprints of the purified labeled proteins, he has demonstrated that different peptides of each protein are exposed on each surface of the membrane. It was concluded that both of these proteins spanned the membrane in an asymmetric fashion, and that neither protein was free to rotate about an axis parallel to the plane of the membrane. Such rotation would randomize the amino groups accessible to FMMP on each side of the membrane.

Labeling experiments have also demonstrated extensive asymmetry of the lipid components of the red cell membrane. Both FMMP (Bretscher, 1972a, b, 1973) and 2,4,6-trinitrobenzenesulfonate (Gordesky and Marinetti, 1973) are found to label the amino-containing phospholipids, phosphatidylethanolamine (PE) and phosphatidylserine (PS), far more readily in the isolated ghost than in the intact cell. This suggests that these lipids are located mainly on the inner surface of the membrane, while PC and sphingomyelin (Sph) are located chiefly on the outer surface. The bilayer of the red cell membrane is thus characterized by a predominance of ionizable primary amino groups on the inner surface, and a predominance of quaternary ammonium groups on the outer surface.

Experiments on the effects of various purified phospholipases on intact red cells and ghosts have recently been reviewed by Zwaal, Roelofson and Colley (1973). A few of the more pertinent observations are as follows:

1. Purified phospholipase A_2 from *Naja naja* hydrolyzes 68 percent of the PC in the intact cell, leaving the other glycerophospholipids untouched. However PC, PE and PS are all completely broken down if the isolated ghost is treated with this enzyme (Verkleij et al., 1973).
2. Intact bovine red cells, which contain large amounts of Sph and little PC, but substantial amounts of PE and PS, are not attacked by phospholipase A_2 (Zwaal, Roelofson and Colley, 1973).
3. Sphingomyelinase cleaves 80–85 percent of the Sph of intact human and pig red cells without lysis (Verkleij et al., 1973).
4. Successive treatment of human red cells with phospholipase A_2 and sphingomyelinase cleaves 48 percent of the total membrane phospholipid without lysis (Verkleij et al., 1973).
5. Evidence exists that the active sites of erythrocyte ATPases are located on the inner membrane surface (Marchesi and Palade, 1967a). Treatment of human erythrocyte ghosts with either phospholipase A_2 or C, both of which degrade all the glycerophospholipids, causes almost complete inactivation of both the Mg^{2+}- and the Na^+,K^+-ATPases. Treatment with sphingomyelinase, causing complete destruction of Sph, did not decrease either activity (Zwaal, Roelofson and Colley, 1973). Substantial reactivation of ghost Na^+,K^+-ATPase after inactivation by phospholipases could be achieved by addition of PS

but not PE or PC (Ohnishi and Kawamura, 1964; Zwaal, Roelofson and Colley, 1973).

These observations demonstrate that PC and Sph are preferentially exposed on the outer surface of the red cell, while PE and PS are exposed little or not at all. In agreement with this, PS specifically activates, and therefore must bind, enzymes which are on the inner surface of the membrane. The lipid asymmetry revealed by phospholipases is thus essentially identical with that shown by labeling studies.

Both the above labeling studies and phospholipase studies have received elegant confirmation through the work of Whiteley and Berg (1974). These workers used two chemically similar radioactive acetimidates, one of which completely penetrated the intact red cell, while the other was essentially nonpenetrating. Both labels reacted with exposed membrane amino groups essentially quantitatively under very mild conditions, to yield identical derivatives, with minimal alteration of membrane function. In addition to the two major proteins previously identified on the outer surface of the red cell, Whiteley and Berg detected four additional minor components using these reagents. The amino-containing phospholipids were found to be located predominantly on the inner surface, in agreement with earlier studies. These workers also provided additional evidence for the close structural similarity between the membrane of the intact red cell and the isolated ghost.

Several reports suggest that lipid asymmetry of the type demonstrated for red cells occurs in other membranes as well. A recent study of influenza virions demonstrated phospholipid asymmetry in the viral bilayer similar to that of the red blood cell (Tsai and Lenard, 1975). After digestion of the viral phospholipid with phospholipase C, the repurified particle was found to have lost over 80 percent of its PC, up to 70 percent of its Sph, but much smaller amounts (30–40 percent) of its PS, PI and PE (Tsai and Lenard, 1975). Since these viral particles form by budding from the host cell plasma membrane (*see* Compans and Choppin, 1975), it seems likely that this asymmetric arrangement of lipids reflects that of the host cell plasma membrane.

There is also evidence suggesting that the myelin bilayer is asymmetric. The asymmetric electron density profile found in X-ray diffraction studies (Caspar and Kirschner, 1971) has been attributed to preferential localization of cholesterol in the outer leaf of the bilayer. Further, the two major polypeptides of myelin are thought to be asymmetrically located in the membrane, and are found to interact with different lipids. The A_1 basic protein, which is the only protein thought to be present in the intraperiod line (extracellular surface) of myelin (Dickinson *et al.*, 1970) interacts specifically with cerebroside sulfate, and to a lesser extent with other acidic lipids, but not with neutral lipids (Demel *et al.*, 1973). These observations are consistent with a stable asymmetric structure for myelin, with acidic lipids preferentially localized on the extracellular surface, and the specific binding of lipids by the different asymmetrically distributed proteins.

The finding that in some cases PC and Sph on the outer membrane surface can be extensively digested by phospholipases with very little digestion of PE and PS suggests that the asymmetry of the lipid is stable, with very little exchange occurring between the inner and outer surfaces. A stable asymmetric bilayer in membranes is consistent with spin label studies by Kornberg and McConnell (1971a), which demonstrated that spin labels incorporated into PC vesicles undergo 'flip-flop' from one surface to the other extremely slowly. A half-time of 'flip-flop' in PC vesicles in excess of 11 days has recently been reported (Rothman and Dawidowicz, 1975).

10.6 ARRANGEMENT AND MOTION OF MEMBRANE PROTEINS

The organization of biological membranes may be considered in terms of two alternative possibilities (Singer, 1971). As one possibility, the membrane might be stabilized by a protein matrix containing interspersed pockets of lipid bilayer. The alternative is a lipid bilayer matrix containing interspersed proteins. Both extremes could satisfy the basic thermodynamic requirements for macromolecular assembly (Singer, 1971), but each would have quite different properties. A protein-matrix membrane would be characterized by long-range order and immobile proteins, while a lipid-matrix membrane should possess no long-range order, and have mobile proteins.

With the emergence of evidence demonstrating the mobility of various membrane proteins (*see* Edidin, 1974), models of membrane structure based on a lipid matrix were proposed (Singer and Nicolson, 1972; Wallach, 1972). The evidence that many proteins do move in functioning, intact membranes is convincing, and will be discussed in Section 10.6.1. Left unexplained by these models, however, are the ways in which integration of function occurs over large portions of a membrane, involving distances of tens or hundreds of nanometers.

As a possible answer to this dilemma, evidence is discussed in Section 10.6.2 which suggests the existence of a second, protein matrix which might provide the long-range order required for many membrane functions. The precise nature of this second matrix is not yet clear, and the different reports suggesting its existence in specific membranes show great variability with regard to structure, function, composition, and even whether or not it is present. This is in contrast to the lipid bilayer matrix, which appears to be a ubiquitous feature of biological membranes.

10.6.1 Motions of proteins in the lipid matrix

It may be useful to distinguish at the outset between three different possible types of diffusional motion which proteins might undergo in membranes: (a) lateral diffusion of the protein molecule or complex in the plane of the membrane; (b) rotational diffusion about an axis perpendicular to the membrane surface, and (c) rotational diffusion about an axis parallel to the membrane surface. Evidence for the widespread occurrence of (a) has come from a variety of approaches. It has also been shown that rhodopsin in rod outer segment membranes undergoes the second type of motion. Interestingly, no evidence has yet been obtained for the third type of motion, which is often assumed to be the basis of carrier-mediated transport phenomena. Rotational and translational diffusion of membrane protein is the subject of an excellent recent review (Edidin, 1974).

Quantitative measurements have been made of the rate of motions (a) and (b) of rhodopsin in the rod outer segment membrane. Rotational diffusion (b) was measured from the rate of decay of photodichroism induced by bleaching with a flash of polarized light (Cone, 1972). The half-life of this decay was around 20 µs. The induced photodichroism did not decay if the proteins were first immobilized by cross-linking with glutaraldehyde (Brown, 1972; Cone, 1972). Lateral diffusion was measured by bleaching one half of a rod and measuring the rate of randomization of the molecules between the bleached and unbleached halves. Randomization occurred with a half-life of 23 s in the mudpuppy rod ($\sim$12 µm across) at 20 °C (Poo and Cone, 1973). Thus the predominant protein of this membrane undergoes extremely rapid rotational (b) and lateral (a) diffusion.

The immunofluorescence experiments of Frye and Edidin (1970) provided a dramatic demonstration of lateral diffusion of proteins in plasma membranes. Antibodies fluorescing with different colors were used to visualize specific mouse and human antigens on the surface of a mouse–human heterokaryon formed by viral fusion. Complete randomization of the two antigens over the cell surface occurred within 40 minutes at 37 °C, but did not occur at 4 °C. The randomization process did not require metabolic energy, and was independent of protein synthesis. These observations demonstrated that a passive lateral diffusion process was occurring. Similar observations have been made on the plasma membranes of cultured muscle fibers (Edidin and Fambrough, 1973).

Several different electron-microscopic techniques have been used to confirm that lateral diffusion of membrane proteins is a quite general phenomenon. Freeze-fracture techniques, and techniques of coupling specific surface proteins to electron-dense markers have been most useful in these studies.

The freeze-fracture technique is known to split many membranes between the two monolayers of the bilayer, thus exposing the membrane interior to view (Pinto da Silva and Branton, 1970; Tillack and Marchesi, 1970). The interior generally contains numerous particles which vary in appearance in different membranes and which are not found in protein-free synthetic bilayers (Deamer et al., 1970). If the red cell membrane is progressively digested away by proteases, the particles first aggregate, then disappear (Branton, 1971; Speth et al., 1972). The number of particles seen in membranes of *Mycoplasma laidlawii* decreases over a period of time if protein synthesis is prevented (Tourtellotte and Zupnik, 1973). In a reconstituted system containing only phospholipids and rhodopsin, the number of particles seen by freeze-fracture is directly proportional to the amount of rhodopsin incorporated into the bilayer (Hong and Hubbell, 1972; Hong, Chen and Hubbell, 1973). The rearrangement of particles seen by freeze-fracture electron microscopy can thus be confidently interpreted in terms of lateral motion of membrane proteins.

The particles in red cell membranes generally appear to be randomly distributed, singly or in small clusters, throughout the membrane under the normal conditions of observation. These particles can undergo reversible aggregation in response to certain (nonphysiological) conditions of pH and ionic strength, thereby moving in the plane of the membrane in response to electrostatic forces acting at the membrane surface (Pinto da Silva, 1972). Aggregation also occurs after proteolytic removal of the accessible (and presumably more hydrophilic) protein on the membrane surface (Branton, 1971; Speth *et al.*, 1972). The state of aggregation of the particles seen in reconstituted rhodopsin-bilayer systems has been shown to depend both upon the state of the protein (photoreactivated or nonphotoreactivated) and upon the physical state of the lipid bilayer (Hong and Hubbell, 1972; Hong, Chen and Hubbell, 1972).

Similar conclusions have been drawn from studies of the distribution of lectin and antibody binding sites on cell surfaces using lectins or antibodies conjugated to ferritin (Nicolson and Singer, 1971). Changes in the cell surface brought about by viral or chemical transformation or trypsinization cause a clustering of these binding sites, a process which requires lateral diffusion of membrane proteins (Singer and Nicolson, 1972; Nicolson, 1971, 1972). The observed clustering was shown to occur in response to an external agent, in this case the lectin itself (Nicolson, 1973a; Rosenblith *et al.*, 1973).

It is clear from the foregoing discussion that many proteins in different membranes can undergo lateral diffusional motions in response to external stimuli. The importance of such motions in naturally occurring membrane processes, however, is not established by these experiments. It should be remembered that only Cone's measurements (1972) of rotational and lateral diffusion of rhodopsin in rod outer segments were carried out on intact, functioning membranes using physiological stimuli. The motions demonstrated in other systems occurred as a response to external agents which do not characterize normal cellular function.

10.6.2 The protein matrix

The observations suggesting the existence of a protein matrix in certain membranes are diverse and varied. If some membranes do indeed possess a level of organization which is characterized by protein–protein interactions over large distances, this level of organization apparently reflects differences of membrane function far more closely than it reflects similarities of membrane structure, such as have been discussed up to this point. It is particularly with regard to a protein matrix that a future model will have to depict a particular membrane, rather than membranes in general.

It was originally demonstrated by Fleischer, Fleischer and Stoeckenius (1967) that the unit membrane appearance of fixed, stained mitochondria was retained after removal of over 90 percent of the mitochondrial lipid. More recently, Demel *et al.* (1973) have reported that after complete destruction of red cell membrane lipids by a combination of phospholipases and lipases, the ghosts retain their vesicular structure as seen by phase contrast microscopy, and less than 1 percent of the membrane protein is released into the medium. Yu, Fischman and Steck (1973) have performed selective extraction experiments on red cell membranes using nonionic detergents. After removal of all glycerolipid, all glycoprotein, and other major protein classes, there remained a 'filamentous meshwork of inner surface polypeptides' which could be seen by electron microscopy.

Speculation regarding a protein matrix in the red cell membrane has centered on the protein spectrin. The filamentous nature of this protein was recognized (Marchesi and Palade, 1967b) even before it had been isolated (Marchesi and Steers, 1968) and characterized (Marchesi et al., 1970). This water-soluble protein of very high molecular weight has been localized on the inner surface of the red cell membrane by electron microscopy using ferritin–antibody conjugates (Nicolson, Marchesi and Singer, 1971). Nicolson (1973b) has shown that specific anti-spectrin antibody can cause rearrangement of glycoprotein, as monitored by use of a sialic-acid-specific label directed against the outer surface. Thus, the filamentous protein spectrin, located on the inner membrane surface, can control, through a glycoprotein which spans the membrane (Bretscher, 1971c), the topography of specific binding sites on the outer membrane surface. Further, Elgsaeter and Branton (1974) have presented evidence that when spectrin is present on the inner surface of red cell ghosts, the movement of the particles seen in freeze-fracture electron microscopy is prevented in a manner similar to that observed in intact red cells.

There is evidence that a protein matrix of quite a different type comprises part of the membrane structure of many enveloped viruses. In several classes of enveloped viruses, specifically myxo-, paramyxo- and rhabdoviruses, the viral polypeptide of lowest molecular weight (ca. 26 000–45 000 daltons) forms a continuous structure which, as suggested by electron-microscope and fluorescence measurements, is located immediately beneath the viral bilayer. The evidence for such a structure has recently been reviewed (Lenard and Compans, 1974). This structure may play an important role in organizing the viral constituents under the plasma membrane of the host cell prior to release of the mature virion by budding. It may also impart mechanical stability to the completed viral particle.

There has been a great deal of recent interest and speculation regarding microfilaments as long-range mediators of membrane function. Microfilaments are visualized in the electron microscope as a network of fibers of thicknesses varying from 4–8 nm located at the interior surface of the membranes of many (but not all) cells and cellular protrusions (Buckley and Porter, 1967; Perdue, 1973). They have been observed to traverse the membrane to the exterior of the cell, and it has been suggested that they participate in a variety of processes, including cell motility, adhesion, cytokinesis, transport, and lateral diffusion of membrane proteins (Jahn and Bovey, 1969).

Microfilaments exhibit contractility and possess ATPase activity. A close structural similarity between microfilaments and actomyosin has been demonstrated by 'decorating' microfilaments with muscle meromyosin to produce arrow-like structures similar to those seen when actin filaments are similarly 'decorated' (Ishikawa, Bischoff and Holtzer, 1969; Nachmias, Huxley and Kessler, 1970; Pollard et al., 1970; Pollard and Korn, 1971; Tilney and Mooseker, 1971). Further, the fungal alkaloid cytochalasin B, which disrupts microfilament structure, interferes also with actomyosin formation by competing with myosin for its specific binding sites on actin molecules (Spudich and Lin, 1972). Recent immunofluorescence experiments have provided strong support for the contention that both actin and myosin are components of extensive microfilamentous structures in several different cells types (Lazarides and Weber, 1974; Weber and Groeschel-Stewart, 1974).

The study of the role of microfilaments in various cell functions has relied heavily on the fact that several fungal alkaloids, e.g. cytochalasin A and B, demecolcine, colchicine and vinblastine, interact with microfilaments and destroy their structure (Wessels *et al.*, 1971), even though these compounds probably exert other effects on the membrane as well (Estensen, Rosenberg and Sheridan, 1971; Wunderlich, Müller and Speth, 1973). The alkaloids do have a profound effect on many membrane functions, including sugar transport (Estensen and Plagemann, 1972; Kletzien, Perdue and Springer, 1972; Mizel and Wilson, 1972; Bloch, 1973), the spatial separation of transport from phagocytic function (Ukena and Berlin, 1972), uptake of particles (Zurier, Hoffstein and Weissmann, 1973), lateral aggregation of surface binding sites (Edidin and Weiss, 1972; Yin, Ukena and Berlin, 1972), cell shape (Puck, Waldren and Hsie, 1972) and lymphocyte-mediated cytotoxicity (Cerottini and Bruner, 1972).

The question arises as to whether these microfilaments, extending in one direction deep into the cytoplasm and in the other perhaps out of the cell altogether, mediating both membrane and nonmembrane functions, are to be considered as authentic 'membrane proteins'. As Steck and Fox (1972) have pointed out, the question may not be a meaningful one 'since the cell itself may not regard the membrane as a discrete structure but rather in organizational continuity with the cytoplasm'. It follows that any membrane model may eventually prove incomplete unless it specifies the structural arrangements of the entire cell.

Acknowledgements

The authors gratefully acknowledge the Marine Biological Laboratory, Woods Hole, Massachusetts, for the use of their library facilities. One of the authors (F.R.L.) was a member of the Department of Chemistry, Indiana University, Bloomington, Indiana, during the preparation of most of this review. This work was supported by National Science Foundation grants GB 43872 and GB 36789. Frank R. Landsberger is an Andrew W. Mellon Foundation Fellow.

REFERENCES

ALTSTIEL, L. D., LANDSBERGER, F. R. and COMPANS, R. W. (1975). *Annual Meeting of the American Society for Microbiology, New York*, p. 238.
ANSELL, G. B., HAWTHORNE, J. N. and DAWSON, R. M. C. (1973). *Form and Function of Phospholipids*, BBA Library, Vol. 3. Amsterdam; Elsevier.
BALDASSARE, J. J., MCAFEE, A. G. and HO, C., (1973). *Biochem. biophys. Res. Commun.*, **53**:617.
BENSON, A. A. (1966). *J. Am. Oil Chem. Soc.*, **43**:265.
BENDER, W. W., GARAN, H. and BERG, H. C. (1971). *J. molec. Biol.*, **58**:783.
BERG, H. C. (1969). *Biochim. biophys. Acta*, **183**:65.
BLOCH, R. (1973). *Biochemistry*, **12**:4799.
BRANTON, D. (1971). *Phil. Trans. R. Soc., Ser. B*, **261**:133.
BRETSCHER, M. S. (1971a). *J. molec. Biol.*, **59**:351.
BRETSCHER, M. S. (1971b). *J. molec. Biol.*, **58**:775.
BRETSCHER, M. S. (1971c). *Nature, New Biol.*, **231**:229.
BRETSCHER, M. S. (1972a). *Nature, New Biol.*, **236**:11.
BRETSCHER, M. S. (1972b). *J. molec. Biol.*, **71**:523.
BRETSCHER, M. S. (1973). *Science, N.Y.*, **181**:622.

BROWN, P. K. (1972). *Nature, New Biol.*, **236**:35.
BUCKLEY, I. K. and PORTER, K. R. (1967). *Protoplasma*, **64**:349.
CARRAWAY, K. L., KOBYLKA, D. and TRIPLETT, R. B. (1971). *Biochim. biophys. Acta*, **241**:934.
CASPAR, D. L. D. and KIRSCHNER, D. A. (1971). *Nature, New Biol.*, **231**:46.
CEROTTINI, J. C. and BRUNER, K. T. (1972). *Nature, New Biol.*, **237**:272.
CHAN, S. I., SEITER, C. H. A. and FEIGENSON, G. W. (1972). *Biochem. biophys. Res. Commun.*, **46**:1488.
CHAPMAN, D. (1973). *Biological Membranes*, Vol. 2, pp. 91–144. Ed. D. CHAPMAN and D. F. H. WALLACH. New York; Academic Press.
COMPANS, R. W. and CHOPPIN, P. W. (1975). *Comprehensive Virology*, Vol. 4, p. 179. Ed. H. FRAENKEL-CONRAT and R. R. WAGNER. New York: Plenum Press.
CONE, R. A. (1972). *Nature, New Biol.*, **236**:39.
DANIELLI, J. F. and DAVSON, H. (1935). *J. cell. Physiol.*, **5**:495.
DARKE, A., FINER, E. G., FLOOK, A. G. and PHILLIPS, M. C. (1972). *J. molec. Biol.*, **63**:265.
DAVIS, D. G. and INESI, G. (1971). *Biochim. biophys. Acta*, **241**:1.
DEAMER, D. W., LEONARD, R., TARDIEU, A. and BRANTON, D. (1970). *Biochim. biophys. Acta*, **219**:47.
DEMEL, R. A., LONDON, Y., GEURTS VAN KESSEL, W. S. M., VOSSENBERG, F. G. A. and VAN DEENEN, L. L. M. (1973). *Biochim. biophys. Acta*, **311**:507.
DEVAUX, P. and MCCONNELL, H. M. (1972). *J. Am. chem. Soc.*, **94**:4475.
DEVAUX, P., SCANDELLA, C. J. and MCCONNELL, H. M. (1973). *J. magn. Res.*, **9**:474.
DICKINSON, J. P., JONES, K. M., APARICIO, S. R. and LUMSDEN, C. E. (1970). *Nature, New Biol.*, **227**:1133.
EDIDIN, M. (1974). *A. Rev. Biophys. Bioengng*, **3**:179.
EDIDIN, M. and FAMBROUGH, D. (1973). *J. Cell Biol.*, **57**:27.
EDIDIN, M. and WEISS, L. (1972). *Proc. natn. Acad. Sci. U.S.A.*, **69**:2456.
ELETR, S. and INESI, G. (1972). *Biochim. biophys. Acta*, **290**:178.
ELETR, S., ZAKIM, D. and VESSEY, D. A. (1973). *J. molec. Biol.*, **78**:351.
ELGSAETER, A. and BRANTON, D. (1974). *J. Cell Biol.*, **63**:1018.
ENGELMAN, D. M. and ROTHMAN, J. E. (1972). *J. biol. Chem.*, **247**:3694.
ESFEHANI, M., BARNES, E. M. and WAKIL, S. J. (1969). *Proc. natn. Acad. Sci. U.S.A.*, **64**:1057.
ESFEHANI, M. and WAKIL, S. J. (1972). *Fedn Proc. Fedn Am. Socs exp. Biol.*, **31**:413.
ESFEHANI, M., LIMBRICK, A. R., KNUTTON, S., OKA, T. and WAKIL, S. J. (1971). *Proc. natn. Acad. Sci. U.S.A.*, **68**:3180.
ESSER, A. F. and LANYI, J. K. (1973). *Biochemistry*, **12**:1933.
ESTENSEN, R. D. and PLAGEMANN, P. G. W. (1972). *Proc. natn. Acad. Sci. U.S.A.*, **69**:1430.
ESTENSEN, R. D., ROSENBERG, M. and SHERIDAN, J. D. (1971). *Science, N.Y.*, **173**:356.
FLEISCHER, S., FLEISCHER, B. and STOECKENIUS, W. (1967). *J. Cell Biol.*, **32**:193.
FRYE, C. D. and EDIDIN, M. (1970). *J. Cell Sci.*, **7**:319.
GODICI, P. E. and LANDSBERGER, F. R. (1974). *Biochemistry*, **13**:362.
GODICI, P. E. and LANDSBERGER, F. R. (1975). *Biochemistry*, **14**:3927.
GORDESKY, S. E. and MARINETTI, G. V. (1973). *Biochem. biophys. Res. Commun.*, **50**:1027.
GORTER, E. and GRENDEL, F. (1925). *J. exp. Med.*, **41**:439.
GRANT, C. W. M. and MCCONNELL, H. M. (1973). *Proc. natn. Acad. Sci. U.S.A.*, **70**:1238.
GRANT, C. W. M. and MCCONNELL, H. M. (1974). *Proc. natn. Acad. Sci. U.S.A.*, **71**:4653.
GREEN, D. E. and PERDUE, J. F. (1966). *Proc. natn. Acad. Sci. U.S.A.*, **55**:1295.
HABERLAND, M. E. and REYNOLDS, J. A. (1973). *Proc. natn. Acad. Sci. U.S.A.*, **70**:2313.
HINZ, H. J. and STURTEVANT, J. M. (1972). *J. biol. Chem.*, **247**:3697.
HONG, K., CHEN, Y. S. and HUBBELL, W. L. (1973). *J. supramolec. Struct.*, **1**:355.
HONG, K. and HUBBELL, W. L. (1972). *Proc. natn. Acad. Sci. U.S.A.*, **69**:2617.
HORWITZ, A. F., HORSLEY, W. J. and KLEIN, M. P. (1972). *Proc. natn. Acad. Sci. U.S.A.*, **69**:590.
HUBBELL, W. L. and MCCONNELL, H. M. (1969). *Proc. natn. Acad. Sci. U.S.A.*, **64**:20.
HUBBELL, W. L. and MCCONNELL, H. M. (1971). *J. Am. chem. Soc.*, **93**:314.
ISHIKAWA, H., BISCHOFF, R. and HOLTZER, H. (1969). *J. Cell Biol.*, **43**:312.
JAHN, T. L. and BOVEY, E. C. (1969). *Physiol. Rev.*, **49**:793.
JOST, P., GRIFFITH, O. H., CAPALDI, R. A. and VANDERKOOI, G. (1973a). *Biochim. biophys. Acta*, **311**:141.
JOST, P., GRIFFITH, O. H., CAPALDI, R. A. and VANDERKOOI, G. (1973b). *Proc. natn. Acad. Sci. U.S.A.*, **70**:480.
JOST, P., LIBERTINI, L. J., HEBERT, V. C. and GRIFFITH, O. H. (1971). *J. molec. Biol.*, **59**:77.
KAVANAU, J. L. (1966). *Fedn Proc. Fedn Am. Socs exp. Biol.*, **25**:1096.

KEOUGH, K. M., OLDFIELD, E. and CHAPMAN, D. (1973). *Chem. Phys. Lipids*, **10**:37.
KIEHN, E. D. and HOLLAND, J. J. (1968). *Proc. natn. Acad. Sci. U.S.A.*, **61**:1370.
KLETZIEN, R. F., PERDUE, J. F. and SPRINGER, A. (1972). *J. biol. Chem.*, **247**:2964.
KORNBERG, R. D. and MCCONNELL, H. M. (1971a). *Biochemistry*, **10**:1111.
KORNBERG, R. D. and MCCONNELL, H. M. (1971b). *Proc. natn. Acad. Sci. U.S.A.*, **68**:2564.
KRAEMER, P. M. (1971a). *Biochemistry*, **10**:1437.
KRAEMER, P. M. (1971b). *Biochemistry*, **10**:1445.
LANDSBERGER, F. R., PAXTON, J. and LENARD, J. (1971). *Biochim. biophys. Acta*, **266**:1.
LAZARIDES, E. and WEBER, K. (1974). *Proc. natn. Acad. Sci. U.S.A.*, **71**:2268.
LEE, A. G., BIRDSALL, N. J. M. and METCALFE, J. C. (1973). *Biochemistry*, **12**:1650.
LEE, A. G., BIRDSALL, N. J. M., LEVINE, Y. K. and METCALFE, J. C. (1972). *Biochim. biophys. Acta*, **255**:43.
LENARD, J. and COMPANS, R. W. (1974). *Biochim. biophys. Acta, Reviews on Biomembranes*, **344**:51.
LEVINE, Y. K. and WILKINS, M. H. F. (1971). *Nature, New Biol.*, **230**:69.
LEVINE, Y. K., BIRDSALL, N. J. M., LEE, A. G. and METCALFE, J. C. (1972). *Biochemistry*, **11**:1416.
LIBERTINI, L. J., WAGGONER, A. S., JOST, P. C. and GRIFFITH, O. H. (1969). *Proc. natn. Acad. Sci. U.S.A.*, **64**:13.
LINDEN, C. D., WRIGHT, K. L., MCCONNELL, H. M. and FOX, C. F. (1973). *Proc. natn. Acad. Sci. U.S.A.*, **70**:2271.
LUCY, J. A. (1964). *J. theor. Biol.*, **7**:360.
MACHTIGER, N. A. and FOX, C. F. (1973). *A. Rev. Biochem.*, **42**:575.
MARCHESI, S. L., STEERS, E., MARCHESI, V. T. and TILLACK, T. W. (1970). *Biochemistry*, **9**:50.
MARCHESI, V. T. and PALADE, G. E. (1967a). *J. Cell Biol.*, **35**:385.
MARCHESI, V. T. and PALADE, G. E. (1967b). *Proc. natn. Acad. Sci. U.S.A.*, **58**:991.
MARCHESI, V. T. and STEERS, E., JR. (1968). *Science, N.Y.*, **159**:203.
MARCHESI, V. T., TILLACK, T. W., JACKSON, R. L., SEGREST, J. P. and SCOTT, R. E. (1972). *Proc. natn. Acad. Sci. U.S.A.*, **69**:1445.
MAVIS, R. D. and VAGELOS, P. R. (1972). *J. biol. Chem.*, **247**:652.
MCCONNELL, H. M. and MCFARLAND, B. G. (1972). *Ann. N.Y. Acad. Sci.*, **195**:207.
MCCONNELL, H. M., WRIGHT, K. L. and MCFARLAND, B. G. (1972). *Biochem. biophys. Res. Commun.*, **47**:273.
MCNAMEE, M. G. and MCCONNELL, H. M. (1973). *Biochemistry*, **12**:2951.
MELCHIOR, D. L., MOROWITZ, H. J., STURTEVANT, J. M. and TSONG, T. Y. (1970). *Biochim. biophys. Acta*, **219**:114.
METCALFE, J. C., BIRDSALL, N. J. M., FEENEY, J., LEE, A. G., LEVINE, Y. K. and PARTINGTON, P. (1971). *Nature, Lond.*, **233**:199.
MIZEL, S. B. and WILSON, L. (1972). *J. biol. Chem.*, **247**:4102.
NACHMIAS, V. T., HUXLEY, H. E. and KESSLER, D. (1970). *J. molec. Biol.*, **50**:83.
NICOLSON, G. L. (1971). *Nature, New Biol.*, **233**:244.
NICOLSON, G. L. (1972). *Nature, New Biol.*, **239**:193.
NICOLSON, G. L. (1973a). *Nature, New Biol.*, **243**:218.
NICOLSON, G. L. (1973b). *J. supramolec. Struct.*, **1**:410.
NICOLSON, G. L., MARCHESI, V. T. and SINGER, S. J. (1971). *J. Cell Biol.*, **51**:265.
NICOLSON, G. L. and SINGER, S. J. (1971). *Proc. natn. Acad. Sci. U.S.A.*, **68**:942.
OHNISHI, T. and KAWAMURA, H. (1964). *J. Biochem., Tokyo*, **56**:377.
OLDFIELD, E. and CHAPMAN, D. (1972). *FEBS Lett.*, **23**:285.
OVERATH, P., SCHAIRER, H. V. and STOFFEL, W. (1970). *Proc. natn. Acad. Sci. U.S.A.*, **67**:606.
PERDUE, J. F. (1973). *J. Cell Biol.*, **58**:265.
PHILLIPS, M. C. and FINER, E. G. (1974). *Biochim. biophys. Acta*, **356**:199.
PHILLIPS, M. C., LADBROOKE, B. D. and CHAPMAN, D. (1970). *Biochim. biophys. Acta*, **196**:35.
PHILLIPS, D. R. and MORRISON, M. (1970). *Biochem. biophys. Res. Commun.*, **40**:284.
PHILLIPS, D. R. and MORRISON, M. (1971a). *Biochemistry*, **10**:1766.
PHILLIPS, D. R. and MORRISON, M. (1971b). *FEBS Lett.*, **18**:95.
PINTO DA SILVA, P. (1972). *J. Cell Biol.*, **53**:777.
PINTO DA SILVA, P. and BRANTON, D. (1970). *J. Cell Biol.*, **45**:598.
POLLARD, T. D. and KORN, E. D. (1971). *J. Cell Biol.*, **48**:216.
POLLARD, T. D., SHELTON, E., WEIHING, R. R. and KORN, E. D. (1970). *J. molec. Biol.*, **50**:91.
POO, M. and CONE, R. A. (1973). *J. supramolec. Struct.*, **1**:354.
PUCK, T. T., WALDREN, C. A. and HSIE, A. W. (1972). *Proc. natn. Acad. Sci. U.S.A.*, **69**:1943.

ROSENBLITH, J. Z., UKENA, T. E., YIN, H. H., BERLIN, R. D. and KARNOVSKY, M. J. (1973). *Proc. natn. Acad. Sci. U.S.A.*, **70**:1625.
ROTHMAN, J. E. and DAWIDOWICZ, E. A. (1975). *Biochemistry*, **14**:2809.
ROTTEM, S., HUBBELL, W. L., HAYFLICK, L. and MCCONNELL, H. M. (1970). *Biochim. biophys. Acta*, **219**:104.
SCANDELLA, C. J., DEVAUX, P. and MCCONNELL, H. M. (1972). *Proc. natn. Acad. Sci. U.S.A.*, **69**:2056.
SCHMIDT-ULLRICH, R., KNÜFERMANN, H. and WALLACH, D. F. H. (1973). *Biochim. biophys. Acta*, **307**:353.
SEELIG, J. and HASSELBACH, W. (1971). *Eur. J. Biochem.*, **21**:17.
SEFTON, B. M. and GAFFNEY, B. J. (1974). *J. molec. Biol.*, **90**:343.
SHEETZ, M. P. and CHAN, S. I. (1972). *Biochemistry*, **11**:548.
SHIMSHICK, E. J. and MCCONNELL, H. M. (1973a). *Biochem. biophys. Res. Commun.*, **53**:446.
SHIMSHICK, E. J. and MCCONNELL, H. M. (1973b). *Biochemistry*, **12**:2351.
SHIMSHICK, E. J., KLEEMAN, W., HUBBELL, W. L. and MCCONNELL, H. M. (1973). *J. supramolec. Struct.*, **1**:285.
SINGER, S. J. (1971). *Structure and Function of Membranes*, p. 145. Ed. L. ROTHFIELD. New York; Academic Press.
SINGER, S. J. and NICOLSON, G. L. (1972). *Science, N.Y.*, **175**:720.
SMITH, R. and TANFORD, C. (1972). *J. molec. Biol.*, **67**:75.
SPETH, V., WALLACH, D. F. H., WEIDEKAMM, E. and KNÜFERMANN, H. (1972). *Biochim. biophys. Acta*, **255**:386.
SPUDICH, A. A. and LIN, S. (1972). *Proc. natn. Acad. Sci. U.S.A.*, **69**:442.
STECK, T. L. and FOX, C. F. (1972). *Membrane Molecular Biology*, pp. 27–75, Ed. C. F. FOX and A. D. KEITH. Stamford, Connecticut; Sinauer Associates.
STEIM, J. M. (1972). *Mitochondrial Biomembranes*, pp. 185–195. Amsterdam; North-Holland.
STEIM, J. M., TOURTELOTTE, M. E., REINERT, J. C., MCELHANEY, R. N. and RADER, R. L. (1969). *Proc. natn. Acad. Sci. U.S.A.*, **63**:104.
TANFORD, C. (1972). *J. molec. Biol.*, **67**:59.
TILLACK, T. W. and MARCHESI, V. T. (1970). *J. Cell Biol.*, **45**:649.
TILNEY, L. G. and MOOSEKER, M. (1971). *Proc. natn. Acad. Sci. U.S.A.*, **68**:2611.
TOURTELLOTTE, M. E., BRANTON, D. and KEITH, A. (1970). *Proc. natn. Acad. Sci. U.S.A.*, **66**:909.
TOURTELLOTTE, M. E. and ZUPNIK, J. S. (1973). *Science, N.Y.*, **179**:84.
TSAI, K. W. and LENARD, J. (1975). *Nature, Lond.*, **253**:554.
UKENA, T. E. and BERLIN, R. D. (1972). *J. exp. Med.*, **136**:1.
VERKLEIJ, A. J., ZWAAL, R. F. A., ROELOFSON, B., COMFURIUS, P., KASTELIJN, D. and VAN DEENEN, L. L. M. (1973). *Biochim. biophys. Acta*, **323**:178.
WALLACH, D. F. H. (1972). *The Plasma Membrane: Dynamic Perspectives, Genetics and Pathology*. New York; Springer-Verlag.
WEBER, K. and GROESCHEL-STEWART, U. (1974). *Proc. natn. Acad. Sci. U.S.A.*, **71**:4561.
WESSELS, N. K., SPOONER, B. S., ASH, J. F., BRADLEY, M. O., LUDUENA, M. A., TAYLOR, E. L., WRENN, J. T. and YAMADA, K. (1971). *Science, N.Y.*, **171**:135.
WHITELEY, N. M. and BERG, H. C. (1974). *J. molec. Biol.*, **87**:541.
WILKINS, M. H. F., BLAUROCK, A. E. and ENGELMAN, D. M. (1971). *Nature, New Biol.*, **230**:72.
WILSON, G., ROSE, S. P. and FOX, C. F. (1970). *Biochem. biophys. Res. Commun.*, **38**:617.
WUNDERLICH, F., MÜLLER, R. and SPETH, V. (1973). *Science, N.Y.*, **182**:1136.
YIN, Y. H., UKENA, T. E. and BERLIN, R. D. (1972). *Science, N.Y.*, **178**:867.
YU, J., FISCHMAN, D. A. and STECK, T. L. (1973). *J. supramolec. Struct.*, **1**:233.
ZURIER, R. B., HOFFSTEIN, S. and WEISSMANN, G. (1973). *Proc. natn. Acad. Sci. U.S.A.*, **70**:844.
ZWAAL, R. F. A., ROELOFSON, B. and COLLEY, C. M. (1973). *Biochim. biophys. Acta*, **300**:159.

Index

Index

Acanthocytes, 156
Acetylcholine, 175
Acetylcholinesterase,
  erythrocyte surface, on, 179
  loss of, 64
  removal of, 40
Acholeplasma membranes,
  lipid bilayers, 82
  physical properties, 133
Acid phosphatase,
  Golgi system, in, 168
  lysosomes, in, 16, 186
Acinar cells, zymogen granules of, 23
Actin,
  filaments, 25
  molecular weight, 67
Actomyosin, 105, 216
  microfilaments and, 260
Acyltransferase, 230
Adenyl cyclase, 174
  aggregating intramembranous particles and, 199
  as membrane marker, 59
Adipose cells, lipid droplets in, 28
Affinity chromatography, in isolation of membranes, 58
Affinity density perturbation method of membrane isolation, 57
Alamethicin, 120
Alkaline phosphatase, 176
Alkaline phosphodiesterase, 186
Amino acid composition of protein in membranes, 104
D-Amino acid oxidase, 19
Aminopeptidase, 166
Amoeba, pinocytosis in, 199, 207, 208, 217
Anchorage-dependent cells, cultivation, 38
Annular diaphragms, 19
Annulate lamellae, 18

Antibody,
  interaction with antigen, 212
  membrane response to, 211
Antigen–antibody interaction, 212
Artificial capillary system in cell cultivation, 39
Aryl sulfatase, 181
Asymmetry of membranes, 253
ATPase in mitochondrial membranes, 107
Attachment plaques, 5

Bacterial membranes, composition of, 98
Bacteriophage coats, protein layers, 88
Basement membrane, 6
Basidiobolus ranarum membrane, 157
Behavior of membranes, 144
Bilayers (*see also* Phospholipid bilayers *and* Lipid bilayers),
  aggregation of particles in, 213
  homogeneity of structure, 87
  phospholipid diffusion along, 112
  protein penetration into, 87, 247
Bile canaliculi, microvilli, 7
Biogenesis of membranes, 224–243
  mitochondria, in, 227
  model systems, 225
Blasia pusilla, 241
Blebbing (blistering), 7
Bone marrow cells, separation of, 34
Brain cell membranes,
  isolation, 52
  lipids in, 102
  polypeptides of, 69
Brush border membranes, 6
  enzymes in, 178
  glucose-6-phosphatase in, 163
  isolation of, 54
  X-ray diffraction studies, 83
Bungarotoxin receptor sites, 60

268 INDEX

Burkitt's lymphoma, 37

Calcium in cell attachment, 33
Cannibalism, 11
Capping response, 211, 215
Carbohydrate,
  content of membranes, 51, 247
  erythrocyte membranes, in, 126
  on surface, 253
  synthesis in Golgi complex, 16
Cardiac muscle cells,
  attachment devices, 4
Catalase, 19
Cell cultivation, 31, 36
  anchorage-dependent cells, 38
  cloning, 41
  flasks, 39
  harvesting, 40
  medium requirements, 36
  permanent cultures, 38
  primary cultures, 37
  single cell isolation, 41
  suspension, in, 40
Cell interactions, 45
Cell junctions, 4, 6, 215
Cell preservation, 41
Cell projections, 6
Cell separation, methods, 32
  centrifugation, 34
  chelating agents, 33
  chemically derivatized surfaces, 35
  collagenase, 33
  enzymatic treatment, 32
  freezing and thawing, 36
  nonenzymatic treatment, 33
  pronase, 32
  sequential enzyme treatment, 33
  trypsin, 32
  using behavior characteristics, 35
  velocity sedimentation, 34
Cells,
  adhering, 35
  anatomy, 1–30 (*see also* under specific components)
    endoplasmic reticulum, 13
    granules, 18
    nonmembranous components, 25
    organelles, 11
    schema, 1
    special organelles, 18
  attachment, 4
    calcium and magnesium in, 33
    plaques, 5
  behavioral characteristics, separation and, 35
  contacts, 4
  membranes (plasmalemmae), 2
    components, 2
  projections, 6
  separation of, 30, 31

Centrifugation,
  cell separation by, 34
  density gradient, in membrane isolation, 55
Centrioles, 26
Centrosomes, 26
Chelating agents in cell separation, 33
Chemical properties of membranes, 125
Chinese hamster cells, cultivation, 40
Chlamydomonas, 225
Cholesterol,
  bilayers, effect on, 249
  erythrocyte membrane, in, 144, 152
  myelin membranes, in, 98, 128
  plasma membranes, in, 246
Cholesterol–lipid–water interaction, 118
Cholesterol–phospholipid ratio, 102, 249
  marker, as, 61, 70
Cholesterol–protein ratio, 98
Choline groups in lecithin layers, 249
Chondroblasts, collagen production by, 21
Chorioallantoic cells, isolation of membranes, 55
Chromatin, 28
Cilia,
  microtubules and, 26
  structure, 27
Close junctions, 215
Collagen, production of, 21
Collagen fibrils, 6
Collagenase in cell separation, 33
Components of membranes, 98, 225 (*see also* under individual components)
  diffusion of, 233
  physical properties of, 106
  synthesis of, 229
Concanavalin A,
  binding sites as membranous particles, 214
  capping surface receptors, 213
  isolation of membranes, in, 57, 58
Contraction, 199
Cristae, 12
Cryoprotective agents in cell preservation, 41
Culture flasks, 39
Cyclic 3′,5′-AMP, 174
Cytochalasin A and B, 214, 261
Cytochrome oxidase, 189
  incorporation into bilayers, 253
  mitochondrial membranes, in, 90
  smooth endoplasmic reticulum, in, 227
Cytochromes, 123
  Golgi apparatus, in, 185
  interaction with phosphatidylinositol, 124
  mitochondrial membranes, in, 107
  nuclear envelope, in, 191
  synthesis, 238
Cytolysosomes, 18

Davson–Danielli model of membrane, 141, 240, 244

Deformation of membranes, 144, 148, 150, 152, 154, 155
Deoxyribonucleic acid, 224
  associated with nuclear envelope, 190
  chromatin and, 28
  mitochondria, in, 13, 241
  synthesis, 240
Desmosomes, 4, 5
  cellular extensions, on, 6
Differential pelleting, 161, 164
Differential thermal analysis, 110
Diplosomes, *see* Centrioles
Drug–lipid interaction, 123
Drugs, affecting membranes, 228, 231, 253

Echinocytes, 156
Ehrlich ascites cells,
  components of surface membranes, 63
  isolation of membranes, 49, 50
  RNA in membranes, 72
  sialic acid content of membranes, 71
Electron density profiles of membranes, 83, 95
  phospholipid bilayers, 87
Electron spin resonance, 78, 81
Electron transport enzymes, 236
Electrophoresis,
  free-flow, 57
  gel, 67
Electrostatic surface potential, permeability and, 199
Elementary particles, 13
Elutriator rotor, 35
Endocytosis, 8, 211
  in leukocytes, 211
Endoplasmic reticulum, 13, 50, 240
  cisternae in, 13
  continuity with mitochondria, 227
  differentiation in, 234, 236
  enzymes in, 166, 175, 180, 191
  function of, 180
  glucose-6-phosphatase in, 162
  granular, 13, 19, 182
    collagen in, 21
    enzymes in, 182
    formation of, 228
    marker enzyme, 169
    protein synthesis on, 23, 27
  lipid–protein interaction in, 252
  lipid synthesis in, 180, 187
  marker enzymes, 169
  membrane proteins in, 105
  protein synthesis in, 15
  smooth, 13, 182
    biogenesis, 225
    effects of drugs on, 228
    enzymes in, 182
    enzyme markers, 169
  unit membranes forming, 13
Endothelial cells,
  fibrous lamina in, 18
  micropinocytosis, 9

Endothermic transitions, 108
Enzymatic treatment for cell separation, 32
Enzymes, 161–197
  active transport, in, 174
  assigning to particular membranes, 165
  concerned with synthesis and renewal, 173
  distribution of, 171
  electron transport, 236
  endoplasmic reticulum, 180, 191
  Golgi apparatus, in, 183, 192
  hormone action, 174
  identification of sites, use of markers, 167
  intramembranous, 177
  lipids and activity of, 251
  lysosomal, 185
  markers for surface membranes, as, 58
  methods of study, 170
  mitochondrial, 187
  nuclear, 191
  periphery, on, 172
  peroxisomal, 190
  plasma, 172
  related to incoming and outgoing molecules, 173
Enzyme treatment of cell separation, sequential, 33
Eosinophilic granulomas, 21
Eosinophils, crystal-containing granules in, 16
Epidermis cells, membrane-coating granules on, 23
Epithelial cells,
  centrosomes in, 26
  desmosomes in, 6
  mitochondria, 12
Epithelium, keratinizing, 29
Erythrocyte ghosts,
  circular dichroism spectra of, 132
  composition of, 103
  intramembranous particles in, 201
  phospholipases affecting, 255
  proteolytic digestion in, 131
Erythrocyte membrane,
  area of, 142
  asymmetry, 254, 255
  behavior, 144
  carbohydrate content, 126
  chemical properties, 126
  cholesterol in, 117, 144, 152
  cholesterol:phospholipid ratio, 70
  composition of, 62, 98
  deformation of, 148, 150, 152
  glycophorin in, 200
  intramembranous particles in, 200
  lipid bilayers, 150
  lipids in, 131, 259
    distribution of, 127
  mechanical properties of, 139
  models of, 141, 155
  NMR studies, 131

Erythrocyte membrane *continued*
 permeability to hemoglobin, 147
 permeability to solutes, 147, 148
 phospholipid in, 144, 152
 physical properties, 131
 polypeptides of, 68
 potential and shape change, 153
 protein matrix in, 259
 proteins in, 88, 103, 126, 131, 147, 254
 resistance to curvature, 155
 second elastic constant from tethering experiments, 149
 shape change, 144, 148, 152, 153, 154
 shaping forces, 155
 sialic acid content of, 71
 size changes, 148
 spiculated, 156
 stretching, 143, 144, 145
 structure of, 141
 thickness of, 140, 153
 two-dimensional elastomer theory, 150
  alternative to, 152
 viscoelastic properties of, 145
 yielding under stress, 148
 X-ray diffraction studies, 83
Erythrocyte stroma, vesiculated, 55
Erythrocytes,
 acetylcholinesterase on surface, 179
 attachment, 149
 behavior of, 139
 carbohydrates on surface, 253
 deformability of, 139, 154
 hemolysis of, 142
 intramembranous particles in, 258, 259
 isolation of membranes in, 46
  loss of components, 64
 movement of, 139
 shape of, 70
 shearing, 150
 X-ray diffraction patterns, 79
Escherichia coli membranes,
 physical properties, 135
 transition temperatures of glucosides in, 252
 X-ray diffraction studies, 79
Euchromatin, 29
Exocytosis, 215
Extracellular basal lamina, 6

Fatty acids, 92, 102
 structure of, 101
 variation in, 106
Ferricytochrome, 124
Fiber fractionation technique, 58
Fibroblasts,
 cisternae in, 13
 collagen production by, 21
 cultivation of, 37
 endoplasmic reticulum in, 13
 fibrous lamina in, 18
 human diploid, 37

 inclusions, 9
 isolation of membranes, 49, 50, 52, 53, 54, 55, 57
 sialic acid content of membranes, 72
 tight junctions, 6
Fibrosarcoma cells, isolation of membranes, 54
Fibrous lamina, 18
Filaments, 25
Fluorescein mercuric acetate, fixation of cells with, 50
Freeze-etch electron microscopy, 112
Freezing and thawing of cells, 36
Fucose as membrane marker, 61
Function of membranes, microfilaments and, 260
Fuzzy layer, 4

Galactosidase in Golgi apparatus, 185
Galactosyltransferase, 173, 184
Gangliosides,
 as membrane marker, 61
 interaction with tetanus toxin, 124
 structure of, 101
Gel electrophoresis, 67
Gingival cells,
 attachment devices, 4
 granular endoplasmic reticulum in, 13
 inclusions in, 11
 micropinocytosis in, 9
 myelin figures in, 24
Glucagon as membrane marker, 60
Glucagon receptors, 58, 175
Glucose-6-phosphatase,
 distribution of, 234
 endoplasmic reticulum, in, 162, 182, 183
 microsomes, in, 252
 plasma membranes, in, 166
Glucuronidase, 168, 181
Glutamyl transpeptidase, 166
Glycerol in cell preservation, 41
Glycerol lysis in isolation of membranes, 54
Glycocalyx, 4, 7
Glycogen, 28
Glycolipids, 230, 247
Glycophorin, 200, 213. 254
Glycoproteins, 105, 247
 arrangement in membranes, 55
 interaction between, 203
 intramembranous particles and, 200
 loss of, 64
 pegs, 153
Glycosyltransferases, 185
Golgi complex, 13, 15
 association with centrioles, 26
 carbohydrate synthesis in, 16
 enzymes in, 166, 168, 183, 192
 granules associated with, 23
 marker enzymes, 169
 role of, 16
Gramicidin A, 120

INDEX

Gramicidin S, 121
Granules, 19
  associated with Golgi complex, 23
  keratohyalin, 29
  mast cell, 24
  membrane-coating, 23
  secretory, 21

Halobacterium halobium,
  cell envelopes, 88
  purple membrane,
    chemical properties, 98
    composition of, 128
    physical properties, 133
    structure, 83
HeLa cells, isolation of membranes, 50
Hemidesmosomes, 4
Hemoglobin, 140, 147
Hemolysis of erythrocytes, 142, 143, 144
Heparin in mast cells, 24
Hepatoma cells,
  gap junctions in, 216
  RNA in membranes, 72
Hepatoma plasma membranes, hexokinase in, 174
Heterochromatin, 29
Hexokinase, 174
Histamine in mast cells, 24
Histiocytosis X, 21
Histocompatibility antigens, 105
Histocytes, Langerhans granules in, 21
Homarus americanus, membrane, 91
Hormone action, enzymes and, 174
Hydrocarbon regions,
  anisotropic motion in, 90, 92
  chain populations, 93
Hydrophilic and hydrophobic interactions, 247
Hypotonic lysis in membrane isolation, 46

Immune-competent cells, separation of, 34
Immunochemical methods of enzyme study, 170, 171
Inclusions, 8, 19
Insulin as membrane marker, 60
Insulin receptors, 58
Intestinal brush border,
  enzymes in, 178
  glucose-6-phosphatase in, 163
  isolation of membranes, 54
Intestinal cells,
  components of surface membranes, 63
  Golgi apparatus, 183
  isolation of membranes, 50
  microvilli, 6
    cholesterol:phospholipid ratio, 70
Intramembranous particles, 198
  aggregation, 199, 207, 208, 215, 217, 241, 258, 259
    adenyl cyclase and, 199
    spontaneous, 216
  charge, 217
  concanavalin A binding sites, as, 214
  distribution of, 206
  electrostatic interaction between, 204, 206, 220
  erythrocytes, in, 200, 258
  forces between, 202
  glycoprotein nature of, 200
  mobility of, 201
  permeability and, 198
Iodination of proteins as membrane markers, 61, 66
Ion permeability changes in *Xenopus* egg membrane, 209
Isolation of surface membranes, 45–77
  activation of endogenous enzymes in, 65
  adsorption of components in, 65
  affinity chromatography, 58
  affinity density perturbation system, 57
  aqueous two-phase system, 55
  contaminants, 61
  density gradient centrifugation, 55
  enucleated cells, of, 46
  free-flow electrophoresis, 57
  glycerol lysis, 54
  hypotonic lysis, by, 46
  hypotonic $Na_2B_4O_7$, 51
  hypotonic $NaHCO_3$, 51
  isolation of specific fragments in, 65, 164
  large fragments, 51
  loss of components, 64
  markers, 58, 66, 70
    biological properties as, 60
    chemical components, 60
    enzymes for, 58
  nitrogen cavitation, 53
  nucleated cells, of, 48
  phagocytosis of latex beads, 57
  procedural artifacts, 64
  rearrangement of proteins in, 65
  sodium chloride in, 54
  Tris method, 48, 49, 50, 52
    loss of components through, 64
  vesicles, 53
  whole, 46
  zinc ion method, 48, 52

Junctional complex, 5

Keratinocytes, melanin in, 19
Keratohyalin granules, 29
Kidney brush border cells, membrane, 51, 71
Kidney cells,
  isolation of membranes, 50, 52, 53, 54, 57
  membrane fragments, 65
  microbodies in, 19
Kidney microvillus, polypeptides of membranes, 69
Kidney tubule cells, mitochondria, 12

272 INDEX

L cells,
  components of surface membranes, 62
  isolation of membranes, 49
  polypeptides of membranes, 68
  sialic acid content of membranes, 71
Lamina densa, 6
Langerhans cells,
  granules, 21
  melanin granules in, 19
Lecithins,
  endothermic transition, 111
  interaction with gramicidin S, 121
Lecithin–cholesterol–water interactions, 118
Lecithin layers, choline groups in, 249
Lectins, membrane response to, 213
1-Leucyl-β-naphthylamidase, 186
Leukemia cells,
  cholesterol:phospholipid ratio of membrane, 70
  isolation of membranes, 49, 51, 53
Leukocytes,
  cultivation of, 37
  endocytosis in, 211
  lysosomes in, 16
  rigidity of, 158
  separation of, 35
Leukokinin, 211
Lipid bilayers, 170
  between protein layers, 87
  cholesterol affecting, 249
  cytochrome incorporation into, 253
  dynamic structure of, 248
  erythrocyte membranes, in, 141, 150
  myelin membrane, in, 128
  phospholipid molecular motion in, 250
  protein motion in, 257
  protein penetration of, 88, 127, 247
  stability of, 198, 247
  structure, 80, 86
  total content, 82
  transition temperature, 81
  translational diffusion, 90
Lipid–cholesterol systems, 117
Lipid droplets, 28
Lipid–drug interactions, 123
Lipid–lipid interactions, 245
Lipid–polypeptide–water systems, 120
Lipid–protein interactions, 78, 198, 251
  bilayers, in, 200
Lipid–protein–water systems, 123
Lipid systems,
  differential thermal analysis of, 108
  DSC heating curves, 113, 114, 115
  endothermic transition, 108
  physical properties of, 106
  water, effects of, 110
Lipids, 3, 98
  arrangement in membranes, 245
  asymmetrical distribution, 253, 255, 256
  'boundary', 253
  classes of, 101

composition of membranes, 246
  enzyme activity and, 251
  erythrocyte membranes, in, 127, 131
  Escherichia coli membranes, in, 135
  fatty acids in, 106
  melting behaviors, 114
  mitochondrial membranes, in, 12, 135
  mobility of, 244
  NMR spectra, 130
  phase transitions, 112, 113, 135
  purple membrane of *Halobacterium halobium*, 133
  separation of, 102
  synthesis, 180, 229, 230
    endoplasmic reticulum, in, 187
    mitochondria in, 230
  thermal transition, 81
  thermotropic mesomorphism, 117
  transport of, 231
  turnover of, 231
Lipofuchsin, 21
Lipoproteins, 105
Liposome systems, permeability in, 125
Liver cells,
  adenyl cyclase, 174
  components of membranes, 63, 98
  cytolysosomes in, 18
  enzyme markers for, 169
  glycogen in, 28
  isolation of membranes, 51, 56
  lipid bilayers, 81
  lipid synthesis in, 180
  loss of protein from membrane, 64
  markers for membranes, 59
  membrane biogenesis in, 225
  microbodies in, 19
  mitochondria, 11
  polypeptides of membranes, 69
  RNA in membranes, 72
  sialic acid content of membranes, 71
Liver plasma membranes,
  composition, 127
  enzymes in, 173, 177
Lymphocytes,
  capping response, 211, 213, 215
  cholesterol:phospholipid ratio of membrane, 70
  isolation of membranes in, 50, 51, 57
  mechanical properties of membranes, 158
  plasma membrane isolation from, 53
  polypeptides of membranes, 68
  RNA in membranes, 72
  separation of, 36
  sialic acid content of membranes, 71
B Lymphocytes, 201
Lyotropic mesomorphism of phospholipids, 119
Lysosomal membranes,
  enzymes in, 163
  preparation of, 185

Lysosomes, 16
  acid phosphatase in, 16
  enzymes in, 168, 181
  marker enzymes, 169
  sialic acid content, 61
  size, 18

Macrophages,
  components of surface membranes, 62
  isolation of membranes, 50
Magnesium in cell attachment, 33
Magnetic resonance studies, 93
  high-resolution, 78
Mammary gland cells, separation of, 33
Markers for surface membranes, 58, 201
  biological properties as, 60
  chemical components, 60
  enzymes, 58
  identifying enzyme sites, for, 165, 167
Mast cells,
  granules, 24
  separation of, 35
Mechanical properties of membranes, 138–159
Melanin, 19
Melanocytes, 18, 19
Melanophages, 19
Melanosome complexes, 11
Membrane-coating granules, 23
Microfilaments, 260, 261
Microkinetospheres, 18
Micropinocytosis, 9
Microscopy, 1
Microsomes,
  enzymes in, 165, 178
  lipid–protein interaction in, 252
  5'-nucleotidase in, 162
Microtubules, 25
  centrioles, of, 27
Microvilli, 6
Mitochondria, 11, 50
  anatomy of, 187
  components, 12
  continuity with endoplasmic reticulum, 227
  cristae, 12
  distribution of, 11
  division of, 239
  DNA in, 13, 224, 241
  enzyme location in, 189
  formation of, 225
  lipid synthesis and, 230
  marker enzymes, 169
  phospholipid synthesis in, 229
  protein synthesis in, 188, 230, 238, 240
  reaction to oxygen deficiency, 239
  size of, 11, 239
    variations in, 13
  structure of, 189
  unit membranes, 12
Mitochondrial membrane,
  biogenesis of, 227
  composition of, 98
  energy generation in, 188
  enzymes in, 187
  inner, function of, 188
  lipid bilayers, 81
  lipids in, 120
  outer,
    enzymes in, 163
    function of, 188
  physical properties, 135
  protein in, 107
  protein solubilization, 103
Models of membranes, 120, 141, 155
  bilayers, 147
  Davson–Danielli, 141, 240, 244
  examples of biogenesis, 225
  permeability studies in, 125
  structural, 244–261
  tennis ball model, 156
Monoamine oxidase as marker, 168
Monolayer studies of lipids, 115
Mosaicism in membranes, 234
Muscle cells,
  components of surface membranes, 63
  cultivation of, 38
  endoplasmic reticulum in, 15, 182 (*see also* Sarcoplasmic reticulum)
  excitable membranes, 80
  filaments, 25
  glycogen in, 28
  lipid synthesis in, 180
Mycoplasma membranes,
  anisotropic motion in, 91
  composition of, 98
  fatty acid composition, 92
  lipid bilayers, 81
  phospholipids in, 250
  protein layers, 88
  proteins in, 133, 258
  thermal phase transitions, 133
  thickness, 246
  X-ray diffraction patterns, 79
Myelin figures, 24, 111
Myelin membranes,
  arrangement of, 85
  asymmetry in, 256
  chemical properties of, 125
  cholesterol in, 98, 128
  composition of, 98
  lipid bilayer, 128
  lipid composition, 99
  Patterson maps, 85
  physical properties, 128
  protein in, 126, 128, 246
  proteolipids in, 126
  X-ray diffraction studies, 79, 80, 82
Myelin sheath,
  cholesterol in, 117
  formation of, 228
Myoblasts, cultivation, 38
Myosin filaments, 25

NADH–cytochrome c reductase, 175, 231, 236
Nerve ending membranes, protein layers, 88
Neuromuscular junctions, electrical conduction at, 216
Neurones,
  isolation of membranes, 50
  polypeptides of membranes, 69
Neurospora, 225, 239
Nissl substance, 15
Nitrogen cavitation, method of membrane isolation, 53
Nonactin, 120
Nuclear envelope, 18
  DNA associated with, 190
  enzymes of, 190
  pores in, 18, 19
  transport system in, 191
Nuclear magnetic resonance, 93, 119, 249
  erythrocyte membranes, of, 131
  mitochondrial membranes, of, 135
  retinal rod membranes, of, 130
Nuclear membranes,
  differentiation, 236
  enzymes of, 191
Nucleolonema, 29
Nucleolus, 29
  RNA in, 29
Nucleoside diphosphatase, 176
5'-Nucleotidase,
  as plasma membrane marker, 59
  Golgi region, in, 166
  microsomes, in, 162
  plasma membranes, in, 166, 176, 179

One-enzyme–one-location hypothesis, 168
Oocytes, mitochondria of, 11
Oral mucosal cells, membrane-coating granules in, 23
Organelles, 11
Osmotic hemolysis, 143
Osteoblasts, 13, 21
Osteoclasts, 12
Oxygen deficiency, reaction of mitochondria to, 239

Pancreatic acinar cells,
  granular endoplasmic reticulum in, 15
  plasma membranes, 234
Paramecium, trichocyst discharge in, 215
Pars amorpha, 29
Particles, intramembranous, *see* Intramembranous particles
Patterson maps, 84, 95
Perinuclear space, 18
Permeability of membranes, 97, 125, 147, 148
  electrostatic control of, 198–223
  pinocytosis and, 209
Permeases, 174

Peroxidase, 19
Peroxisomes (microbodies), 19
  enzyme markers, 169
  enzymes in, 190
Phagocytosis, 8, 174
  cannibalism, 11
  latex beads, of, 57
Phagolysosomal membranes, isolation, 57
Phosphatases, 176
Phosphatides, synthesis, 173
Phosphatidylcholine, 229
Phosphatidylethanolamine, 229
Phosphodiesterases, 177
Phospholipase C, 24
Phospholipases, effect on erythrocyte ghosts, 255
Phospholipid bilayers, 78
  electron density profiles of, 87
  physical properties, 108
  physical studies, 86
  polarity in, 92
  structure of, 87
Phospholipids,
  conformation, 247
  differential thermal analysis, 109
  diffusion along bilayer, 112
  erythrocyte membranes, in, 144, 152
  fatty acid structure, 102
  lyotropic mesomorphism, 119
  model membranes, as, 120
  monolayers, 115
  motions between molecules, 248
  myelin figure formation with, 111
  physical properties, 108
  ratio to cholesterol, 102, 249
  role of as cement, 24
  structures of, 100
  synthesis of, 187, 229
  thermal data for, 111
  thermotropic phase transitions, 112
  translational motion of molecules, 250
Physical studies of membranes, 78–96, 128
  calorimetry, 129
  differential thermal analysis, 109, 110
  ESR, 78, 81
  freeze-etch electron microscopy, 112
  NMR, 93, 119, 130, 131, 135, 249
  Patterson maps, 84, 95
  spin labeling studies, 248, 252
  thermotropic phase transitions, 112
  X-ray diffraction studies, 78, 79, 82, 87
Physicochemical studies of membranes, 97–137
Phytohemagglutinin, 38
Pinocytosis, 8, 214
  amoeba, in, 199, 207, 208, 217
  permeability of membranes and, 209
  vertebrate cells, in, 211
Pinosomes, 18
Plant lectins, 38
Plasma cells, endoplasmic reticulum in, 15

INDEX

Plasma membranes, 46
  chemical properties of, 127
  composition of, 98, 234
  enzymes in, 161, 172, 178
  marker enzymes, 169
  proteins in, 105, 257, 258
  synthesis and renewal, 173
Plasmocytoma cells, membranes, 51
Platelet membranes,
  composition of, 62, 98, 127
  isolation of, 54, 55
  polypeptides of, 68
  protein synthesis in, 73
  sialic acid content, 71
Pokeweed mitogen, 38
Polypeptide chains in myelin membranes, 125
Polypeptide–lipid–water systems, 120
Polypeptides of surface membranes, 68, 69
Pores in nuclear envelope, 18, 144
Premelanosomes, 21
Production of membranes, 157
Projections, 6
Promitochondria, 225
Pronase, cell separation with, 32, 41
Pronase digestion, 131
Protein in membranes, 3, 103
  amino acid composition, 104
  arrangement in, 55, 65, 133, 257
  asymmetrical distribution, 253, 255
  content of membranes, 246
  diffusion in membranes, 258
  distribution of, 104
  erythrocyte membranes, in, 103, 126, 131, 147
  extraction of, 106
  interaction with bilayers, 200
  interaction with lipid, *see* Lipid–protein interaction
  iodination of, 61, 66
  layers covering lipid bilayers, 87
  marker for membranes, as, 60
  matrix, 259
  microfilaments as, 261
  mitochondrial, 107
  mobility in membranes, 244, 257
  Mycoplasma membranes, in, 133
  myelin membranes, in, 126, 128
  organization of movement, 184
  penetration into bilayers, 87, 127, 247
  permeability to ions, effect on, 124
  solubilization of, 103
  surface membranes, in, 70, 89
  technique of isolation, 102
  transport of, 231
  transversing membranes, 245
  turnover of, 231
Protein layers, 88
  mobile state of, 89
Protein–lipid interactions, 88, 124
Protein–lipid–water systems, 123

Protein synthesis, 23, 239
  endoplasmic reticulum involved in, 15
  mitochondria, in, 188, 230, 238, 241
  ribosomes, in, 27
Proteolipids, 126

Rashgeldi threshold system, 35
Residual bodies, 24
Retinal rod membranes,
  lipids in, 102
  physical properties, 130
  protein in, 246
  rhodopsin in, 90, 202, 257, 258
  X-ray diffraction patterns, 79, 82
Rhodopsin, 202, 253, 258
Rhodopsin molecules, organization of, 90
Ribonucleic acid,
  content of membranes, 72
  nucleolus, in, 29
  ribosomal, 27
Ribosomes, 27
  annulate lamellae and, 19
  attached to endoplasmic reticulum, 236
  attached to surface membranes, 72
  enzyme synthesis in, 181
  protein synthesis in, 27
Ricinus communis, 201

Sarcoplasmic reticulum, 15
  enzyme markers, 59
  lipid–protein interactions, 93
  physical properties, 132
Sarcoplasmic vesicles,
  excitable membranes, 80
  lipid bilayers, 90
  phospholipids in, 201
  polarity in, 92
  structure, 95
Second elastic constant, 149
Secretory granules, 21
Shape changes in membranes, *see* Deformation of membranes
Sialic acid,
  content of lysosomes, 61
  membranes, in, 70
Size of membranes, 142
Skeletal muscle cells, glycogen in, 28
Sodium chloride in membrane isolation, 54
Sodium pump, 174
Solubilization in study of enzymes, 170
Spectrin, 259, 260
Sphingolipids,
  structures, 100
  variation in, 106
Spiculated red cell, 156
Spleen, erythrocyte movement in, 139
Steroids, conjugation of, 181
Stokes' law, 34
Storage granules, enzymes in, 190
Stretching of membranes, 143, 144, 145

Structure of membranes, 78, 93, 95, 97
  anisotropy, 90
  asymmetry, 253
  erythrocytes, in, 141
  exocytosis and, 215
  forces holding together, 247
  models of, 244–261
  protein arrangement and motion, 257
  protein matrix, 259
Surface membranes, isolation of, *see* Isolation of surface membranes
Synaptosomes,
  components of surface membranes, 63
  isolation of membranes, 54
  polypeptides of membranes, 69

3T3 cells,
  polypeptides of membranes, 68
  SV_{40}-transformed, 35
Tennis ball model of membranes, 156
Thiamine pyrophosphatase, 184
Thickness of membranes, 140
  erythrocyte, in, 153
Thymocytes,
  components of surface membranes, 62
  isolation of membranes, 53
Thyroid acinar cells, microvilli, 7
Thyroid plasma membranes, isolation, 52
Tight junctions, 6
Tonofilaments, 5, 27
Transport within membranes, 224
Tris procedure in isolation of membranes, 48, 49, 50, 52, 64
Tritosomes, 186
Tuftsin, 211
Two-dimensional elastomer theory, 150, 157
  alternative to, 152

UDP–galactose-splitting galactosidase, 175
Unit membranes, 3
  annulate lamellae, forming, 19
  endoplasmic reticulum, forming, 13
  mitochondria, of, 12
Urate oxidase, 19

Valinomycin, 120
Velocity sedimentation, in cell separation, 34
Vesicles, isolation of, 53
Viral bilayers, asymmetry in, 256
Viral membranes, protein matrix in, 260
Viscoelastic properties of membranes, 145

Watch-band model, 156
Water, lipid systems affected by, 110
Water–lipid–cholesterol interactions, 118
Water–lipid–polypeptide systems, 120
Water–lipid–protein systems, 123
Wolfgram protein, 126

X-ray diffraction studies, 78, 79, 82, 87
Xenopus egg membrane, ion permeability change in, 209

Zeiosis, 7
Zinc ion method of membrane isolation, 48, 52
Zona pellucida, removal of, 32
Zymogen granules, 15, 23

THE LIBRARY
UNIVERSITY OF CALIFORNIA
San Francisco

THIS BOOK IS DUE ON THE LAST DATE STAMPED BELOW

Books not returned on time are subject to fines according to the Library Lending Code. A renewal may be made on certain materials. For details consult Lending Code.

14 DAY
DEC 17 1985

14 DAY
AUG - 8 1988

RETURNED
AUG - 9 1988

28 DAY
AUG - 4 1996

Series 4128